DYNAMICAL SYSTEMS & DIFFERENTIAL EQUATIONS

DYNAMICAL SYSTEMS & DIFFERENTIAL EQUATIONS

Editor

Torsten Dohman

scitus
academics

Dynamical Systems & Differential Equations
Edited by **Torsten Dohman**

This Book is Printed on Acid-Free Paper. ♾

ISBN: 9781681179599
Library of Congress Control Number: 2017949094

SCITUS Academics LLC,
616, Corporate Way, Suite 2, 4766,
Valley Cottage, NY 10989

www.scitusacademics.com

Notice

Printed in United States of America.

Preface

Dynamical systems theory, one of the rapidly developing areas of modern mathematics, deals with powerful theoretical base for exploring a variety of models that arise in natural and social sciences, engineering and technology. The blend of the internal wealth and beauty of results with the outstanding practical importance motivates a growing number of specialists to study into dynamical systems. Differential equations have a remarkable ability to predict the world around us. They are used in a wide variety of disciplines, from biology, economics, physics, chemistry and engineering. They can describe exponential growth and decay, the population growth of species or the change in investment return over time. In applications, the functions usually represent physical quantities, the derivatives represent their rates of change, and the equation defines a relationship between the two. Because such relations are extremely common, differential equations play a prominent role in many disciplines including engineering, physics, economics, and biology. In pure mathematics, differential equations are studied from several different perspectives, mostly concerned with their solutions—the set of functions that satisfy the equation.

This book 'Dynamical Systems & Differential Equations' covers all areas related to dynamical systems and differential equations and their applications in biology, economics, engineering, physics, and other related areas of science. It provides dynamical aspects of ordinary differential equations and the relations between dynamical systems and certain fields outside pure mathematics with an emphasis on systems of ordinary differential equations and will be most appropriate for upper level undergraduate and research students as well as serve scientists through significant advances in any branch of science and technology and the discussion of new scientific developments.

Table of Contents

CHAPTER 1

Similarity Reduction of Nonlinear Partial Differential Equations

Amnah S. Al-Johani 1,2

[1]*Department of Applied Mathematics, College of Science, Northern Borders University, Arar, Saudi Arabia*

[2]*College of Home Economics, Northern Borders University, Arar, Saudi Arabia*

ABSTRACT

In this work, the HB method is extended to search for similarity reduction of nonlinear partial differential equations. This method is generalized and will apply for a (2 + 1)-dimensional higher order Broer-Kaup System. Some new exact solutions of Broer-Kaup System are found.

Keywords: Similarity Reduction; Exact Solutions; Nonlinear Partial Differential Equations

1. INTRODUCTION

In the past few decades, there has been the noticeable progress in the construction of the exact solutions for nonlinear partial differential equations, which has long been a major concern for both mathematicians and physicists. The effort in finding exact solutions to nonlinear differential equation, when they exist, is very important for the understanding of most nonlinear physical phenomena. For instances, the nonlinear wave phenomena observed in fluid dynamics, plasma and optical fibers are often modelled by the bell shaped sech solutions and the kink shaped tanh solutions.

We consider the following a (2 + 1)-dimensional higher order Broer-Kaup system:

$$u_t + 4\left(u_{xx} + u^3 - 3uu_x + 3uw + 3p\right)_x = 0, \tag{1}$$

$$v_t + 4\left(v_{xx} + 3vu^2 + uv_x + 3vw\right)_x = 0, \tag{2}$$

$$w_y - v_x = 0, \tag{3}$$

$$p_y - (uv)_x = 0. \tag{4}$$

which is obtained from the Kadomtsev-Petviashvili (KP) equation by the symmetry constraint [1].

The systems (1)-(4) were given by Li et al [2] solving it via a transformation and tanh-function method to obtain many new exact solutions. Jain et al. [3] reduced a system to a simple (1 + 1)-dimensional nonlinear evolution equation through a simple transformation, and by using the new generally projective Riccati equation expansion method to explore many families of soliton-like and periodic solutions for it. Recently, Li et al. [4] have obtained some new types of multisoliton solutions for the systems (1)-(4) by using some simple transformations as $v = u_y, w = u_x + \alpha(x,t), p = uu_x + \beta$ and homogenous balance method.

The homogenous balance (HB) method is a powerful tool to find solitary wave solutions of nonlinear partial differential equations. Fan et al. [5] presented an improved HB method to obtain more other kinds of exact solutions and introduced a continuation of [5] in [6]. The traditional method for finding similarity reduction of nonlinear partial differential equations is to use classical Lie approach [7,8]. However, the method involves tedious algebraic calculations and still can not be used to find all similarity solutions. Recently,Clarkson and Kruskal devloped a direct and simple method to find more similarity solutions of nonlinear PDEs.

In this work, the HB method is extended to search for similarity reduction of nonlinear partial differential equations. So, more solutions can be obtained by the improved HB method. This method is generalized and can be applied to other nonlinear partial differential equations [9-15].

2. SIMILARITY REDUCTION OF NONLINEAR PARTIAL DIFFERENTIAL EQUATIONS

We describe the main steps of our method. For a given PDE, say in three variables, say x, y, t

$$u_t = K\left(u, u_x, u_y, u_{xx}, u_{yy}, \cdots\right) \tag{5}$$

we seek its similarity reductions in the form

$$u(x,y,t) = \frac{\partial^{\alpha}}{\partial x^{\alpha}} f(s) + u_0, \tag{6}$$

where α is a constant to determine by balancing between the highest order derivative of the linear terms of u and the nonlinear terms of u, where s, u_0 are regarded as undetermined functions.

Substituting from Equation (6) into Equation (5) and collecting all terms of f with the same derivative and power. To make the associated equation be an ordinary equations of f and s, requiring ratios of their coefficients being functions of s, we obtain a set of determining equations for s, u_0 and other undermined functions, from which s and u_0 will be obtained.

To explain this method, we will apply for a (2 + 1)-dimensional higher order Broer-Kaup system (1)-(4), we suppose their similarity solutions are of the form

$$u(x,y,t) = \frac{\partial^{\alpha_1}}{\partial x^{\alpha_1}} f(s) + u_0, \quad v(x,y,t) = \frac{\partial^{\alpha_2}}{\partial x^{\alpha_2}} g(s) + v_0,$$
$$w(x,y,t) = \frac{\partial^{\alpha_3}}{\partial x^{\alpha_3}} h(s) + w_0, \quad p(x,y,t) = \frac{\partial^{\alpha_4}}{\partial x^{\alpha_4}} k(s) + p_0. \tag{7}$$

where

$$s = s(x,y,t), u_0 = u_0(x,y,t), v_0 = v_0(x,y,t), w_0 = w_0(x,y,t), p_0 = p(x,y,t).$$

are determined functions. Balancing the highest order of linear term with the nonlinear terms in every equations (1)-(4) to determining $\alpha_1, \alpha_2, \alpha_3, \alpha_4$, we obtain

$$\alpha_1 = 1, \alpha_2 = \alpha_3 = 2, \alpha_4 = 3 \tag{8}$$

the Equation (64) take the following form

$$u(x,y,t) = \frac{\partial}{\partial x} f(s) + u_0 = f' s_x + u_0, \tag{9}$$

$$v(x,y,t) = \frac{\partial^2}{\partial x^2} g(s) + v_0 = g'' s_x^2 + g' s_{xx} + v_0, \tag{10}$$

$$w(x,y,t) = \frac{\partial^2}{\partial x^2} h(s) + w_0 = h'' s_x^2 + h' s_{xx} + w_0, \tag{11}$$

$$p(x,y,t) = \frac{\partial^3}{\partial x^3} k(s) + p_0 = k''' s_x^3 + 3k'' s_x s_{xx} + k' s_{xxx} + p_0. \tag{12}$$

Substituting Equations (9)-(12) into the original system (1)-(4) and collecting all terms of f, g, h, k with the same derivative and power leads to

$$\begin{aligned}&(-12 s_x s_{xxx}) f'^2 + (12 u_{0x} s_x^2 + 36 u_0 s_x s_{xx}) h'' + (36 s_{xx}^2 + 48 s_x s_{xx}) k'' + 12 f h' s_x s_{xxx} \\ &+ (24 u_0 u_{0x} s_x + 12 w_{0x} s_x + s_{xt} + 12 u_0^2 s_{xx} + 12 w_0 s_{xx} - 24 u_{0x} s_{xx} - 12 u_0 s_{xxx} + 4 s_{xxxx}) f' \\ &+ (12 u_{0x} s_{xx} + 12 u_0 s_{xxx}) h' + 12 k' s_{xxxx} + (12 p_{0x} + u_{0t} + 12 u_0^2 u_{0x} + 12 w_0 u_{0x} - 12 u_{0x}^2 - 12 u_0 u_{0xx} + 4 u_{0xxx}) = 0,\end{aligned} \tag{13}$$

$$\begin{aligned}
&\left(24ff''g''+12f'2g'''+4f''g'''+12h''g'''+12g''h''+4f'g''''+4g'''''\right)s_x^5\\
&+\left(4u_0s_x^4+40s_x^3s_{xx}\right)g'''+\left(24u_0s_x^4+28s_x^3s_{xx}\right)f'g''+\left(24u_0s_x^4+12s_x^3s_{xx}\right)f''g''\\
&+\left(24f'g'f''+60f'^2g''+72g''h''+12h'g'''+12g'h'''\right)s_x^3s_{xx}\\
&+\left(s_x^2s_t+12u_0^2s_x^3+12w_0s_x^3+4u_{0x}s_x^3+24u_0s_x^2s_{xx}+60s_xs_{xx}^2+40s_x^2s_{xxx}\right)g'''\\
&+\left(24u_{0x}s_x^3+96u_0s_x^2s_{xx}+24s_xs_{xx}^2+16s_x^2s_{xxx}\right)f'g''+\left(24s_xs_{xx}^2+12s_x^2s_{xx}\right)f'^2g'\\
&+\left(24u_0s_x^2s_{xx}+4s_x^2s_{xxx}\right)f''g'+\left(36s_xs_{xx}^2+12s_x^2s_{xxx}\right)h'g''\\
&+\left(36s_xs_{xx}^2+12s_x^2s_{xxx}\right)g'h''+\left(24ff'''+12h'''\right)v_0s_x^3\left(24u_0v_0s_x^2+4v_{0x}s_x^2\right)f''\\
&+\left(24u_0u_{0x}s_x^2+12w_{0x}s_x^2+2s_xs_{xt}+36u_0^2s_xs_{xx}+36w_0s_xs_{xx}+12u_{0x}s_xs_{xx}+12u_0s_{xx}^2+16u_0s_xs_{xxx}\right.\\
&\left.+40s_{xx}s_{xxx}+20s_xs_{xxxx}\right)g''+\left(12v_{0x}s_x^2+36v_0s_xs_{xx}\right)h''+\left(12v_{0x}s_x^2+24v_0s_xs_{xx}\right)f'^2\\
&+\left(24u_0s_xs_{xx}+24u_0s_{xx}^2+24u_0s_xs_{xxx}+4s_{xx}s_{xxx}+4s_xs_{xxxx}\right)f'g'+24s_{xx}s_{xxxx}g'h'\\
&+\left(24v_0u_{0x}s_x+24u_0v_0s_{xx}+4v_{0x}s_{xx}\right)f'\\
&+\left(24u_0u_{0x}s_{xx}+12w_{0x}s_{xx}+s_{xxt}+12u_0^2s_{xxx}+12w_0s_{xxx}+4u_{0x}s_{xxx}+4u_0s_{xxxx}+4s_{xxxxx}\right)g'\\
&+\left(12v_{0x}s_{xx}+12v_0s_{xxx}\right)h'\\
&+\left(24u_0v_0u_{0x}+v_{0t}+12u_0^2v_{0x}+12w_0v_{0x}+4u_{0x}v_{0x}+12v_0w_{0x}+4u_0v_{0xx}+4v_{0xxx}\right)=0,
\end{aligned}\tag{14}$$

$$-g'''s_x^3+h'''s_x^2s_y+\left(s_{xx}s_y+2s_xs_{xy}\right)h''-3g's_xs_{xx}+s_{xxy}h'-s_{xxx}g'+\left(w_{0y}-v_{0x}\right)=0,\tag{15}$$

$$\begin{aligned}
&-\left(f''g''+f'g'''\right)s_x^4-\left(u_{0x}s_x^2+3u_0s_xs_{xx}\right)g''-\left(s_{xx}^2+s_xs_{xxx}\right)f'g'\\
&+\left(s_ys_{xxx}+3s_{xx}s_{xy}+3s_xs_{xxy}\right)k''-v_0s_x^2f''+\left(3s_xs_ys_{xx}+3s_x^2s_{xy}+s_x^3s_y\right)k'''\\
&-\left(g'f''+4f'g''\right)s_x^2s_{xx}-u_0s_x^3g'''-\left(v_{0x}s_x+v_0s_{xx}\right)f'-\left(u_{0x}s_{xx}+u_0s_{xxx}\right)g'\\
&+s_{xxxy}k'+\left(p_{0y}-v_0u_{0x}-u_0v_{0x}\right)=0.
\end{aligned}\tag{16}$$

To make Equations (13)-(16) be an ordinary differential equations of f,g,h and k only for s, the ratios of the coefficients of different derivative and power of f,g,h,k must be functions of s. That is to say, the following constrained conditions are satisfied

$$s_x^2s_{xx}=s_x^4\Gamma_1(s)\tag{17}$$

$$-12u_0s_x^3+24s_x^2s_{xx}=s_x^4\Gamma_2(s)\tag{18}$$

$$24u_0 s_x^3 - 60 s_x^2 s_{xx} = s_x^4 \Gamma_3(s) \tag{19}$$

$$s_x s_t + 12u_0 s_x^2 + 12w_0 s_x^2 + 12 s_{xx}^2 - 24u_{0x} s_x^2 - 36u_0 s_x s_{xx} + 16 s_x s_{xxx} = s_x^4 \Gamma_4(s) \tag{20}$$

$$12u_{0x} s_x^2 + 36u_0 s_x s_{xx} = s_x^4 \Gamma_5(s) \tag{21}$$

$$36 s_{xx}^2 + 48 s_x s_{xx} = s_x^4 \Gamma_6(s) \tag{22}$$

$$12u_{0x} s_x^2 + 24u_0 s_x s_{xx} - 12 s_{xx}^2 - 12 s_x s_{xxx} = s_x^4 \Gamma_7(s) \tag{23}$$

$$s_x s_{xxx} = s_x^4 \Gamma_8(s) \tag{24}$$

$$12(w_0 s_x)_x + 12(u_0^2 s_x)_x + s_{xt} - 24u_{0x} s_{xx} - 12u_0 s_{xxx} + 4 s_{xxxx} = s_x^4 \Gamma_9(s) \tag{25}$$

$$12u_{0x} s_{xx} + 12u_0 s_{xxx} = s_x^4 \Gamma_{10}(s) \tag{26}$$

$$s_{xxxx} = s_x^4 \Gamma_{11}(s) \tag{27}$$

$$12 p_{0x} + u_{0t} + 12u_0^2 u_{0x} + 12w_0 u_{0x} - 12u_{0x}^2 - 12u_0 u_{0xx} + 4u_{0xxx} = s_x^4 \Gamma_{12}(s) \tag{28}$$

$$s_{xx} s_x^3 = s_x^4 \Gamma_{13}(s) \tag{29}$$

$$4u_0 s_x^4 + 40 s_x^3 s_{xx} = s_x^5 \Gamma_{14}(s) \tag{30}$$

$$24u_0 s_x^4 + 28 s_x^3 s_{xx} = s_x^5 \Gamma_{15}(s) \tag{31}$$

$$24u_0 s_x^4 + 12 s_x^3 s_{xx} = s_x^5 \Gamma_{16}(s) \tag{32}$$

$$s_x^2 s_t + 12u_0^2 s_x^3 + 12 w_0 s_x^3 + 4u_{0x} s_x^3 + 24u_0 s_x^2 s_{xx} + 60 s_x s_{xx}^2 \tag{33}$$

$$s_x^2 s_{xx} = s_x^4 \Gamma_1(s) \tag{34}$$

$$-12u_0 s_x^3 + 24 s_x^2 s_{xx} = s_x^4 \Gamma_2(s) \tag{35}$$

$$24u_0 s_x^3 - 60 s_x^2 s_{xx} = s_x^4 \Gamma_3(s) \tag{36}$$

$$s_x s_t + 12u_0 s_x^2 + 12 w_0 s_x^2 + 12 s_{xx}^2 - 24u_{0x} s_x^2 - 36u_0 s_x s_{xx} + 16 s_x s_{xxx} = s_x^4 \Gamma_4(s) \tag{37}$$

$$12u_{0x} s_x^2 + 36u_0 s_x s_{xx} = s_x^4 \Gamma_5(s) \tag{38}$$

$$36 s_{xx}^2 + 48 s_x s_{xx} = s_x^4 \Gamma_6(s) \tag{39}$$

$$12u_{0x} s_x^2 + 24u_0 s_x s_{xx} - 12 s_{xx}^2 - 12 s_x s_{xxx} = s_x^4 \Gamma_7(s) \tag{40}$$

$$s_x s_{xxx} = s_x^4 \Gamma_8(s) \tag{41}$$

$$12(w_0 s_x)_x + 12(u_0^2 s_x)_x + s_{xt} - 24u_{0x} s_{xx} - 12u_0 s_{xxx} + 4 s_{xxxx} = s_x^4 \Gamma_9(s) \tag{42}$$

$$12u_{0x} s_{xx} + 12u_0 s_{xxx} = s_x^4 \Gamma_{10}(s) \tag{43}$$

$$s_{xxxx} = s_x^4 \Gamma_{11}(s) \tag{44}$$

$$12p_{0x} + u_{0t} + 12u_0^2 u_{0x} + 12w_0 u_{0x} - 12u_{0x}^2 - 12u_0 u_{0xx} + 4u_{0xxx} = s_x^4 \Gamma_{12}(s) \quad (45)$$

$$s_{xx} s_x^3 = s_x^4 \Gamma_{13}(s) \quad (46)$$

$$4u_0 s_x^4 + 40 s_x^3 s_{xx} = s_x^5 \Gamma_{14}(s) \quad (47)$$

$$24u_0 s_x^4 + 28 s_x^3 s_{xx} = s_x^5 \Gamma_{15}(s) \quad (48)$$

$$24u_0 s_x^4 + 12 s_x^3 s_{xx} = s_x^5 \Gamma_{16}(s) \quad (49)$$

$$s_x^2 s_t + 12u_0^2 s_x^3 + 12w_0 s_x^3 + 4u_{0x} s_x^3 + 24u_0 s_x^2 s_{xx} + 60 s_x s_{xx}^2 \quad (50)$$

$$+40 s_x^2 s_{xxx} s_x^5 = \Gamma_{17}(s) \quad (51)$$

$$24u_{0x} s_x^3 + 96u_0 s_x^2 s_{xx} + 24 s_x s_{xx}^2 + 16 s_x^2 s_{xxx} = s_x^5 \Gamma_{18}(s) \quad (52)$$

$$24 s_x s_{xx}^2 + 12 s_x^2 s_{xx} = s_x^5 \Gamma_{19}(s) \quad (53)$$

$$24u_0 s_x^2 s_{xx} + 4 s_x^2 s_{xxx} = s_x^5 \Gamma_{20}(s) \quad (54)$$

$$36 s_x s_{xx}^2 + 12 s_x^2 s_{xxx} = s_x^5 \Gamma_{21}(s) \quad (55)$$

$$36 s_x s_{xx}^2 + 12 s_x^2 s_{xxx} = s_x^5 \Gamma_{22}(s) \quad (56)$$

$$v_0 s_x^3 = s_x^5 \Gamma_{23}(s) \tag{57}$$

$$24u_0 v_0 s_x^2 + 4v_{0x} s_x^2 = s_x^5 \Gamma_{24}(s) \tag{58}$$

$$24u_0 u_{0x} s_x^2 + 12w_{0x} s_x^2 + 2s_x s_{xt} + 36u_0^2 s_x s_{xx} + 36w_0 s_x s_{xx} + 12u_{0x} s_x s_{xx} \tag{59}$$

$$+12u_0 s_{xx}^2 + 16u_0 s_x s_{xxx} + 40s_{xx} s_{xxx} + 20s_x s_{xxxx} = s_x^5 \Gamma_{25}(s) \tag{60}$$

$$12v_{0x} s_x^2 + 36v_0 s_x s_{xx} = s_x^5 \Gamma_{26}(s) \tag{61}$$

$$12v_{0x} s_x^2 + 24v_0 s_x s_{xx} = s_x^5 \Gamma_{27}(s) \tag{62}$$

$$24u_0 s_x s_{xx} + 24u_0 s_{xx}^2 + 24u_0 s_x s_{xxx} + 4s_{xx} s_{xxx} + 4s_x s_{xxxx} = s_x^5 \Gamma_{28}(s) \tag{63}$$

$$24s_{xx} s_{xxxx} = s_x^5 \Gamma_{29}(s) \tag{64}$$

$$24v_0 u_{0x} s_x + 24u_0 v_0 s_{xx} + 4v_{0x} s_{xx} = s_x^5 \Gamma_{30}(s) \tag{65}$$

$$12\left(u_0^2 s_{xx}\right)_x + s_{xxt} + 12\left(w_0 s_{xx}\right)_x + \left(4u_0 s_{xxx}\right)_x + 4s_{xxxxx} s_x^5 = \Gamma_{31}(s) \tag{66}$$

$$12v_{0x} s_{xx} + 12v_0 s_{xxx} = s_x^5 \Gamma_{32}(s) \tag{67}$$

$$24u_0 v_0 u_{0x} + v_{0t} + 12u_0^2 v_{0x} + 12\left(w_0 v_0\right)_x + 4\left(u_0 v_{0x}\right)_x + 4v_{0xxx} = s_x^5 \Gamma_{33}(s) \tag{68}$$

$$-s_x^2 s_y = s_x^3 \Gamma_{34}(s) \tag{69}$$

$$-\left(s_{xx}s_y + 2s_x s_{xy}\right) = s_x^3 \Gamma_{35}(s) \tag{70}$$

$$3s_x s_{xx} = s_x^3 \Gamma_{36}(s) \tag{71}$$

$$-s_{xxy} = s_x^3 \Gamma_{37}(s) \tag{72}$$

$$s_{xxx} = s_x^3 \Gamma_{38}(s) \tag{73}$$

$$v_{0x} - w_{0y} = s_x^3 \Gamma_{39}(s) \tag{74}$$

$$u_{0x}s_x^2 + 3u_0 s_x s_{xx} = s_x^4 \Gamma_{40}(s) \tag{75}$$

$$s_{xx}^2 + s_x s_{xxx} = s_x^4 \Gamma_{41}(s) \tag{76}$$

$$-\left(s_y s_{xxx} + 3s_{xx}s_{xy} + 3s_x s_{xxy}\right) = s_x^4 \Gamma_{42}(s) \tag{77}$$

$$v_0 s_x^2 = s_x^4 \Gamma_{43}(s) \tag{78}$$

$$-\left(3s_x s_y s_{xx} + 3s_x^2 s_{xy} + s_x^3 s_y\right) = s_x^4 \Gamma_{44}(s) \tag{79}$$

$$s_x^2 s_{xx} = s_x^4 \Gamma_{45}(s) \tag{80}$$

$$u_0 s_x^3 = s_x^4 \Gamma_{46}(s) \tag{81}$$

$$v_{0x}s_x + v_0 s_{xx} = s_x^4 \Gamma_{47}(s) \tag{82}$$

$$u_{0x}s_{xx} + u_0 s_{xxx} = s_x^4 \Gamma_{48}(s) \tag{83}$$

$$-s_{xxxy} = s_x^4 \Gamma_{49}(s) \tag{84}$$

$$-\left(p_{0y} - v_0 u_{0x} - u_0 v_{0x}\right) = s_x^4 \Gamma_{50}(s) \tag{85}$$

where $\Gamma_i\,(i = 1,\cdots,50)$ are some arbitrary functions of s to be determined and take the following form

$$\Gamma_1(s) = \frac{12f'^2 f'' - 12f''^2 + 12fh'' - 12ff''' + 12fh''' + 4f'''' + 12k''''}{12\left(f'^3 + h'f'' + 4fh''\right)}$$

$$\Gamma_2(s) = \frac{12f'^2 f'' - 12f''^2 + 12fh'' - 12ff''' + 12fh''' + 4f'''' + 12k''''}{f'''}$$

$$\Gamma_3(s) = \frac{12f'^2 f'' - 12f''^2 + 12fh'' - 12ff''' + 12fh''' + 4f'''' + 12k''''}{ff''}$$

$$\Gamma_4(s) = \frac{12f'^2 f'' - 12f''^2 + 12fh'' - 12ff''' + 12fh''' + 4f'''' + 12k''''}{f''}$$

$$\Gamma_5(s) = \frac{12f'^2 f'' - 12f''^2 + 12fh'' - 12ff''' + 12fh''' + 4f'''' + 12k''''}{h''}$$

$$\Gamma_6(s) = \frac{12f'^2 f'' - 12f''^2 + 12fh'' - 12ff''' + 12fh''' + 4f'''' + 12k''''}{k''}$$

$$\Gamma_7(s) = \frac{12f'^2 f'' - 12f''^2 + 12fh'' - 12ff''' + 12fh''' + 4f'''' + 12k''''}{f'^2}$$

$$\Gamma_8(s)=\frac{12f'^2f''-12f''^2+12f''h''-12ff'''+12f'h'''+4f''''+12k''''}{12f'h'}$$

$$\Gamma_9(s)=\frac{12f'^2f''-12f''^2+12f''h''-12ff'''+12f'h'''+4f''''+12k''''}{f'}$$

$$\Gamma_{10}(s)=\frac{12f'^2f''-12f''^2+12f''h''-12ff'''+12f'h'''+4f''''+12k''''}{h'}$$

$$\Gamma_{11}(s)=\frac{12f'^2f''-12f''^2+12f''h''-12ff'''+12f'h'''+4f''''+12k''''}{k'}$$

$$\Gamma_{12}(s)=12f'^2f''-12f''^2+12f''h''-12ff'''+12f'h'''+4f''''+12k''''$$

$$\Gamma_{13}(s)=\frac{24ff''g''+12f'2g'''+4f''g'''+12h''g'''+12g''h''+4f'g''''+4g''''}{24f'g'f''+60f'^2g''+72g''h''+12h'g'''+12g'h'''}$$

$$\Gamma_{14}(s)=\frac{24ff''g''+12f'2g'''+4f''g'''+12h''g'''+12g''h''+4f'g''''+4g''''}{g''''}$$

$$\Gamma_{15}(s)=\frac{24ff''g''+12f'2g'''+4f''g'''+12h''g'''+12g''h''+4f'g''''+4g''''}{f'g'''}$$

$$\Gamma_{16}(s)=\frac{24ff''g''+12f'2g'''+4f''g'''+12h''g'''+12g''h''+4f'g''''+4g''''}{f''g''}$$

$$\Gamma_{17}(s)=\frac{24ff''g''+12f'2g'''+4f''g'''+12h''g'''+12g''h''+4f'g''''+4g''''}{g'''}$$

$$\Gamma_{18}(s)=\frac{24ff''g''+12f'2g'''+4f''g'''+12h''g'''+12g''h''+4f'g''''+4g''''}{f'g''}$$

$$\Gamma_{19}(s)=\frac{24ff''g''+12f'2g'''+4f''g'''+12h''g'''+12g''h''+4f'g''''+4g''''}{f'^2g'}$$

$$\Gamma_{20}(s)=\frac{24ff'f''g''+12f'2g'''+4f''g'''+12h''g'''+12g''h''+4f'g''''+4g''''}{f''g'}$$

$$\Gamma_{21}(s)=\frac{24ff'f''g''+12f'2g'''+4f''g'''+12h''g'''+12g''h''+4f'g''''+4g''''}{g''h'}$$

$$\Gamma_{22}(s)=\frac{24ff'f''g''+12f'2g'''+4f''g'''+12h''g'''+12g''h''+4f'g''''+4g''''}{g'h''}$$

$$\Gamma_{23}(s)=\frac{24ff'f''g''+12f'2g'''+4f''g'''+12h''g'''+12g''h''+4f'g''''+4g''''}{\left(24ff''+12h'''\right)}$$

$$\Gamma_{24}(s)=\frac{24ff'f''g''+12f'2g'''+4f''g'''+12h''g'''+12g''h''+4f'g''''+4g''''}{f''}$$

$$\Gamma_{25}(s)=\frac{24ff'f''g''+12f'2g'''+4f''g'''+12h''g'''+12g''h''+4f'g''''+4g''''}{g''}$$

$$\Gamma_{26}(s)=\frac{24ff'f''g''+12f'2g'''+4f''g'''+12h''g'''+12g''h''+4f'g''''+4g''''}{h''}$$

$$\Gamma_{27}(s)=\frac{24ff'f''g''+12f'2g'''+4f''g'''+12h''g'''+12g''h''+4f'g''''+4g''''}{f'^{2}}$$

$$\Gamma_{28}(s)=\frac{24ff'f''g''+12f'2g'''+4f''g'''+12h''g'''+12g''h''+4f'g''''+4g''''}{f'g'}$$

$$\Gamma_{29}(s)=\frac{24ff'f''g''+12f'2g'''+4f''g'''+12h''g'''+12g''h''+4f'g''''+4g''''}{g'h'}$$

$$\Gamma_{30}(s)=\frac{24ff'f''g''+12f'2g'''+4f''g'''+12h''g'''+12g''h''+4f'g''''+4g''''}{f'}$$

$$\Gamma_{31}(s)=\frac{24ff'f''g''+12f'2g'''+4f''g'''+12h''g'''+12g''h''+4f'g''''+4g''''}{g'}$$

$$\Gamma_{32}(s)=\frac{24ff''g''+12f'2g'''+4f''g'''+12h''g'''+12g''h''+4f'g''''+4g''''}{h'}$$

$$\Gamma_{33}(s)=24ff''g''+12f'^{2}g'''+4f''g'''+12h''g'''+12g''h''+4f'g''''+4g''''$$

$$\Gamma_{39}(s)=g''' \quad \Gamma_{38}(s)=\frac{g'''}{g'} \quad \Gamma_{37}(s)=\frac{g'''}{h'} \quad \Gamma_{36}(s)=\frac{g'''}{g''} \quad \Gamma_{35}(s)=\frac{g'''}{h''}$$

$$\Gamma_{34}(s)=\frac{g'''}{h'''}$$

$$\Gamma_{43}(s)=\frac{f''g''+f'g'''}{f''} \quad \Gamma_{42}(s)=\frac{f''g''+f'g'''}{k''} \quad \Gamma_{41}(s)=\frac{f''g''+f'g'''}{f'g'}$$

$$\Gamma_{40}(s)=\frac{f''g''+f'g'''}{g''}$$

$$\Gamma_{47}(s)=\frac{f''g''+f'g'''}{f'} \quad \Gamma_{46}(s)=\frac{f''g''+f'g'''}{g'''} \quad \Gamma_{45}(s)=\frac{f''g''+f'g'''}{g'f''+4f'g''}$$

$$\Gamma_{44}(s)=\frac{f''g''+f'g'''}{k'''}$$

$$\Gamma_{50}(s)=f''g''+f'g''' \quad \Gamma_{49}(s)=\frac{f''g''+f'g'''}{k'} \quad \Gamma_{48}(s)=\frac{f''g''+f'g'''}{g'} \tag{86}$$

There are freedoms in the determination of u_0, v_0, w_0, p_0, s which can exploit the following rules, without loss of generality:

If u_0 has the form $u_0=u'(x,y,t)+\frac{\partial}{\partial x}\Omega$ then we can assume that $\Omega=0$

$\left(\text{make the transformation } f(s)\rightarrow f(s)-\Omega\right)$,

If v_0 has the form $v_0=v'(x,y,t)+\frac{\partial^2}{\partial x^2}\Omega$ then we can assume that $\Omega=0$

$\left(\text{make the transformation } f(s)\rightarrow f(s)-\Omega\right)$,

If w_0 has the form $w_0 = w'(x,y,t)+\frac{\partial}{\partial x}\Omega$ then we can assume that $\Omega=0$ (make the transformation $f(s)\to f(s)-\Omega$),

If p_0 has the form $p_0 = p'(x,y,t)+\frac{\partial^3}{\partial x^3}\Omega$ then we can assume that $\Omega=0$ (make the transformation $f(s)\to f(s)-\Omega$).

If $s(x,y,t)$ is defined by an equation of the form $\Omega(s)=s_0(x,y,t)$, we can also assume that $\Omega=s$ (make the transformation $s\to\Omega^{-1}(s)$) .

From Equation (17), we get

$$\frac{s_{xx}}{s_x}=s_x\Gamma_1(s) \tag{87}$$

integrating Equation (87) with respect to x, we get

$$Lns_x+Ln\Gamma(s)=Ln\theta(y,t), \tag{88}$$

$$s_x\Gamma(s)=\theta(y,t). \tag{89}$$

integrating Equation (89) with respect to x, we obtain

$$\gamma(s)=\theta(y,t)x+\sigma(y,t). \tag{90}$$

By using the rule (e) into Equation (90), we obtain a function $s(x,y,t)$ in the form

$$s(x,y,t)=\theta(y,t)x+\sigma(y,t) \tag{91}$$

substituting from Equation (90) into Equations (17), (20), (24), (25), (26), (53), (54), (55), (56), (63), (64), (71), (72), (73), (76), (77), (80), (83)and (84) we obtain

$$\Gamma_1(s)=\Gamma_6(s)=\Gamma_8(s)=\Gamma_{10}(s)=\Gamma_{11}(s)=\Gamma_{19}(s)=\Gamma_{20}(s)=\Gamma_{21}(s)=\Gamma_{22}(s)$$
$$\Gamma_{28}(s)=\Gamma_{29}(s)=\Gamma_{31}(s)=\Gamma_{32}(s)=\Gamma_{36}(s)=\Gamma_{37}(s)=\Gamma_{38}(s)=\Gamma_{41}(s)=\Gamma_{42}(s)$$
$$\Gamma_{45}(s)=\Gamma_{48}(s)=\Gamma_{49}(s)=0, \tag{92}$$

where $s_{xx}=0$ which clear in equation (91) .

By using Equation (91) into Equation (18), we get

$$u_0=-\frac{1}{12}\theta\Gamma_2(s)=-\frac{1}{12}\frac{\partial}{\partial x}\left[\gamma_2(s)\right] \tag{93}$$

where $\Gamma_2(s)=\frac{\mathrm{d}}{\mathrm{d}s}\gamma_2(s),\ \theta=\frac{\partial s}{\partial x}$.

By apply the rule (a) on Equation (93), we obtain

$$u_0=0.\ \gamma_2(s)=0\Rightarrow\Gamma_2(s)=0, \tag{94}$$

substituting from Equations (91), (94) into Equation (20), we obtain

$$\Gamma_3(s)=0. \tag{95}$$

using Equations (91)and (94) into Equation (19), we get

$$w_0=\frac{1}{12}\theta^2\Gamma_4(s)-\frac{\theta_t}{12\theta}x-\frac{\sigma_t}{12\theta}=\frac{1}{12}\frac{\partial^2}{\partial x^2}\left[\gamma_4(s)\right]-\frac{\theta_t}{12\theta}x-\frac{\sigma_t}{12\theta}, \tag{96}$$

where $\Gamma_4(s)=\frac{\mathrm{d}^2}{\mathrm{d}s^2}\gamma_2(s),\ \theta^2=\frac{\partial^2 s}{\partial x^2}$.

By using the rule (c) into the above equation (96) to become in the form

$$\gamma_4(s)=0\Rightarrow\Gamma_4(s)=0,$$

then Equation (96) take the form

$$w_0 = -\frac{1}{12}\left[\theta_t x + \sigma_t\right]. \tag{97}$$

substituting from Eqs.(91), (94), (97) into Equations (20), (23), (25), (29), (30), (31), (32), (33)and (34), we obtain

$$\Gamma_5(s) = \Gamma_7(s) = \Gamma_9(s) = \Gamma_{13}(s) = \Gamma_{14}(s) = \Gamma_{15}(s) = \Gamma_{16}(s) = \Gamma_{17}(s) = \Gamma_{18}(s) = 0 \tag{98}$$

from Equation (91) into Equation (57), we get

$$v_0 = \theta^2 \Gamma_{23}(s) = \frac{\partial^2}{\partial x^2}\left[\gamma_{23}(s)\right] \tag{99}$$

where $\Gamma_{23}(s) = \frac{d^2}{ds^2}\gamma_{23}(s),\ \theta^2 = \frac{\partial^2 s}{\partial x^2}.$

By using the rule (b) into the above equation (99)

$$v_0 = 0.\ \gamma_{23}(s) = 0 \Rightarrow \Gamma_{23}(s) = 0, \tag{100}$$

substituting from Equations.(91), (94) and (97), into Equations.(61), (62), (65), (68), (75), (78), (81)and (82) then,we obtain

$$\Gamma_{26}(s) = \Gamma_{27}(s) = \Gamma_{30}(s) = \Gamma_{33}(s) = \Gamma_{40}(s) = \Gamma_{43}(s) = \Gamma_{46}(s) = \Gamma_{47}(s) = 0. \tag{101}$$

Substituting from Equations (91), (94), (97) and (101) into Equations (60), (69), (70), we obtain

$$\Gamma_{25}(s) = \frac{A}{\theta^2} = C, \tag{102}$$

$$\Gamma_{34}(s) = -(Ds + F), \tag{103}$$

$$\Gamma_{35}(s) = -2D. \tag{104}$$

where $\theta = \theta(y,t), \sigma = \sigma(y,t)$.

and

$$A = \frac{\theta_t}{\theta^2}, C = \frac{A}{\theta^2}, \tag{105}$$

$$D = \frac{\theta_y}{\theta^2}, F = \frac{\sigma_y}{\theta} - D\sigma, \tag{106}$$

$$B = \frac{\sigma_t}{\theta} - A\sigma. \tag{107}$$

Using this notation, the equation (97) take the following form

$$w_0 = -\frac{1}{12}(As + B), \qquad ((108))$$

substituting Equations (97) and (100) into Equation (74), we obtain

$$\Gamma_{39}(s) = \frac{1}{12}C(Ds + F), \tag{109}$$

using any equation which we need into Equation (28), we obtain

$$12p_{0x} = \theta^4\Gamma_{12}(s) = \frac{\partial^4}{\partial x^4}\gamma_{12}(s) \tag{110}$$

where $\Gamma_{12}(s) = \frac{d^4}{ds^4}\gamma_{12}(s)$, $\theta^4 = \frac{\partial^4 s}{\partial x^4}$.

By using the rule (d) after differential with respect to x into Equations (110), we obtain

$$p_0 = 0.\ \gamma_{12}(s) = 0 \Rightarrow \Gamma_{12}(s) = 0, \tag{111}$$

Substituting into Equations (9)-(12), we obtain the similarity solutions of the Broer-Kaup system Equations (1)-(4) in the form

$$p(x,y,t) = \theta^3 H(s).\ w(x,y,t) = \theta^2 R(s) - \frac{1}{12}(As+B),\ v(x,y,t) = \theta^2 Q(s),$$
$$u(x,y,t) = \theta P(s), \tag{112}$$

where

$$P(s) = f', Q(s) = g'', R(s) = h'' \text{ and } H(s) = k'''. \tag{113}$$

with $s(x,y,t) = \theta(y,t)x + \sigma(y,t)$.

Substituting from Equation (112) to obtain an ordinary differential equations from the origin system (1)-(4), we get

$$P''' - 3PP'' + 3PR' - 3P'^2 + 3RP' + 3P^2P' + 3H' = 0, \tag{114}$$

$$4Q''' + 24QPP' + 12P^2Q' + 12RQ' + 4P'Q' + 12QR' + 4PQ'' - CQ = 0, \tag{115}$$

$$-12Q' - ADs + 12R'Ds + 12FR' + 24DR - FC = 0, \tag{116}$$

$$(PQ)' - 3DH - FH' - DsH' = 0. \tag{117}$$

where $'' ' '' = \frac{d}{ds}.$

The general solution for the variable $\theta(y,t), \sigma(y,t)$ which satisfy Equations (105)-(107) are

$$\sigma(y,t)=\frac{1}{A}\left[\frac{(BD-AF)y+c_2}{At+Dy+c_1}-B\right]\theta(y,t)=-\frac{1}{At+Dy+c_1}, \tag{118}$$

where c_1, c_2 are arbitrary constants.

There some subcases for the constants A, B, D, F

$D=F=0, A\neq 0,$ the solutions of Equations (105)-(107) are

$$\sigma(y,t)=\sigma(t)=\frac{1}{A}\left[\frac{c_5}{At+c_4}-B\right]\theta(y,t)=\theta(t)=-\frac{1}{At+c_4}, \tag{119}$$

where c_4, c_5 are arbitrary constants. In this case the equations (114)-(117) take the form

$$P'''-3PP''+3PR'-3P'^2+3RP'+3P^2P'+3H'=0, \tag{120}$$

$$4Q'''+24QPP'+12P^2Q'+12(RQ)'+4P'Q'+4PQ''-CQ=0, \tag{121}$$

$$Q'=0, \tag{122}$$

$$(PQ)'=0. \tag{123}$$

the solutions for equations (120)-(123) are

$$Q(s)=c_7, \quad P(s)=\frac{c_6}{c_7}, \tag{124}$$

$$H(s)=-\frac{c_6c}{c_7}s^2-\frac{c_6c_8}{c_7}s+c_9. \quad R(s)=\frac{1}{12}cs+c_8,$$

To obtain the solutions for the original system (1)-(4), we substituting from the equations (119), (124) into Equations (112), we get

$$u(x,y,t)=-\frac{c_6}{c_7(At+c_4)},\ v(x,y,t)=-\frac{c_7}{(At+c_4)^2},$$

$$w(x,y,t)=-\frac{cs+12c_8}{12(At+c_4)^2}-\frac{1}{12}(As+B),\ p(x,y,t)=\frac{(c_6cs^2+c_6c_8s-c_7c_9)}{c_7(At+c_4)^3}. \tag{125}$$

REFERENCES

1. H. A. Zedan, "Exact Solutions for the Generalized KdV Equation by Using Backlund Transformations," Journal of the Franklin Institute, Vol. 348, No. 8, 2011, pp. 1751-1768.
2. D.-S. Li, et al., "Solving the (2 + 1)-Dimensional Higher Order Broer-Kaup System via a Transformation and Tanh-Function mEthod," Chaos, Solitons & Fractals, Vol. 20, No. 5, 2004, pp. 1021-1025.
3. J. Mei, et al., "New Soliton-Like and Periodic Solution of (2 + 1)-Dimensional Higher Order Broer-Kaup System," Chaos, Solitons & Fractals, Vol. 22, No. 3, 2004, pp. 669-674.
4. D.-S. Li, et al., "Some New Types of Multisoliton Solutions for the (2 + 1)-Dimensional Higher-Order Broer-Kaup System," Applied Mathematics and Computation, Vol. 152, No. 3, 2004, pp. 847-853.
5. E. G. Fan and H. Q. Zhang, "A Note on the Homogeneous Balance Method," Physics Letters A, Vol. 246, No. 5, 1998, pp. 403-406.
6. E. G. Fan, "Two New Applications of the Homogeneous Balance Method," Physics Letters A, Vol. 265, No. 5-6, 2000, pp. 353-357.
7. G. Bluman, "Symmetries and Differential Equations," Springer-Verlag, New York, 1989.
8. P. J. Olver, "Applications of Lie Group to Differential Equation," Springer-Verlag, New York, 1986.
9. P. A. Clarkson, "New Similarity Solutions for the Modified Boussinesq Equation," Journal of Physics A: Mathematical and General, Vol. 22, No. 13, 1989, pp. 2355-2365.
10. Y. Yu, Q. Wang and H. Q. Zhang, "The Extended Jacobi Elliptic Function Method to Solve a Generalized Hirota-Satsuma Coupled KdV Equations," Chaos, Solitons & Fractals, Vol. 26, No. 5, 2005, pp. 1415-1421.
11. J. L. Zhang, M. L. Wang, Y. M. Wang, Z. D. Fang, "The Improved F-Expansion Method and Its Applications," Physics Letters A, Vol. 350, No. 1-2, 2006, pp. 103-109.
12. E. M. E. Zayed and H. Zedan, "On the Solitary Wave Solutions for Nonlinear Hirota-Satsuma Coupled KdV of Equations," Chaos, Solitons & Fractals, Vol. 22, No. 2, 2004, pp. 285-303.
13. M. Wadati, "Introduction to Solitons," Pramana: Journal of Physics, Vol. 57, No. 5-6, 2001, pp. 841-847.
14. Z. Chen, D. H. Zhao and J. Ruan, "Dynamic Analysis of High-Order Cohen-Grossberg Neural Networks with Time Delay," Chaos, Solitons & Fractals, Vol. 32, No. 4, 2007, pp. 1538-1546.
15. Hassan A. Zedan, "Solution of (3 + 1) - Dimensional Nonlinear Cubic Schrodinger Equation by Differential Transform Method," Mathematical Problems in Engineering, Vol. 2012, 2012, 14 p.

CHAPTER 2

Hausdorff Dimension of Multi-Layer Neural Networks

Jung-Chao Ban[1*], Chih-Hung Chang[2#]

[1]*Department of Applied Mathematics, National Dong Hwa University, Hualien, Taiwan*

[2]*Department of Applied Mathematics, Feng Chia University, Taichung, Taiwan*

ABSTRACT

This elucidation investigates the Hausdorff dimension of the output space of multi-layer neural networks. When the factor map from the covering space of the output space to the output space has a synchronizing word, the Hausdorff dimension of the output space relates to its topological entropy. This clarifies the geometrical structure of the output space in more details.

Keywords: Multi-layer neural networks; Hausdorff dimension; sofic shift; output space

1. INTRODUCTION

The multi-layer neural networks (MNN, [1,2]) have received considerable attention and were successfully applied to many areas such as signal processing, pattern recognition ([3,4]) and combinatorial optimization ([5,6]) in the past few decades. The investigation of mosaic solution is the most essential in MNN models due to the learning algorithm and training processing. In [7-9], the authors proved that the output solutions space $\mathbf{Y}^{(2)}$ of a 2-layer MNN forms a so-called sofic shift space, which is a factor of a classical subshift of finite type. Thus, MNN model indeed produces abundant output patterns and makes learning algorithm more efficient. A useful quantity to classify the output solution space is the topological entropy $h(\mathbf{Y}^{(2)})$ ([10]). We call the output solution space pattern formation if $h(\mathbf{Y}^{(2)})=0$, and call it spatial chaos if $h(\mathbf{Y}^{(2)})>0$. The $h(\mathbf{Y}^{(2)})=0$ indicates that the output patterns grow subexponentially and exponentially for $h(\mathbf{Y}^{(2)})=0$. For positive entropy systems, the explicit value of $h(\mathbf{Y}^{(2)})$ presents how chaotic the system is. In [7], Ban and Chang provided a method to compute explicit values of $h(\mathbf{Y}^{(2)})$ for a 2-layer MNN. The method is quite general and it makes the computation of $h(\mathbf{Y}^{(N)})$ possible for arbitrary $2 < N \in \mathbb{N}$, i.e., N-layer MNN.

From the dynamical system (DS) point of view, the topological entropy reveals the complexity of the global patterns. However, it provides less information of the inner structure of a given DS, e.g., self-similarity or recurrent properties. The possible quantity reveals that such properties are the Hausdorff dimension (HD, [11]) since the Hausdorff dimension is an indicator of the geometrical structure. For most DS, the computation of Hausdorff dimension is not an easy task, and the box dimension (BD) is usually computed first to give the upper bound for HD. Due to the relationship of topological entropy and BD ([12]) of a symbolic DS[1], the previous work ([7]) for topological entropy gives the upper bound for HD of N-layer MNN. Nature question arises: Given a MNN, how to compute the explicit value for HD? The aim of this paper is to establish the HD formula for N-layer MNNs. Using the tool of symbolic DS, the HD formula will be established for N-layer MNNs which possesses a synchronizing word (Theorem 2.4). The result leads us to exploit the inner structure for a N-layer MNN. We believe that further interesting applications of the results presented (or of the generalizations) can be obtained.

This paper is organized as follows. Section 2 contains a brief disscussion for the computation of topological entropy in [7]. The main result is stated and proved therein. Section 3 presents an MNN model for which we can compute its HD.

2. PRELIMINARIES AND MAIN RESULTS

A one-dimensional multi-layer neural network (MNN) is realized as

$$\begin{cases} \frac{\mathrm{d}}{\mathrm{d}t} x_i^{(k)}(t) = -x_i^{(k)}(t) + z^{(k)} \\ \qquad + a^{(k)} f\left(x_i^{(k)}(t)\right) + \sum_{\ell \in \mathrm{N}} b_\ell^{(k)} f\left(x_{i+\ell}^{(k-1)}(t)\right), \\ \frac{\mathrm{d}}{\mathrm{d}t} x_i^{(1)}(t) = -x_i^{(1)}(t) + z^{(1)} \\ \qquad + a^{(1)} f\left(x_i^{(1)}(t)\right) + \sum_{\ell \in \mathrm{N}} a_\ell^{(1)} f\left(x_{i+\ell}^{(1)}(t)\right), \end{cases} \tag{1}$$

for some $N \in \square$, $k = 2,\cdots,N$ and $i \in \mathrm{Z}$. The finite subset $\mathrm{N} \subset \mathrm{Z}$ indicates the neighborhood, and the piecewise linear map

$$f(x) = \frac{1}{2}\left(|x+1| - |x-1|\right)$$

is called the output function. The template $\mathrm{T} = [\boldsymbol{A}, \boldsymbol{B}, \boldsymbol{z}]$ is composed of feedback template $\boldsymbol{A} = (A_1, A_2)$ with

$A_2 = \left(a_\ell^{(1)}\right)_{\ell \in \mathrm{N}}$ $A_1 = \left(a^{(1)},\cdots,a^{(N)}\right)$, controlling template

$\boldsymbol{B} = (B_2,\cdots,B_N)$ and threshold

$\boldsymbol{z} = \left(z^{(1)},\cdots,z^{(N)}\right)$ where

$B_k = \left(b_\ell^{(k)}\right)_{\ell \in \mathrm{N}}$ for $k \geq 2$.

A solution

$$\boldsymbol{x}(t) = \left(x_i^{(1)}(t),\cdots,x_i^{(N)}(t)\right)_{i \in \mathrm{Z}} \in \mathrm{R}^{\mathrm{Z}_{\infty} \times N}$$

of (1) is called mosaic if

$$\left|x_i^{(k)}(t)\right| > 1$$

for $1 \leq k \leq N, i \in \mathrm{Z}$, and $t \geq T_0$ for some T_0.

$$y = \left(y_i^{(1)} \cdots y_i^{(N)}\right)_{i \in Z} \in \{-1,1\}^{Z_\infty \times N}$$

of a mosaic solution is called a mosaic pattern, where

$$y_i^{(k)} = f\left(x_i^{(k)}\right).$$

The solution space $\mathbf{Y}$ of (1) stores the patterns y, and the output space $\mathbf{Y}^{(N)}$ of (1) is the collection of the output patterns; more precisely,

$$\mathbf{Y}^{(N)} = \left\{\left(y_i^{(N)}\right)_{i \in Z} : \left(y_i^{(1)} \cdots y_i^{(N)}\right)_{i \in Z} \in \mathbf{Y}\right\}.$$

A neighborhood N is called the nearest neighborhood if $\mathrm{N} = \{-1,1\}$. In [7], the authors showed that 2-layer MNNs with nearest neighborhood are essential for the investigation of MNNs. In the rest of this manuscript, we refer MNNs to 2-layer MNNs with nearest neighborhood unless otherwise stated.

2.1. Topological Entropy and Hausdorff Dimension

Since the neighborhood N is finite and is invariant for each i, the output space is determined by the so-called basic set of admissible local patterns. Replace the pattern -1 and 1 by - and +, respectively; the basic set of admissible local patterns of the first and second layer is a subset of

$$\{---,--+,-+-,-++,+--,+-+,++-,+++\}; \tag{2}$$

and $\{p_1,\cdots,p_8\}$, respectively, where $\{p_1,\cdots,p_8\}$ denotes

$$\left\{\begin{smallmatrix} & - & \\ - & & - \end{smallmatrix}, \begin{smallmatrix} & - & \\ - & & + \end{smallmatrix}, \begin{smallmatrix} & - & \\ + & & - \end{smallmatrix}, \begin{smallmatrix} & - & \\ + & & + \end{smallmatrix}, \begin{smallmatrix} & + & \\ - & & - \end{smallmatrix}, \begin{smallmatrix} & + & \\ - & & + \end{smallmatrix}, \begin{smallmatrix} & + & \\ + & & - \end{smallmatrix}, \begin{smallmatrix} & + & \\ + & & + \end{smallmatrix}\right\}. \tag{3}$$

To ease the notation, we denote

$$\begin{matrix} & \alpha_1 & \\ \alpha_2 & & \alpha_3 \end{matrix}$$

by $\alpha_1 \Diamond \alpha_2 \alpha_3$. Given a template T, the basic set of admissible local pattern

$$\mathrm{B}(\mathrm{T}) = \left(\mathrm{B}^{(1)}, \mathrm{B}^{(2)}\right)$$

is determined, where $\mathrm{B}^{(1)}$ and $\mathrm{B}^{(2)}$ are the basic set of admissible local patterns of the first and second layer, respectively. Let

$$\mathrm{P}_8 = \left\{\left(a^{(2)}, b_{-1}^{(2)}, b_1^{(2)}, z^{(2)}, a^{(1)}, a_{-1}^{(1)}, a_1^{(1)}, z^{(1)}\right)\right\}$$

denote the parameter space of (1). Theorem 2.1 asserts that P_8 can be partitioned into finitely many subregions so that two templates in the same partition exhibit the same basic set of admissible local patterns.

THEOREM 2.1. (See [7]) There is a positive integer K and unique set of open subregions $\{P_k\}_{k=1}^{K}$ satisfying [(i)]

1) $\mathrm{P}_8 = \bigcup_{k=1}^{K} \overline{P}_k$.

2) $P_i \cap P_j = \varnothing$ if $i \neq j$.

3) Templates $\mathrm{T}, \mathrm{T}' \in P_k$ for some K if and only if $\mathrm{B}(\mathrm{T}) = \mathrm{B}(\mathrm{T}')$.

Since the template of MNNs is spatially invariant, the so-called transition matrix is used to investigate the complexity of MNNs. The transition matrix $\boldsymbol{T}$ is defined by

$$\boldsymbol{T}(i,j) = \begin{cases} 1, & p_i, p_j \in \mathrm{B}^{(2)} \quad \text{and} \quad \alpha_{i-1}\alpha_{j-1}\alpha_{i+1}, \alpha_{j-1}\alpha_{i+1}\alpha_{j+1} \in \mathrm{B}^{(1)}; \\ 0, & \text{otherwise}; \end{cases} \tag{4}$$

herein p_k is presented as $\alpha_k \Diamond \alpha_{k-1}\alpha_{k+1}$ for $k = 1, \cdots, 8$. Furthermore, the transition matrix of the second layer

$$T_2 \in \mathrm{M}_{8\times 8}\left(\{0,1\}\right)$$

is defined by

$$T_2(i,j)=1 \quad \text{if and only if } p_i, p_j \in \mathrm{B}^{(2)}; \tag{5}$$

the transition matrix of the first layer

$$T_1 \in \mathrm{M}_{4\times 4}(\{0,1\})$$

is defined by

$$T_1(i,j)=1 \quad \text{if and only if } \mathrm{X}(i,j)\in \mathrm{B}^{(1)}, \tag{6}$$

where

$$\mathbb{X} = \begin{array}{c} \\ \boxed{--} \\ \boxed{-+} \\ \boxed{+-} \\ \boxed{++} \end{array} \begin{array}{cccc} \boxed{--} & \boxed{-+} & \boxed{+-} & \boxed{++} \\ \end{array} \! \begin{pmatrix} \boxed{---} & \boxed{--+} & \varnothing & \varnothing \\ \varnothing & \varnothing & \boxed{-+-} & \boxed{-++} \\ \boxed{+--} & \boxed{+-+} & \varnothing & \varnothing \\ \varnothing & \varnothing & \boxed{++-} & \boxed{+++} \end{pmatrix}.$$

Write

$$T_1^2 = \left(T_{i,j}\right)_{i,j=1}^{2}$$

as four smaller 2×2 matrices. Define $\overline{T}_1$ by

$$\overline{T}_1(p,q)=T_{i,j}(k,l), \quad \text{where} \quad p=2i+j-2, q=2k+l-2. \tag{7}$$

Ban and Chang [7] decomposed $\boldsymbol{T}$ as the product of T_1 and T_2.

THEOREM 2.2. (See [7]) Suppose $\boldsymbol{T}$ is the transition matrix of (1), and T_1 and T_2 are the transition matrices of the first and second layer, respectively. Let $\overline{T}_1$ be defined as in (7). Then

$$\boldsymbol{T} = T_2 \circ \left(E_2 \otimes \overline{T}_1 \right), \tag{8}$$

where E_k is a $k \times k$ matrix with all entries being 1's; $\circ$ and $\otimes$ are the Hadamard and Kronecker product, respectively.

As being demonstrated in [7-9,13], the solution space $\mathbf{Y}$ is a so-called shift of finite type (SFT, also known as a topological Markov shift) and the output space $\mathbf{Y}^{(2)}$ is a sofic shift. More specifically, a SFT can be represented as a directed graph $G = (\mathrm{V}, \mathrm{E})$ and a sofic shift can be represented as a labeled graph $\mathrm{G} = (G, \mathrm{L})$ for some labeling $\mathrm{L} : \mathrm{E} \to \mathrm{A}$ and finite alphabet A. A labeled graph $\mathrm{G} = (G, \mathrm{L})$ is called right-resolving if the restriction of L to E_I is one-to-one for all $I \in \mathrm{V}$, where E_I consists of those edges starting from I. If G is not right-solving, there exists a labeled graph H, derived by applying subset construction method (SCM) to G, such that the sofic shift represented by H is identical to the original space. A detailed instruction is referred to [14].

One of the most frequently used quantum for the measure of the spatial complexity is the topological entropy. Let X be a symbolic space and let $\Sigma_n(X)$ denote the collection of the patterns of length n in X. The topological entropy of X is defined by

$$h(X) = \lim_{n \to \infty} \frac{\log \Gamma_n(X)}{n}, \quad \text{provided the limit exists.}$$

Herein

$$\Gamma_n(X) = \left| \Sigma_n(X) \right|$$

indicates the cardinality of $\Sigma_n(X)$.

THEOREM 2.3. (See [7,9,13]) Let G be the labeled graph obtained from the transition matrix $\boldsymbol{T}$ of (1). The topological entropies of $\mathbf{Y}$ and $\mathbf{Y}^{(2)}$ are $h(\mathbf{Y}) = \log \rho_T$ and

$$h\left(\mathbf{Y}^{(2)}\right) = \begin{cases} \log \rho_T, & \text{if G is right-resolving;} \\ \log \rho_H, & \text{otherwise;} \end{cases} \tag{9}$$

respectively, where H is the transition matrix of the labeled graph $\mathcal{H}$ which is obtained by applying SCM to $\mathcal{G}$.

Aside from the topological entropy, the Hausdorff dimension characterizes its geometrical structure. The concept of the Hausdorff dimension generalizes the notion of the dimension of a real vector space and helps to distinguish the difference of measure zero sets. Let $\mathcal{A}$ be a finite set with cardinality $|\mathcal{A}| = n$, which we consider to be an alphabet of symbols. Without the loss of generality, we usually take

$$\mathcal{A} = \{0,1,\cdots,n-1\}.$$

The full $\mathcal{A}$-shift $\mathcal{A}^{\mathbb{Z}}$ is the collection of all biinfinite sequences with entries from $\mathcal{A}$. More precisely,

$$\mathcal{A}^{\mathbb{Z}} = \{\alpha = (\alpha_i)_{i\in\mathbb{Z}} : \alpha_i \in \mathcal{A} \text{ for all } i \in \mathbb{Z}\}.$$

It is well-known that $\mathcal{A}^{\mathbb{Z}}$ is a compact metric space endowed with the metric

$$d(x,y) = \sum_{i\in\mathbb{Z}} \frac{|x_i - y_i|}{n^{|i|+1}}, \quad x,y \in \mathcal{A}^{\mathbb{Z}}.$$

Suppose X is a subspace of $\mathcal{A}^{\mathbb{Z}}$. Set

$$X^+ = \{(x_i)_{i\ge 0} : (x_i)_{i\in\mathbb{Z}} \in X\}$$
$$\text{and } X^- = \{(x_i)_{i\le 0} : (x_i)_{i\in\mathbb{Z}} \in X\}.$$

It follows that X^+ and X^- can be embedded in the close interval $[0,1]$ separately. Moreover, $\mathcal{A}^{\mathbb{Z}+}$ and $\mathcal{A}^{\mathbb{Z}-}$ can be mapped onto the close interval $[0,1]$, and X is identified with the direct product $X^+ \times X^-$. This makes the elucidation of the Hausdorff dimension of the output space $\mathbf{Y}^{(2)}$ comprehensible. (Recall that the alphabet $\mathcal{A}$ of $\mathbf{Y}^{(2)}$ is $\mathcal{A} = \{-,+\} := \{0,1\}$).

2.2. Main Result

Suppose X,Y are shift spaces and $\phi : X \to Y$ is a factor map. We say that ϕ has a synchronizing word if there is a finite word $y_1 y_2 \cdots y_n \in \Sigma_n(Y)$ such that each element in $\phi^{-1}(y_1 y_2 \cdots y_n)$ admits the same terminal entry. More precisely, for any

$$x_1 x_2 \cdots x_m, x_1' x_2' \cdots x_m' \in \Sigma_m(X)$$

satisfying

$$\phi(x_1 x_2 \cdots x_m) = \phi(x_1' x_2' \cdots x_m') = y_1 y_2 \cdots y_n$$

we have $x_m = x_m'$.

Suppose $\mathrm{G} = (G, \mathrm{L})$ is a labeled graph representation of the output space $\mathbf{Y}^{(2)}$ of (1). Denote by W the SFT represented by the graph G if G is right-resolving; otherwise, denote by W the SFT represented by the graph H, where $\mathrm{H} = (H, \mathrm{L}')$ is obtained by applying SCM to G. It follows that W is a covering space of $\mathbf{Y}^{(2)}$ and there is a factor map $\phi : W \to \mathbf{Y}^{(2)}$ which is represented by the labeling L (or L'). Theorem 2.4 asserts that the Hausdorff dimension of the output space $\dim \mathbf{Y}^{(2)}$ relates to the topological entropy of its covering space $h(W)$ if ϕ has a synchronizing word.

THEOREM 2.4. Along with the same assumption of Theorem 2.3. Let W, which is represented by G if G is right-resolving and is represented by H otherwise, be the covering space of $\mathbf{Y}^{(2)}$. Suppose the factor map $\phi : W \to \mathbf{Y}^{(2)}$, which is represented by the labeling L (or L'), has a synchronizing word. Then

$$\dim \mathbf{Y}^{(2)} = \begin{cases} \dfrac{\log \rho_T^2}{\log 2}, & \text{if G is right-resolving;} \\[2ex] \dfrac{\log \rho_H^2}{\log 2}, & \text{otherwise.} \end{cases} \tag{10}$$

Restated,

$$\dim \mathbf{Y}^{(2)} = \frac{2h\left(\mathbf{Y}^{(2)}\right)}{\log 2}. \tag{11}$$

Proof. Suppose X is a SFT and μ is an invariant probability measure on X. The Variational Principle indicates that the topological entropy of X is the supremum of the measure-theoretic entropy of X; more precisely,

$$h(X) = \sup\{h_\mu(X):\ \mu \text{ is an invariant probability measure on } X\}$$

A measure μ is called maximal if $h_\mu(X)$ attains the supremum. Let μ be a Markov measure which is derived from the transition matrix of X. Then μ is the unique measure that satisfies

$$h(X) = h_\mu(X)$$

if X is topologically transitive (cf. [15]). Ban and Chang showed that, if $\phi: W \to \mathbf{Y}^{(2)}$ has a synchronizing word, then the Hausdorff dimension of the output space is

$$\dim \mathbf{Y}^{(2)} = \frac{2h_{\mu^+}\left(W^+\right)}{\log 2} = \frac{2h_{\mu^-}\left(W^-\right)}{\log 2},$$

where μ^+/μ^- is a maximal measure of W^+/W^- (see [16], Theorem 2.6). Since Π is right-resolving, the factor map $\phi: W \to \mathbf{Y}^{(2)}$ is finite-to-one. It follows that

$$h(W) = h\left(\mathbf{Y}^{(2)}\right).$$

Theorem 2.3 demonstrates that the topological entropy of the output space

$$h\left(\mathbf{Y}^{(2)}\right) = \log \rho_H$$

(respectively $h\left(\mathbf{Y}^{(2)}\right)=\log\rho_T$) if G is not rightresolving (respectively G is right-resolving). A straightforward examination infers that

$$h(W)=h\left(W^{+}\right)=h\left(W^{-}\right).$$

Hence we have

$$\dim \mathbf{Y}^{(2)}=\frac{2h\left(\mathbf{Y}^{(2)}\right)}{\log 2}=\begin{cases}\dfrac{\log\rho_T^2}{\log 2}, & \text{if G is right-resolving;}\\[2mm] \dfrac{\log\rho_H^2}{\log 2}, & \text{otherwise.}\end{cases}$$

this completes the proof. □

3. EXAMPLE

Suppose $\mathrm{T}=(\boldsymbol{A},\boldsymbol{B},z)$ with

$\boldsymbol{B}=(-2.6,-1.4)$ $A_2=(-4,-2)$, $A_1=(2.2,1.7)$, and $z=(-1.2,0.3)$. The transition matrices for the first and second layer are

$$T_1=\begin{pmatrix}0&1&0&0\\0&0&1&1\\1&1&0&0\\0&0&0&0\end{pmatrix}$$

$$\text{and } T_2=\begin{pmatrix}0&0&0&0&0&0&0&0\\0&0&0&0&0&0&0&0\\0&0&1&1&1&1&0&0\\0&0&1&1&1&1&0&0\\0&0&1&1&1&1&0&0\\0&0&1&1&1&1&0&0\\0&0&0&0&0&0&0&0\\0&0&0&0&0&0&0&0\end{pmatrix}$$

respectively. Therefore, the transition matrix and the symbolic transition matrix of the MNN are

$$\boldsymbol{T}=\begin{pmatrix} 0&0&0&0&0&0&0&0\\ 0&0&0&0&0&0&0&0\\ 0&0&0&0&0&1&0&0\\ 0&0&0&0&1&1&0&0\\ 0&0&1&0&0&1&0&0\\ 0&0&1&0&0&1&0&0\\ 0&0&0&0&0&0&0&0\\ 0&0&0&0&0&0&0&0 \end{pmatrix}$$

$$\text{and }\boldsymbol{S}=\begin{pmatrix} \varnothing&\varnothing&\varnothing&\varnothing&\varnothing&\varnothing&\varnothing&\varnothing\\ \varnothing&\varnothing&\varnothing&\varnothing&\varnothing&\varnothing&\varnothing&\varnothing\\ \varnothing&\varnothing&\varnothing&\varnothing&\varnothing&a_{01}&\varnothing&\varnothing\\ \varnothing&\varnothing&\varnothing&\varnothing&a_{01}&a_{01}&\varnothing&\varnothing\\ \varnothing&\varnothing&a_{10}&\varnothing&\varnothing&a_{11}&\varnothing&\varnothing\\ \varnothing&\varnothing&a_{10}&\varnothing&\varnothing&a_{11}&\varnothing&\varnothing\\ \varnothing&\varnothing&\varnothing&\varnothing&\varnothing&\varnothing&\varnothing&\varnothing\\ \varnothing&\varnothing&\varnothing&\varnothing&\varnothing&\varnothing&\varnothing&\varnothing \end{pmatrix}$$

respectively, where

$$a_{00}=--,\quad a_{01}=-+,$$
$$a_{10}=+-,\quad a_{11}=++.$$

It is seen from the symbolic transition matrix S that the labeled graph G is not right-resolving, and applying SMC to G derives a right-resolving labeled graph H (cf. **Figure 1**). The transition matrix of H, indexed by $p_3, p_4, p_5, p_6, \{p_5, p_6\}$, is

$$\boldsymbol{H}=\begin{pmatrix} 0&0&0&1&0\\ 0&0&0&0&1\\ 1&0&0&1&0\\ 1&0&0&1&0\\ 0&0&1&1&0 \end{pmatrix}.$$

Theorem 2.3 indicates that

$$h\left(\mathbf{Y}^{(2)}\right) = \log \rho_H = \log g$$

where

$$g = \frac{1+\sqrt{5}}{2}$$

is the golden mean.

The symbolic transition matrix of II is

$$\boldsymbol{S}' = \begin{pmatrix} \varnothing & \varnothing & \varnothing & a_{01} & \varnothing \\ \varnothing & \varnothing & \varnothing & \varnothing & a_{01} \\ a_{10} & \varnothing & \varnothing & a_{11} & \varnothing \\ a_{10} & \varnothing & \varnothing & a_{11} & \varnothing \\ \varnothing & \varnothing & a_{10} & a_{11} & \varnothing \end{pmatrix}.$$

It is seen that both $a_{10} = +-$ and $a_{11} = ++$ are synchronizing words of ϕ. Theorem 2.4 demonstrates that

$$\dim \mathbf{Y}^{(2)} = \frac{\log g^2}{\log 2} \approx 1.3885.$$

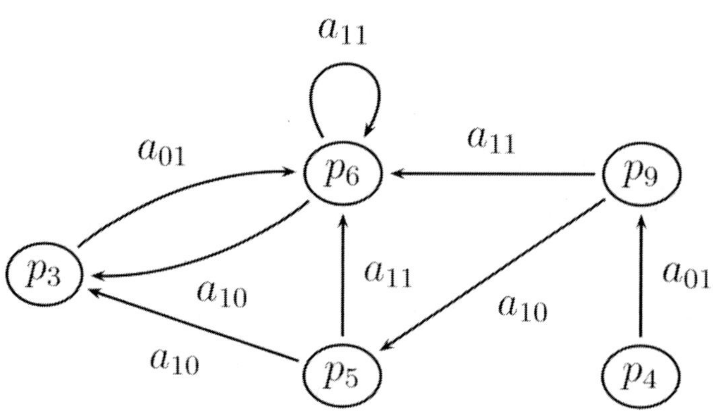

Figure 1. The right-resolving labeled graph $\mathbf{H}$ obtained by applying SCM to $\mathbf{G}$ in example. Here $p_9 = \{p_5, p_6\}$.

The fractal set of the output space $\mathbf{Y}^{(2)}$ is seen in **Figure 2**.

Jung-Chao Ban, Chih-Hung Chang#

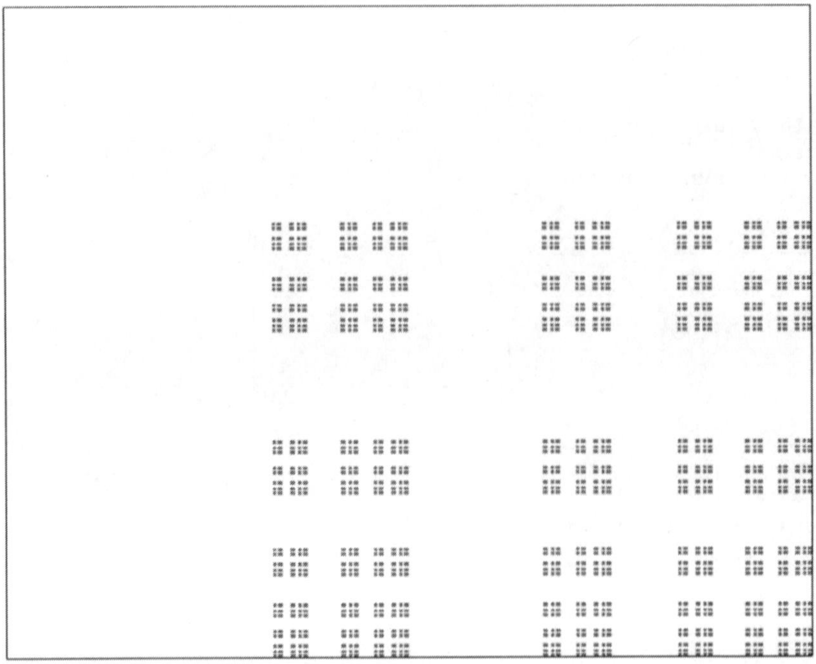

Figure 2. The fractal set of the output space $\mathbf{Y}^{(2)}$.

REFERENCES

1. K. Hornik, M. Stinchcombe and H. White, "Multilayer Feedforward Networks Are Universal Approximators," Neural Networks, Vol. 2, No. 5, 1989, pp. 359-366.
2. B. Widrow and M. Lehr, "30 Years of Adaptive Neural Networks: Perceptron Madaline, and Backpropagation," Proceedings of the IEEE, Vol. 78, No. 9, 1990, pp. 1415- 1442.
3. Y. A. Alsultanny and M. M. Aqul, "Pattern Recognition Using Multilayer Neural-Genetic Algorithm," Neurocomputing, Vol. 51, 2003, pp. 237-247.
4. B. Widrow, "Layered Neural Nets for Pattern Recognition," IEEE Transactions on Acoustics, Speech, and Signal Processing, Vol. 36, No. 7, 1962, pp. 1109-1118.
5. J. J. Hopeld and D. W. Tank, "Neural Computation of Decisions in Optimization Problems," Biological Cybernetics, Vol. 52, No. 3, 1985, pp. 141-152.
6. C. Peterson and B. Soderberg, "A New Method for Mapping Optimization Problems onto Neural Network," International Journal of Neural Systems, Vol. 1, 1989, pp. 3-22.
7. J.-C. Ban and C.-H. Chang, "The Learning Problem of Multi-Layer Neural Networks," Neural Networks, Vol. 46, 2013, pp. 116-123.

8. J.-C. Ban, C.-H. Chang and S.-S. Lin, "The Structure of Multi-Layer Cellular Neural Networks," Journal of Differential Equations, Vol. 252, No. 8, 2012, pp. 4563-4597.
9. J.-C. Ban, C.-H. Chang, S.-S. Lin and Y.-H. Lin, "Spatial Complexity in Multi-Layer Cellular Neural Networks," Journal of Differential Equations, Vol. 246, No. 2, 2009, pp. 552-580.
10. S.-N. Chow and J. Mallet-Paret, "Pattern Formation and Spatial Chaos in Lattice Dynamical Systems: I and II," IEEE Transactions on Circuits and Systems I: Fundamental Theory and Applications, Vol. 42, No. 10, 1995, pp. 746-756.
11. K. Falconer, "Fractal Geometry: Mathematical Foundations and Application," 2nd Edition, John Wilet & Sons, New York, London, Sydney, 2003.
12. Y. Pesin, "Dimension Theory in Dynamical Systems: Contemporary Views and Application," The University of Chicago Press, Chicago, 1997.
13. J. Juang and S.-S. Lin, "Cellular Neural Networks: Mosaic Pattern and Spatial Chaos," SIAM Journal on Applied Mathematics, Vol. 60, No. 3, 2000, pp. 891-915.
14. D. Lind and B. Marcus, "An Introduction to Symbolic Dynamics and Coding," Cambridge University Press, Cambridge, 1995.
15. B. Kitchens, "Symbolic Dynamics. One-Sided, Two-Sided and Countable State Markov Shifts," Springer-Verlag, New York, 1998.
16. J.-C. Ban and C.-H. Chang, "On the Structure of MultiLayer Cellular Neural Networks. Part II: The Complexity between Two Layers," Submitted, 2012.

NOTES

*Ban is partially supported by the National Science Council (Contract No NSC 100-2115-M-259-009-MY2).

#Chang is grateful for the partial support of the National Science Council (Contract No NSC 101-2115-M-035-002-).

[1]Pesin showed that the box dimension of a shift space X is the quotient of the topological entropy and the metric on X. More precisely, if the metric is defined by $d(x,y)=\sum_{i\in\mathbb{Z}}\frac{|x_i-y_i|}{\beta^{|i|+1}}$ for $x,y\in X,\beta>1$, then the box dimension is $\dim_B X = h(X)/\log\beta$.

CHAPTER 3

Time-Optimal Control Problem for n × n Co-Operative Parabolic Systems with Control in Initial Conditions

Mohammed A. Shehata

Department of Mathematics, Faculty of Science, Jazan University, Jazan, KSA

ABSTRACT

In this paper, time-optimal control problem for a liner $n \times n$ co-operative parabolic system involving Laplace operator is considered. This problem is, steering an initial state $y(0) = u$, with control u, so that an observation $y(t)$ hits a given target set in minimum time. First, the existence and uniqueness of solutions of such system under conditions on the coefficients are proved. Afterwards necessary and sufficient conditions of optimality are obtained. Finally a scaler case is given.

Keywords: Time-optimal control problems; bang-bang controls; parabolic system; co-operative systems

1. INTRODUCTION

The "time optimal" control problem is one of the most important problems in the field of control theory. The simple version is that steering the initial state y_0 in a Hilbert space H to hit a target set $K \subset H$ in minimum time, with control subject to constraints $(u \in U \subset H)$.

In this paper, we will focus our attention on some special aspects of minimum time problems for co-operative parabolic system involving Laplace operator with control acts in the initial conditions. In order to explain the results we have in mind, it is convenient to consider the abstract form.

Let V and H be two real Hilbert spaces such that V is a dense subspace of H Identifying the dual of H with H we may consider $V \subset H \subset V'$, where the embedding is dense in the following space. Let $A(t)$ $(t \in]0,T[)$ be a family of continuous operators associated with a bilinear forms $\pi(t;.,.)$ defined on $V \times V$ which are satisfied with Gårding's inequality

$$\pi(t;y,y) + c_0 \|y\|_H^2 \geq c_1 \|y\|_V^2 \tag{1}$$

for $y \in V, t \in [0,T]$ and $c_0 \geq 0, c_1 > 0.$

Then, from [1] and [2], for $t \in]0,T[$ and B being a bounded linear operator on H the following abstract systems:

$$\left.\begin{aligned} &\frac{d}{dt} y(t) + A(t) y(t) = f, f \in L^2(0,T;V'), \\ &y(0) = Bu \end{aligned}\right\} \tag{2}$$

have a unique weak solution y such that $y \in C([0,T];H).$ We shall denote by $y(t;u)$ the unique solution of the Equation (2) corresponding to the control u. The time optimal control problem that we shall concern reads:

$$\min\{\tau : y(\tau;u) \in K, u \in U\} \tag{3}$$

where K is a given subset of H which is called the target set of the Problem (3). A control u^0 is called a time optimal control if $u^0 \in U$ and if there is a number $\tau^0 > 0$ such that $y(\tau^0; u^0) \in K$ and

$$\tau^0 = \min\{\tau : y(\tau; u) \in K, u \in U\}. \tag{4}$$

We call the number τ^0 as the optimal time for the time optimal control Problem (3).

Three questions (problems) arise naturally in connection with this problem:

1) Is there a control u, and $\tau > 0$ such that $y(\tau; u) \in K$? (this is an approximate controllability problem).

2) Assume that the answer to 1) is in the affirmative and

$$\tau^0 = \min\{\tau : y(\tau; u) \in K, u \in U\}.$$

Is there a control u^0 which steers $y(\tau^0)$ to hit a target set K in minimum time?

3) If u^0 exists, is it unique? What additional properties does it have?

Let $\Omega \subset R^N$ be a bounded open domain with smooth boundary Γ. and set $Q = \Omega \times]0, T[$ $\Sigma = \Gamma \times]0, T[$. In the works [1] and [3], the existence of time optimal controls of the following controlled linear parabolic equations with distributed control u was obtained:

$$\left.\begin{array}{ll} \dfrac{\partial y}{\partial t} = \Delta y + u & \text{in } Q, \\ y(x,0) = y_0(x) & \text{in } \Omega, \\ y(x,t) = 0 & \text{on } \Sigma, \end{array}\right\} \tag{5}$$

where $y_0(x)$ is a given function in $L^2(\Omega)$, $u \in U$ and U is a closed bounded set in $L^2(\Omega)$. The results in [3] partly overlap with results in [1] and they were shown that if the system (5) is controllable and if $K = \{0\}$ then the corresponding time optimal control problem has at least one solution and it is bangbang.

In the work [4], the authors gave a sufficient and necessary condition for the existence of time optimal control for the problem with the target set $K=\{0\}$ and certain controlled systems. These results will be stated as follows. Consider the following controlled system

$$\left.\begin{array}{ll}\dfrac{\partial y}{\partial t}=\Delta y+ay+u & \text{in } Q,\\ y(x,0)=y_0(x) & \text{in } \Omega,\\ y(x,t)=0 & \text{on } \Sigma,\end{array}\right\} \tag{6}$$

where a is a real number. Let $\{\lambda_i\}_{i\geq 1}, \lambda_1 \leq \lambda_2 \leq \lambda_3 \leq \ldots$, be the eigenvalues of $-\Delta$ with the Dirichlet boundary condition and $\{e_i\}_{i\geq 1}$ be the corresponding eigenfunctions, which forms an orthogonal basis of $L^2(\Omega)$. We take the target set K to be the origin $\{0\}$ in $L^2(\Omega)$ and the control set U to be the set

$$U_\varepsilon=\left\{u(.,t)\in L^2(\Omega):\|u\|_{L^2(\Omega)}\leq\varepsilon\right\}$$

where ε is a positive number, namely, $U_\varepsilon=B(0,\varepsilon)$, the closed ball in $L^2(\Omega)$ centered at 0 and of radius ε It was proved that if $K=\{0\}$ and $U=U_\varepsilon$, then the corresponding time optimal control problem has at least one solution if and only if $a\leq\lambda_1$.

More early, in the works [5-7], the time optimal controls problem for globally controlled linear and semilinear parabolic equations was considered.

In our papers [8,9], the time optimal control problem of $n\times n$ co-operative hyperbolic systems with different cases of the observation and distributed or boundary controls constraints was considered.

In [10], optimal control of infinite order hyperbolic equation with control via initial conditions was considered.

In the present paper, the above results for the time optimal control of systems governed by parabolic equations are extended to the case of $n\times n$ co-operative parabolic systems as well as control via initial conditions. First, the existence and uniqueness of solutions for $n\times n$ co-operative parabolic system are proved under conditions on the coefficients stated by the principal eigenvalue of the Laplace eigenvalue problem, then the time optimal control problem is formulated and the existence of a time optimal control is proved. Then the necessary and sufficient conditions which the optimal controls must satisfy are derived in terms of the adjoint. Finally, the scaler case is given.

2. N × N CO-OPERATIVE PARABOLIC SYSTEMS

Let $H_0^1(\Omega)$, be the usual Sobolev space of order one which consists of all $\phi \in L^2(\Omega)$ whose distributional derivatives $\frac{\partial \phi}{\partial x_i} \in L^2(\Omega)$ and $\phi_\Gamma = 0$ with the scalar product norm

$$\langle y, \phi \rangle_{H_0^1(\Omega)} = \langle y, \phi \rangle_{L^2(\Omega)} + \langle \nabla y, \nabla \phi \rangle_{L^2(\Omega)},$$

$$\text{where } \nabla = \sum_{k=1}^{N} \frac{\partial}{\partial x_k}.$$

We have the following dense embedding chain [11]

$$\left(H_0^1(\Omega)\right)^n \subseteq \left(L^2(\Omega)\right)^n \subseteq \left(H_0^{-1}(\Omega)\right)^n.$$

where $H_0^{-1}(\Omega)$ is the dual of $H_0^{-1}(\Omega)$

Here and everywhere below the vectors are denoted by bold letters. For $\boldsymbol{y} = (y_i)_{i=1}^n, \boldsymbol{\phi} = (\phi_i)_{i=1}^n \in \left(H_0^1(\Omega)\right)^n$ and $t \in]0, T[$, let us define a family of continues bilinear forms

$$\pi(t;.,.) : \left(H_0^1(\Omega)\right)^n \times \left(H_0^1(\Omega)\right)^n \to \Re \text{ by}$$

$$\pi(t; \boldsymbol{y}, \boldsymbol{\phi}) = \sum_{i=1}^{n} \int_\Omega \left[(\nabla y_i)(\nabla \phi_i) - a_i(x,t) y_i \phi_i \right] \mathrm{d}x - \sum_{i,j=1}^{n} \int_\Omega a_{ij}(x,t) y_j \phi_i \mathrm{d}x \tag{7}$$

where

$$\left.\begin{array}{l} a_i(x,t) \text{ and } a_{ij}(x,t) \text{ are positive functions in } L^\infty(Q), \\ a_{ij}=0 \text{ when } i=j \quad \text{and } a_{ij} \le \sqrt{a_i a_j} \text{ when } i \ne j \end{array}\right\} \tag{8}$$

The bilinear form (7) can be but in the operator form:

$$\begin{aligned} \pi(t;y,\phi) &= \sum_{i=1}^{n}\int_\Omega\left[(-\Delta y_i)-a_i(x,t)y_i\right]\phi_i dx \\ &- \sum_{i,j=1}^{n}\int_\Omega a_{ij}(x,t)y_j\phi_i dx \\ &= \sum_{i=1}^{n}\left\langle -(A(t)y)_i,\phi\right\rangle_{L^2(\Omega)} \end{aligned}$$

where $A(t)$ is $n\times n$ matrix operator which maps $(H_0^1(\Omega))^n$ onto $(H_0^{-1}(\Omega))^n$ and takes the form

$$A(t)y = \begin{pmatrix} \Delta+a_1 & a_{12} & \text{L} & a_{1n} \\ a_{21} & \Delta+a_2 & \text{L} & a_{2n} \\ \text{M} & \text{M} & \text{O} & \text{M} \\ a_{n1} & a_{n2} & \text{L} & \Delta+a_n \end{pmatrix}\begin{pmatrix} y_1 \\ y_2 \\ \text{M} \\ y_n \end{pmatrix}.$$

LEMMA 2.1. If Ω is a regular bounded domain in R^N, with boundary Γ, and if m is positive on Ω and smooth enough (in particular $m \in L^\infty(\Omega)$,) then the eigenvalue problem:

$$\left.\begin{array}{ll} -\Delta y = \lambda m(x) y & \text{in } \Omega, \\ y=0 & \text{on } \Gamma \end{array}\right\}$$

possesses an infinite sequence of positive eigenvalues:

$$0<\lambda_1(m)<\lambda_2(m)\le \text{L} \le \lambda_k(m) \le \text{L} ;$$
$$\lambda_k(m)\to\infty, \text{ as } k\to\infty.$$

Moreover $\lambda_1(m)$ is simple, its associate eigenfunction e_m is positive, and $\lambda_1(m)$ is characterized by:

$$\lambda_1(m)\int_\Omega my^2\mathrm{d}x \le \int_\Omega |\nabla y|^2\,\mathrm{d}x \tag{9}$$

PROOF. See [12]. ,.

Now, let

$$\lambda_1(a_i) \ge n, \quad i = 1,2,\mathrm{L}\ ,n \tag{10}$$

LEMMA 2.2. If (8) and (10) hold then, the bilinear form (7) satisfy the Gårding inequality

$$\pi(t;y,y) + c_0\|y\|^2_{(L^2(\Omega))^n} \ge c_1\|y\|^2_{(H^1_0(\Omega))^n},$$

$$c_0, c_1 > 0$$

PROOF. In fact

$$\begin{aligned}\pi(t;y,y) &= \sum_{i=1}^{n}\int_\Omega\left[|\nabla y_i|^2 - a_i(x,t)y_i^2\right]\mathrm{d}x\\ &\quad -\sum_{i,j=1}^{n}\int_\Omega a_{ij}(x,t)y_iy_j\mathrm{d}x\\ &\ge \sum_{i=1}^{n}\int_\Omega\left[|\nabla y_i|^2 - a_i(x,t)y_i^2\right]\mathrm{d}x\\ &\quad -2\sum_{i>j}^{n}\int_\Omega\sqrt{a_i(x,t)a_j(x,t)}\,y_iy_j\mathrm{d}x\end{aligned}$$

By Cauchy Schwarz inequality and (9), we obtain

$$\pi(t;y,y) \geq \sum_{i=1}^{n}\left(1-\frac{1}{\lambda_1(a_i)}\right)\int_\Omega |\nabla y_i|^2\, dx$$

$$-2\sum_{i>j}^{n}\int_\Omega \frac{1}{\sqrt{\lambda_1(a_i)\lambda_1(a_j)}}\left(\int_\Omega |\nabla y_i|^2\, dx\right)^{\frac{1}{2}}\left(\int_\Omega |\nabla y_j|^2\, dx\right)^{\frac{1}{2}}$$

$$\geq \sum_{i=1}^{n}\left(\frac{\lambda_1(a_i)-n}{\lambda_1(a_i)}\right)\int_\Omega |\nabla y_i|^2\, dx$$

From (10) we have

$$\pi(t;y,y) \geq \alpha\left[\sum_{i=1}^{n}\int_\Omega |\nabla y_i|^2\, dx\right] \quad \alpha > 0$$

Add $\|y\|_{(L^2(\Omega))^n}$ to two sides, then we have the result. ,.

We can now apply Theorem 1.1 and Theorem 1.2 Chapter 3 in [1] (with $V=\left(H_0^1(\Omega)\right)^n$ and $H=\left(L^2(\Omega)\right)^n$) to obtain the following theorem:

THEOREM 2.3. If (8) and (10) hold, then there exist a unique solution

$$y \in W(0,T) = \left\{y : y \in L^2\left(0,T;\left(H_0^1(\Omega)\right)^n\right), \frac{\partial y}{\partial t} \in \left(H_0^1(\Omega)\right)^n\right\}$$

satisfying the following $n\times n$ system: $i=1,2,\ldots,n$

$$\left.\begin{aligned} &\frac{\partial y_i}{\partial t} = \left(A(t)y\right)_i + f_i, \quad f_i \in L^2\left(0,T;H_0^{-1}(\Omega)\right) && \text{in } Q, \\ &y_i(x,0) = u_i(x), \quad u_i(x) \in L^2(\Omega) && \text{in } \Omega, \\ &y_i(x,t) = 0 && \text{on } \Sigma. \end{aligned}\right\} \tag{11}$$

Moreover y is continuous from $[0,T] \to \left(L^2(\Omega)\right)^n$.

3. MINIMUM TIME AND CONTROLLABILITY

We denote the unique solution of (11), at time t for each control $\boldsymbol{u}=(u_1,u_2,\ldots,u_n)$ by $\boldsymbol{y}(t;\boldsymbol{u})$. Occasionally, we write $\boldsymbol{y}(x,t;\boldsymbol{u})$ when the explicit dependence on x is required. We can now formulate the time optimal control problem corresponding to the $n\times n$ cooperative parabolic system (11):

$$\min\left\{t:\boldsymbol{y}(t;\boldsymbol{u})\in K_\varepsilon^n,\ \boldsymbol{u}\in U_\varepsilon^n\right\}, \tag{12}$$

with constraints

$$\left.\begin{aligned}&\boldsymbol{y}(t;\boldsymbol{u})\text{ is the solution of }(11),\\ &U_\varepsilon^n=\left\{\boldsymbol{u}=(u_1,\ldots,u_n)\in\left(L^2(\Omega)\right)^n:\|u_i\|_{L^2(\Omega)}\le\varepsilon\right\},\\ &K_\varepsilon^n=\left\{z=(z_1,\ldots,z_n)\in\left(L^2(\Omega)\right)^n:\|z_i-z_{id}\|_{L^2(\Omega)}\le\varepsilon\right\},\end{aligned}\right\} \tag{13}$$

and $\varepsilon,\varepsilon>0$ and $z_{id}\in L^2(\Omega)$ are given.

THEOREM 3.1. If (8) and (10) are hold, then the system whose state is given by (11) is controllable, i.e.,

$$\text{there exists a }\tau\in\,]0,T]\text{ and }\boldsymbol{u}\in U_\varepsilon^n\text{ with }\boldsymbol{y}(\tau;\boldsymbol{u})\in K_\varepsilon^n \tag{14}$$

PROOF. Let us first remark that by translation we may always reduce the problem of controllability to the case were the system (11) with $f_i=0$. We can show quit easily that (11) is approximately controllable in $(L^2(\Omega))^n$ in any finite time $\tau>0$, if and only if, $\left\{\boldsymbol{y}(\tau;\boldsymbol{u}):\boldsymbol{u}\in\left(L^2(\Omega)\right)^n\right\}$ is dense in $(L^2(\Omega))^n$. By the Hahn-Banach theorem, this will be the case if

$$\int_\Omega \bar{z}_i(x)\,y_i(x,\tau;\boldsymbol{u})\,\mathrm{d}x=0,\quad \bar{z}_i\in L^2(\Omega), \tag{15}$$

for all $u \in \left(L^2(\Omega)\right)^n$, implies that $\overline{z}_i(x) = 0, \quad i = 1, 2,$

L $,n.$

Let us introduce the adjoint state $p(t;u)$ by the solution of the following system

$$\left.\begin{array}{ll} -\dfrac{\partial p_i}{\partial t}(t;u) - \left(A^*(t)\,p(t;u)\right)_i = 0 & \text{in } \Omega \times]0,\tau[, \\ p_i(x,\tau) = \overline{z}_i(x) & \text{in } \Omega, \\ p_i(x,t) = 0 & \text{on } \Gamma \times]0,\tau[, \end{array}\right\} \tag{16}$$

where $A^*(t)$ is the adjoint of $A(t)$ which is defined by

$$\left\langle A^*(t)\phi, \psi \right\rangle = \left\langle \phi, A(t)\psi \right\rangle, \quad \phi, \psi \in \left(H_0^1(\Omega)\right)^n.$$

The existence of a unique solution for the Problem (16) can be proved using Theorem 2.3, with an obvious change of variables.

Multiply the first equation in (16) by $y_i(t;u)$ and integrate by parts from 0 to τ, we obtain the following identity:

$$\begin{aligned} 0 &= \int_0^\tau \int_\Omega \left[-\frac{\partial p_i}{\partial t} - \left(A^*(t)\,p(t;u)\right)_i \right] y_i(t;u)\,\mathrm{d}x\mathrm{d}t \\ &= -\int_\Omega p_i(t;u)\, y_i(t;u)\Big|_0^\tau \mathrm{d}x \\ &\quad + \int_0^\tau \int_\Omega p_i(t;u) \left[\frac{\partial}{\partial t} y_i(t;u) - \left(A(t)\,y(t;u)\right)_i \right] \mathrm{d}x\mathrm{d}t \\ &= \int_\Omega \overline{z}_i(x)\, y_i(x,\tau;u)\,\mathrm{d}x - \int_\Omega p_i(0;u)\, u_i \mathrm{d}x. \end{aligned}$$

and so, if (15) holds, then

$$\int_\Omega p_i(x,0;u)\, u_i \mathrm{d}x = 0 \quad \forall u_i \in L^2(\Omega)$$

hence $p_i(x,0;\boldsymbol{u})=0.$ But from the backward uniqueness property, $\boldsymbol{p}=(p_i)_{i=1}^n\equiv 0$ and hence $\bar{z}_i(x)=0.$,.

Now set

$$\tau^0=\inf\left\{\tau:\boldsymbol{y}(\tau;\boldsymbol{u})\in K_\varepsilon^n \text{ fore some } \boldsymbol{u}\in U_\varepsilon^n\right\}. \tag{17}$$

Then , the following result holds.

THEOREM 3.2. If (8) and (10) are hold, then there exist an admissible control $\boldsymbol{u}^0$ to the problem (12)-(17), which steering $\boldsymbol{y}(t;\boldsymbol{u}^0)$ to hitting a target set K_ε^n in minimum time τ^0 (defined by (17)). Moreover

$$\sum_{i=1}^{n}\int_\Omega\left(y_i\left(\tau^0;\boldsymbol{u}^0\right)-z_{id}\right)\left(y_i\left(\tau^0;\boldsymbol{u}\right)-y_i\left(\tau^0;\boldsymbol{u}^0\right)\right)\mathrm{d}x\geq 0$$
$$\forall\boldsymbol{u}\in U_\varepsilon^n \tag{18}$$

PROOF. Fixe x, we can choose $\tau^m\to\tau^0$ and admissible controls $\{\boldsymbol{u}^m\}$ such that

$$\boldsymbol{y}\left(\tau^m;\boldsymbol{u}^m\right)\in K_\varepsilon^n,\quad m=1,2,\mathrm{L}\ .$$

Set $\boldsymbol{y}^m=\boldsymbol{y}(\boldsymbol{u}^m).$ Since U_ε^n is bounded, we may verify that $\boldsymbol{y}^m$ ranges in a bounded set in

$$\left(L^2\left(0,T;\left(L^2(\Omega)\right)^n\right)=\left(L^2(Q)\right)^n\right).$$

We may then extract a subsequence, again denoted by $\{\boldsymbol{u}^m,\boldsymbol{y}^m\}$ such that

$$\left.\begin{aligned}&\boldsymbol{u}^m\to\boldsymbol{u}^0\quad\text{weakly in }\left(L^2(\Omega)\right)^n,\quad u^0\in U_\varepsilon^n,\\&\boldsymbol{y}^m\to\boldsymbol{y}\quad\text{weakly in }L^2\left(0,T;\left(H_0^1(\Omega)\right)^n\right).\end{aligned}\right\} \tag{19}$$

We deduce from the equality

$$\frac{\mathrm{d}\boldsymbol{y}^m}{\mathrm{d}t} = f - A(t)\boldsymbol{y}^m$$

that

$$\frac{\mathrm{d}\boldsymbol{y}^m}{\mathrm{d}t} \to \frac{\mathrm{d}\boldsymbol{y}}{\mathrm{d}t} = f - A(t)\boldsymbol{y} \quad \text{in } L^2\left(0,T;\left(H^{-1}(\Omega)\right)^n\right),$$

and

$$\boldsymbol{y}^m(0) \to \boldsymbol{y}(0) = \boldsymbol{u}^0 \quad \text{in } U_\varepsilon^n.$$

But

$$\begin{aligned}
&\boldsymbol{y}\left(\tau^m;\boldsymbol{u}^m\right) - \boldsymbol{y}\left(\tau^0;\boldsymbol{u}^0\right) \\
&= \boldsymbol{y}\left(\tau^m;\boldsymbol{u}^m\right) - \boldsymbol{y}\left(\tau^0;\boldsymbol{u}^m\right) + \boldsymbol{y}\left(\tau^0;\boldsymbol{u}^m\right) - \boldsymbol{y}\left(\tau^0;\boldsymbol{u}^0\right)
\end{aligned}$$

Now from (19)

$$\boldsymbol{y}\left(\tau^0;\boldsymbol{u}^m\right) \to \boldsymbol{y}\left(\tau^0;\boldsymbol{u}^0\right) \quad \text{weakly in } \left(L^2(\Omega)\right)^n \tag{20}$$

and

$$\begin{aligned}
&\left\|\boldsymbol{y}\left(\tau^m;\boldsymbol{u}^m\right) - \boldsymbol{y}\left(\tau^0;\boldsymbol{u}^m\right)\right\|_{\left(H^{-1}(\Omega)\right)^n} \\
&= \left\|\int_{\tau^0}^{\tau^m} \frac{\mathrm{d}}{\mathrm{d}t}\boldsymbol{y}\left(t;\boldsymbol{u}^m\right)\mathrm{d}t\right\|_{\left(H^{-1}(\Omega)\right)^n} \\
&\le \sqrt{\tau^m - \tau^0}\left(\int_{\tau^0}^{\tau^m} \left\|\frac{\mathrm{d}}{\mathrm{d}t}\boldsymbol{y}\left(t;\boldsymbol{u}^m\right)\mathrm{d}t\right\|_{\left(H^{-1}(\Omega)\right)^n}^2 \mathrm{d}t\right)^{\frac{1}{2}} \\
&\le c\sqrt{\tau_n - \tau^0}
\end{aligned} \tag{21}$$

Combine (20) and (21) show that

$$y\left(\tau^{m};u^{m}\right)-y\left(\tau^{0};u^{0}\right)\rightarrow 0 \quad \text{weakly in } \left(H_0^{-1}(\Omega)\right)^n . \tag{22}$$

and so, $y\left(\tau^{0};u^{0}\right)\in K_{\varepsilon}^{n}$ as K_{ε}^{n} is closed and convex, hence weakly closed. This shows that K_{ε}^{n} is reached in time τ^{0} by admissible control u^{0}.

For the second part of the theorem, really, from Theorem 2.3, the mapping $t\rightarrow y(t;u)$ from $[0,T]\rightarrow\left(L^{2}(\Omega)\right)^{n}$ is continuous for each fixed u and so $y\left(\tau^{0};u\right)\notin \operatorname{int} K_{\varepsilon}^{n}$, for any $u\in U_{\varepsilon}^{n}$, by minimality of τ^{0}.

Using Theorem 2.3, it is easy to verify that the mapping $u\rightarrow y\left(\tau^{0};u\right)$, from $\left(L^{2}(\Omega)\right)^{n}\rightarrow\left(L^{2}(\Omega)\right)^{n}$, is continuous and linear. then, the set

$$\mathscr{A}\left(\tau^{0}\right)=\left\{y\left(\tau^{0};u\right):u\in U_{\varepsilon}^{n}\right\}$$

is the image under a linear mapping of a convex set hence $\mathscr{A}\left(\tau^{0}\right)$ is convex. Thus we have

$\mathscr{A}\left(\tau^{0}\right)\cap \operatorname{int} K_{\varepsilon}^{n}=\varnothing$ and $y\left(\tau^{0};u^{0}\right)\in\partial K_{\varepsilon}^{n}$ (boundary of K_{ε}^{n}) . Since $\operatorname{int} K_{\varepsilon}^{n}\neq\varnothing$ (from (14)) so there exists a closed hyperplane separating $A\left(\tau^{0}\right)$ and K_{ε}^{n} containing $y\left(\tau^{0};u^{0}\right)$, i.e. there is a nonzero $g\in\left(L^{2}(\Omega)\right)^{n}$ such as

$$\begin{aligned}\int_{\Omega}\left\langle g,y\left(\tau^{0};u\right)\right\rangle_{\left(L^{2}(\Omega)\right)^{n}} &\leq\left\langle g,y\left(\tau^{0};u^{0}\right)\right\rangle_{\left(L^{2}(\Omega)\right)^{n}}\\ &\leq \inf_{y\in K_{\varepsilon}^{n}}\left\langle g,y\left(\tau^{0};u\right)\right\rangle_{\left(L^{2}(\Omega)\right)^{n}}\end{aligned} \tag{23}$$

From the second inequality in (23), g must support the set K_{ε}^{n} at $y\left(\tau^{0};u^{0}\right)$ i.e.

$$\left\langle g,\left(y\left(\tau^0;u\right)-y\left(\tau^0;y^0\right)\right)\right\rangle_{\left(L^2(\Omega)\right)^n}\geq 0$$
$$\forall u\in U_\varepsilon^n$$

and since $\left(L^2(\Omega)\right)^n$ is a Hilbert space, g must be of the form

$$g=\lambda\left(y\left(\tau^0;u^0\right)-z_{id}\right)$$
for some $\lambda>0$.

Dividing the inequality (23) by λ gives the desired result. ,.

Now Inequality (18) can be interpreted as follows: let us introduce the adjoint state $p\left(t;u^0\right)$ by the solution of the following system

$$\left.\begin{array}{ll}-\dfrac{\partial p_i}{\partial t}\left(t;u^0\right)+\left(A^*(t)\,p\left(t;u^0\right)\right)_i=0 & \text{in } \Omega\times\left]0,\tau^0\right[,\\ p_i\left(x,\tau^0\right)=\left(y_i\left(x,\tau^0\right)-z_{id}\right) & \text{in}\Omega,\\ p_i\left(x,t\right)=0 & \text{on } \Gamma\times\left]0,\tau^0\right[.\end{array}\right\}\qquad (24)$$

As the proof of Theorem 3.1, we multiply the first equation int (24) by $y_i\left(t;u\right)-y_i\left(t;u^0\right)$ and integrate by parts from 0 to τ^0, we obtain the following identity:

$$\int_\Omega\left(y_i\left(\tau^0;u^0\right)-z_{id}\right)\left(y_i\left(x,\tau^0;u\right)-y_i\left(x,\tau^0;u^0\right)\right)\mathrm{d}x$$
$$=\int_\Omega p_i\left(0;u^0\right)\left(u-u^0\right)\mathrm{d}x.$$

hence condition (18) becomes

$$\sum_{i=1}^n\int_\Omega p_i\left(x,0;u^0\right)\left(u-u^0\right)\mathrm{d}x\geq 0$$
$$\forall u\in U_\varepsilon^n.\qquad (25)$$

Using controllability condition (14), the backward uniqueness property implies $p_i\left(x,0;\boldsymbol{u}^0\right)=0.$ then the optimal control is bang-bang, i.e., $\left\|u_i^0\right\|_{L^2(\Omega)}=\varepsilon$ and since

U_ε^n is strictly convex, then the optimal control is unique. We have thus proved:

THEOREM 3.3. If (8) and (10) are hold, then there exist the adjoint state

$$\boldsymbol{p}\in W\left(0,\tau^0\right)$$
$$=\left\{\boldsymbol{p}:\boldsymbol{p}\in L^2\left(0,\tau^0;\left(H_0^1(\Omega)\right)^n\right),\frac{\partial \boldsymbol{p}}{\partial t}\in\left(H_0^{-1}(\Omega)\right)^n\right\}$$

such that the optimal control $\boldsymbol{u}^0$ of problem (12)-(17) is bang-bang unique and it is determined by (24), (25) together with (11) (with $u_i=u_i^0, i=1,2,\mathrm{L}\ ,n$).

4. SCALER CASE

Here, we take the case where $n=2,$ in this case, the time optimal problem therefore is

$$\min\left\{t:y\left(x,t;\boldsymbol{u}\right)\in K_\varepsilon^2,\ \boldsymbol{u}=\left(u_1,u_2\right)\in U_\varepsilon^2\right\}$$

The state $\boldsymbol{y}=\left(y_1,y_2\right)$ is solution of the following equations

$$\left.\begin{array}{l}\dfrac{\partial y_1}{\partial t}-\Delta y_1=a_{11}(x,t)\,y_1+a_{12}(x,t)\,y_2\\ \qquad\qquad +f_1,\quad x\in\Omega,t\in\left]0,\tau^0\right[,\\ \dfrac{\partial y_2}{\partial t}-\Delta y_2=a_{21}(x,t)\,y_1+a_{22}(x,t)\,y_2\\ \qquad\qquad +f_2,\quad x\in\Omega,t\in\left]0,\tau^0\right[,\\ y_1(x,0)=u_1^0(x),\quad y_2(x,0)=u_2^0(x),\\ x\in\Omega,\quad y_1(x,t)=y_2(x,t)=0,\\ \quad x\in\Gamma,t\in\left]0,\tau^0\right[,\end{array}\right\}$$

with

$$\left.\begin{array}{l}a_{ij}(x,t),i,j=1,2\text{ are positive functions in }L^\infty(Q),\\ \lambda_1(a_{11})\geq 2,\quad \lambda_1(a_{22})\geq 2.\end{array}\right\}$$

The adjoint is solution of the following equations

$$\left.\begin{array}{l}-\dfrac{\partial p_1}{\partial t}-\Delta p_1=a_{22}(x,t)\,p_1-a_{12}(x,t)\,p_2\\ \qquad\qquad +f_1,\quad x\in\Omega,t\in\left]0,\tau^0\right[,\\ -\dfrac{\partial p_2}{\partial t}-\Delta p_2=-a_{21}(x,t)\,p_1+a_{11}(x,t)\,p_2\\ \qquad\qquad +f_2,\quad x\in\Omega,t\in\left]0,\tau^0\right[,\\ p_1(x,\tau^0)=\left(y_1(x,\tau^0)-z_{1d}\right),\quad x\in\Omega,\\ p_2(x,\tau^0)=\left(y_2(x,\tau^0)-z_{2d}\right),\quad x\in\Omega,\\ p_1(x,t)=p_2(x,t)=0,\quad x\in\Gamma,t\in\left]0,\tau^0\right[.\end{array}\right\}$$

The maximum condition is

$$\int_0^{\tau^0}\int_\Omega\left[p_1(x,0)(u_1-u_1^0)+p_2(x,0)(u_2-u_2^0)\right]\mathrm{d}x\mathrm{d}t\ge 0$$
$$\forall \boldsymbol{u}=(u_1,u_2)\in U_\varepsilon^2.$$

5. COMMENTS

We note that, in this paper, we have chosen to treat a special systems involving Laplace operator just for simplicity. Most of the results we described in this paper apply without any change on the results to more general parabolic systems involving the following second order operator:

$$L(x,.)=\sum_{i,j=1}^{n}b_{ij}(x,.)\frac{\partial^2}{\partial x_i\partial x_j}+\sum_{j=1}^{n}b_j(x,.)\frac{\partial}{\partial x_j}+b_0(x,.)$$

with sufficiently smooth coefficients (in particular, $b_0\in L^\infty(Q), b_j, b_0>0\ b_j,\ b_{ij},$) and under the LegendreHadamard ellipticity condition

$$\sum_{i,j=1}^{n}\eta_i\eta_j\ge\sigma\sum_{i=1}^{n}\eta_i\quad \forall(x,t)\in Q,$$

for all $\eta_i\in\Re$ and some constants $\sigma>0.$

In this case, we replace the first eigenvalue of the Laplace operator by the first eigenvalue of the operator L (see [12]).

In this paper, we have chosen to treat a co-operative parabolic system with Dirichlet boundary conditions. The results can be extended to the case of $n\times n$ cooperative parabolic system with Neumann boundary conditions: if we take $H^1(\Omega)$ instead of $H_0^1(\Omega),$ we have to replace the Dirichlet boundary conditions $y_i=0, p_i=0$ on the boundary by Neumann boundary conditions $\frac{\partial y_i}{\partial \nu}=0, \frac{\partial p_i}{\partial \nu}=0,$ where ν is the outward normal.

The results in this paper carry over to the fixed-time problem ([1] chapter 3).

$$\text{minimize}\sum_{i=1}^{n}\int_\Omega\left|y_i(x,T;\boldsymbol{u})-z_{id}(x)\right|^2\mathrm{d}x,\quad T\text{ fixed,}$$

Mohammed A. Shehata (11) [except in the trivial case where $z_{id}(x)=y_i(x,T;\boldsymbol{u})\ \forall i=1,2,\ldots,n$ for some admissible control $\boldsymbol{u}$]. This can be proven in an analogous manner, as the necessary and sufficient conditions for optimality for this problem coincide with (11), (16) and (25) (with $u_i=u_i^0, i=1,2,\ldots,n$).

REFERENCES

1. J. L. Lions, "Optimal Control of Systems Governed by Partial Differential Equations," Springer-Verlag, New York, 1971.
2. J. L. Lions and E. Magenes, "Non-Homogeneous Boundary Value Problem and Applications, I, II," Spring-Verlage, New York, 1972.
3. H. O. Fattorini, "Infinite Dimensional Optimization Theory and Opotimal Control," Cambridge University Press, Cambridge, 1998.
4. K. D. Phung, G. Wang and X. Zhang, "On the Existence of Time Optimal Controls for Linear Evolution Equations," Discrete and Continuous Dynamical Systems, Series B, Vol. 8, No. 4, 2007, pp. 925-941.
5. G. Knowles, "Time Optimal Control of Parabolic Systems with Boundary Condition Involving Time Delays," Journal of Optimization Theory and Applications, Vol. 25, No. 1, 1978, pp. 563-574.
6. H. O. Fattorini, "Infinite Dimensional Linear Control Systems: The Time Optimal and Norm Optimal Problems," North-Holland Mathematics Studies, Vol. 201, Elsevier, Amsterdam, 2005.
7. X. Li and J. Yong, "Optimal Control Theory for Infinite Dimensional Systems," Birkhauser, Boston, 1995.
8. H. A. El-Saify, H. M. Serag and M. A. Shehata, "TimeOptimal Control for Co-Operative Hyperbolic Systems Involving Laplace Operator," Journal of Dynamical and Control Systems, Vol. 15, No. 3, 2009, pp. 405-423.
9. M. A. Shehata, "Some Time-Optimal Control Problems for n × n Co-Operative Hyperbolic Systems with Distributed or Boundary Controls," Journal of Mathematical Sciences: Advances and Applications, Vol. 18, No. 1-2, 2012, pp. 63-83.
10. Kowalewski, "Optimal Control via Initial State of an Infinite Order Time Delay Hyperbolic System," 15th International Conference on Methods and Models in Automation and Robotics (MMAR), 23-26 August 2010, pp. 275-277.
11. R. A. Adams, "Sobolev Spaces," Academic Press, New York, 1975.
12. J. Fleckinger, J. Hernández and F. de Thélin, "On the Existence of Multiple Principal Eigenvalues for Some Indefinite Linear Eigenvalue Problems," Revista de la Real Academia de Ciencias Exactas, Físicas y Naturales. Serie A: Matemáticas, Vol. 97, No. 2, 2003, pp. 461-466.

CHAPTER 4

On Decay of Solutions and Spectral Property for a Class of Linear Parabolic Feedback Control Systems

Takao Nambu

Department of Applied Mathematics, Graduate School of System Informatics, Kobe University, Kobe, Japan

ABSTRACT

Unlike regular stabilizations, we construct in the paper a specific feedback control system such that $u(t)$ decays exponentially with the designated decay rate, and that some non-trivial linear functionals of u decay exactly faster than $\|u(t)\|$. The system contains a dynamic compensator with another state v in the feedback loop, and consists of two states u and v. This problem entirely differs from the one with static feedback scheme in which the system consists only of a single state u. To show the essential difference, some specific property of the spectral subspaces associated with our control system is studied.

Keywords: Stabilization of linear parabolic systems; decay of functionals; dynamic feedback scheme; spectral structures of composite systems; complete observability of systems

1. INTRODUCTION

Stabilization problems for linear parabolic control systems have the history of more than three decades. Although some difficult problems are left unresolved, it seems that the study has reached a degree of maturity in a sense. The so-called dynamic compensators are introduced in the feedback loop to cope with the most difficult case such as the scheme of boundary observation/boundary input (see the literature, e.g., [1-6]). In [2-4,6], no Riesz basis is assumed, corresponding to the coefficient elliptic operators with complicated boundary operators (see (1) below). Let H be a separable Hilbert space with the inner product $\langle \cdot,\cdot \rangle$ and the norm $\|\cdot\|$. A standard control system with state $u(t) \in H$, $t \geq 0$ consists of a finite number of inputs $g_k(t)$, $1 \leq k \leq M$, and outputs $c_k(u)$, $1 \leq k \leq N$, and is described by the following linear differential equation in H:

$$\frac{du}{dt} + Lu = \sum_{k=1}^{M} g_k(t) h_k, \quad t > 0, \quad u(0) = u_0.$$

Here, L denotes a linear closed operator with dense domain $\mathscr{D}(L)$ such that the resolvent $(\lambda - L)^{-1}$ is compact; $h_k \in H$ actuators through which the scalarvalued inputs $g_k(t)$ are inserted in the equation; and $c_k(u)$ linear functionals of u which allow unboundedness but are subordinate to L. The control system also reflects boundary feedback schemes by interpreting h_k and the differential equation in weaker topologies. In general stabilization studies, the inputs $g_k(t)$ are designed as a suitable feedback of the outputs $c_k(u)$, so that the state $u(t)$ could be stabilized as $t \to \infty$. Then every linear functional of u also decays at least with the same decay rate. This is true in the case where the functional is unbounded and subordinate to L.

We then raise a question: can we find a nontrivial linear functional which decays faster than $\|u(t)\|$? The purpose of the paper is to construct a specific feedback control system such that $\|u(t)\|$ decays exponentially with the designated decay rate, and that some nontrivial linear functionals of $u(t)$, say $\langle u(t), f \rangle$, decay definitely faster than $\|u(t)\|$ for any initial state. To achieve this property, our control scheme contains a dynamic compensator with state $v(t)$ in another separable Hilbert space $\mathscr{H}$ in the feedback loop to connect $c_k(u(t))$ and $g_k(t)$. Thus the control system has state $(u(t), v(t))$ in the product space $H \times \mathscr{H}$. We note that the above decay property is achieved in a straightforward manner in the static feedback control scheme in which we set $M = N$ and $g_k(t) = c_k(u(t))$, $1 \le k \le M$. In fact, the static feedback system contains a single state $u(t)$ only, and the so-called spectral decomposition of H associated with the elliptic operator enables us to find such an $f \neq 0$ in some spectral subspace. Such typical examples are the Fourier coefficients corresponding to higher frequencies. In our control system, however, it is indispensable in the spectral decomposition method to ensure a vector of the form $\boldsymbol{f} = (f, 0)$ in the spectral subspace of $H \times \mathscr{H}$ to achieve a faster decay of $\langle \boldsymbol{x}, \boldsymbol{f} \rangle_{H \times \mathscr{H}} = \langle u, f \rangle$, where $\boldsymbol{x} = (u, v)$1. It is very unlikely and almost denied to find such a vector $\boldsymbol{f} = (f, 0)$ in our control system with state (u, v). To make the paper clearer and more readable, unlikeliness of the above vector $\boldsymbol{f} = (f, 0)$ is discussed in detail in Section 3, which turns out to be a new spectral feature of the control system, and has never appeared in the literature; this spectral property also justifies the relevance of our problem setting.

Let us begin with the characterization of the controlled plant. Let Ω denote a bounded domain in $\mathbb{R}^m$ with the boundary Γ which consists of a finite number of smooth components of $(m-1)$-dimension. Let $(\mathscr{L}, \tau)$ be a pair of linear operators defined by

$$\mathscr{L} u = -\sum_{i,j=1}^{m} \frac{\partial}{\partial x_i}\left(a_{ij}(x) \frac{\partial u}{\partial x_j} \right) + \sum_{i=1}^{m} b_i(x) \frac{\partial u}{\partial x_i} + c(x) u,$$

$$\tau u = \alpha(\xi) u + (1 - \alpha(\xi)) \frac{\partial u}{\partial \nu}, \tag{1}$$

where $a_{ij}(x) = a_{ji}(x)$ for $1 \le i, j \le m$, $x \in \bar{\Omega}$; $\sum_{i,j=1}^{m} a_{ij}(x) \xi_i \xi_j \ge \delta |\xi|^2$, $\forall \xi = (\xi_1, \cdots, \xi_m) \in \mathbb{R}^m$, $\forall x \in \bar{\Omega}$ for some positive δ and

$$0 \le \alpha(\xi) \le 1, \qquad \frac{\partial u}{\partial \nu} = \sum_{i,j=1}^{m} a_{ij}(\xi) \nu_i(\xi) \left. \frac{\partial u}{\partial x_j} \right|_{\Gamma},$$

$\nu(\xi) = (\nu_1(\xi), \cdots, \nu_m(\xi))$ being the unit outer normal at $\xi \in \Gamma$. Henceforth set $H = L^2(\Omega)$. The pair $(\mathscr{L}, \tau)$ defines an operator $\hat{L}$ closable in H as $\hat{L} u = \mathscr{L} u$ for

$$u \in \mathscr{D}(\hat{L}) = \{u \in C^2(\Omega) \cap C^1(\bar{\Omega});\ \mathscr{L}u \in L^2(\Omega),\ \tau u = 0\}.$$

The closure of $\hat{L}$ is denoted as L. It is well known (see [7]) that L has a compact resolvent $(\lambda - L)^{-1}$; that the spectrum $\sigma(L)$ lies in the complement $(\bar{\Sigma} - b)^c$ of some sector $\bar{\Sigma} - b$, where $\bar{\Sigma} = \{\lambda \in \mathrm{C}; \theta_0 \le |\arg \lambda| \le \pi\},\ 0 < \theta_0 < \pi/2, b \in \mathrm{R}^1$; and that the following estimates hold:

$$\left\|(\lambda - L)^{-1}\right\| \le \frac{\text{const}}{1 + |\lambda|}, \quad \lambda \in \bar{\Sigma} - b, \tag{2}$$

where the symbol $\|\cdot\|$ also denotes the $\mathscr{L}(H)$-norm. Thus $-L$ is an infinitesimal generator of an analytic semigroup $e^{-tL}, t > 0$. The fractional powers L_c^{ω}, $\omega > 0$ are defined in a standard manner, where $L_c = L + c$ and $c > 0$ is sufficiently large. It is not very clear on how the domain $\mathscr{D}(L_c^{\omega})$ is characterized by the fractional Sobolev spaces, since the Dirichlet boundary is continuously connected by the Robin boundary. Such a characterization is, however, neither essential nor necessary in our study. There is a set of generalized eigenpairs $\{\lambda_i, \varphi_{ij}\}$

such that

1) $\sigma(L) = \{\lambda_i\}_{i=1}^{\infty}, \quad \mathrm{Re}\,\lambda_1 \le \cdots \le \mathrm{Re}\,\lambda_i \le \cdots \to \infty$;

2) $\lambda_i \ne \lambda_j$ for $i \ne j$; and

3) $L\varphi_{ij} = \lambda_i \varphi_{ij} + \sum_{k<j} \alpha_{jk}^i \varphi_{ik}, \quad i \ge 1,\ 1 \le j \le m_i (< \infty)$.

It is well known (see, e.g., page 285 of [8]) that the set $\{\varphi_{ij}\}$ spans $H = L^2(\Omega)$, but does not necessarily form a Riesz basis for H. Let P_{λ_i} be the (not necessarily orthogonal) projector corresponding to the eigenvalue λ_i. The restriction of L onto the invariant subspace $P_{\lambda_i}H$ is, according to the basis $\{\varphi_{ij}; 1 \le j \le m_i\}$, equivalent to the $m_i \times m_i$ upper triangular matrix Λ_i:

$$\Lambda_i|_{(j,k)} = \begin{cases} \alpha^i_{kj}, & j<k, \\ \lambda_i, & j=k, \\ 0, & j>k. \end{cases} \tag{3}$$

By setting $\Lambda_i = \lambda_i + N_i$, the matrix N_i is nilpotent, that is, $N_i^{m_i} = 0$. The minimum integer $n = l_i$ such that $\ker N_i^n = \ker N_i^{n+1}$ is called the ascent of $\lambda_i - L$. It is well known that the ascent l_i coincides with the order of the pole λ_i of $(\lambda - L)^{-1}$ (see Theorem 5.8-A of [9] for more details). Let $(\mathscr{L}^*, \tau^*)$ be the formal adjoint of $(\mathscr{L}, \tau)$:

$$\begin{aligned} \mathscr{L}^*\varphi &= -\sum_{i,j=1}^m \frac{\partial}{\partial x_i}\left(a_{ij}(x)\frac{\partial \varphi}{\partial x_j}\right) - \operatorname{div}(\boldsymbol{b}(x)\varphi) + c(x)\varphi, \\ \tau^*\varphi &= \alpha(\xi)\varphi \\ &\quad + (1-\alpha(\xi))\left(\frac{\partial \varphi}{\partial \nu} + (\boldsymbol{b}(\xi)\cdot \boldsymbol{\nu}(\xi))\varphi\right), \end{aligned} \tag{4}$$

where $\boldsymbol{b}(x) = (b_1(x), \cdots, b_m(x))$. The pair $(\mathscr{L}^*, \tau^*)$ defines an operator $\hat{L}^*$ just as the above $\hat{L}$. Then the adjoint of L, denoted by L^*, is given as the closure of $\hat{L}^*$ in H. There is a set of generalized eigenpairs $\{\overline{\lambda}_i, \psi_{ij}\}$ such that

1) $\sigma(L^*) = \{\overline{\lambda}_1, \overline{\lambda}_2, \cdots, \overline{\lambda}_i, \cdots\}$; and

2) $L^*\psi_{ij} = \overline{\lambda}_i \psi_{ij} + \sum_{k<j} \beta^i_{jk}\psi_{ik}, \; i \ge 1, \; 1 \le j \le m_i (< \infty)$.

Similarly, the set $\{\psi_{ij}\}$ spans H. Setting $P_{\lambda_i} u = \sum_{j=1}^{m_i} u_{ij}\varphi_{ij}$, we have the relationship:

$$\begin{pmatrix} u_{i1} \\ \vdots \\ u_{im_i} \end{pmatrix} = \Pi_{\lambda_i}^{-1} \begin{pmatrix} \langle u, \psi_{i1} \rangle \\ \vdots \\ \langle u, \psi_{im_i} \rangle \end{pmatrix},$$
$$\Pi_{\lambda_i} = \left(\langle \varphi_{ij}, \psi_{il} \rangle ; \begin{matrix} j \to 1, \cdots, m_i \\ l \downarrow 1, \cdots, m_i \end{matrix} \right). \tag{5}$$

Let us turn to the characterization of a dynamic compensator. Let $\mathscr{H}$ be any separable Hilbert space with inner product $\langle \cdot, \cdot \rangle_{\mathscr{H}}$ and norm $\|\cdot\|_{\mathscr{H}}$. Relabelling an orthonormal basis for $\mathscr{H}$, let $\{\eta_{ij}^{\pm}; i \geq 1, \quad 1 \leq j \leq n_i (< \infty)\}$ be a new orthonomal basis for $\mathscr{H}$. Every vector $v \in \mathscr{H}$ is then expressed in terms of the basis $\{\eta_{ij}^{+}\}$ as a Fourier series: $\sum_{i,j} v_{ij}^{\pm} \eta_{ij}^{\pm}$ $v = \sum_{i,j} \left(v_{ij}^{+} \eta_{ij}^{+} + v_{ij}^{-} \eta_{ij}^{-} \right) =$, $v_{ij}^{\pm} = \langle v, \eta_{ij}^{\pm} \rangle_{\mathscr{H}}$. Let $\{\mu_i\}$ be a sequence of increasing positive numbers: $0 < \mu_1 < \mu_2 < \cdots \to \infty$, and set $\omega^{\pm} = a \pm \sqrt{-1}\sqrt{1-a^2}$, where $0 < a < 1$. Let B be a linear closed operator defined as

$$Bv = \sum_{i,j} \mu_i \omega^{\pm} v_{ij}^{\pm} \eta_{ij}^{\pm} \tag{6}$$

with dense domain $\mathscr{D}(B) = \left\{ v; \sum_i \left| \mu_i v_{ij}^{\pm} \right|^2 < \infty \right\}$. Then, 1) $\sigma(B) = \{\mu_i \omega^{\pm}; i \geq 1\}$; and 2) $(\mu_i \omega^{\pm} - B)\eta_{ij}^{\pm} = 0$, $i \geq 1$, $1 \leq j \leq n_i$. Thus $-B$ is the infinitesimal generator of an analytic semigroup $e^{-tB}, t > 0$, which is expressed by $e^{-tB} v = \sum_{i,j} e^{-\mu_i \omega^{\pm} t} v_{ij}^{\pm} \eta_{ij}^{\pm}$. The semigroup e^{-tB} satisfies the decay estimate

$$\left\| e^{-tB} \right\|_{\mathscr{H}} \leq e^{-a\mu_1 t}, \quad t \geq 0. \tag{7}$$

The adjoint operator B^* of B is described as

$$B^* v = \sum_{i,j} \mu_i \omega^{\mp} v_{ij}^{\pm} \eta_{ij}^{\pm}, \quad v \in \mathscr{D}(B^*) = \mathscr{D}(B), \tag{8}$$

and thus $B^* \eta_{ij}^{\pm} = \mu_i \omega^{\mp} \eta_{ij}^{\pm}$. Let $\mathscr{P}_n$, $n \geq 1$ be the projector in $\mathscr{H}$ such that $\mathscr{P}_n v = \sum_{i(\leq n), j} v_{ij}^{\pm} \eta_{ij}^{\pm}$.

Our control system has state $(u,v) \in H \times \mathscr{H}$, and is described as a differential equation in $H \times \mathscr{H}$:

$$\begin{aligned}
&\frac{du}{dt} + Lu = -\left\langle v, \left(\alpha - B^*\right) f \right\rangle_{\mathscr{H}} \eta + \sum_{k=1}^{M} \left\langle v, \rho_k \right\rangle_{\mathscr{H}} \zeta_k, \\
&\frac{dv}{dt} + Bv = \sum_{k=1}^{N} c_k(u) \xi_k - \left\langle v, \left(\alpha - B^*\right) f \right\rangle_{\mathscr{H}} X\eta \\
&\qquad + \sum_{k=1}^{M} \left\langle v, \rho_k \right\rangle_{\mathscr{H}} X\zeta_k.
\end{aligned} \tag{9}$$

The equation with state v means a dynamic compensator equipped with a set of outputs $\left\langle v, \left(\alpha - B^*\right) f \right\rangle_{\mathscr{H}}$ and $\left\langle v, \rho_k \right\rangle_{\mathscr{H}}$, $1 \le k \le M$, where α > 0. The parameters η and ζ_k denote given actuators of the controlled plant, and ξ_k actuators of the compensator to be designed. The outputs $c_k(u)$ of the controlled plant are considered on the boundary, and defined as

$$c_k(u) = \begin{cases} \left\langle u, w_k \right\rangle_{\Gamma}, & \text{when } \alpha(\xi) \not\equiv 1, \\ \left\langle \dfrac{\partial u}{\partial \nu}, w_k \right\rangle_{\Gamma}, & \text{when } \alpha(\xi) \equiv 1, \end{cases} \tag{10}$$

where $w_k \in L^2(\Gamma)$ denotes observation weights. The operator $X \in \mathscr{L}(H; \mathscr{H})$, specified later, denotes a unique solution to the operator equation on $\mathscr{D}(L)$,

$$XL - BX = C, \quad \text{where } C = -\sum_{k=1}^{N} c_k(\cdot)\xi_k. \tag{11}$$

Given a suitable vector $f \in \mathscr{H}$, the control law is to construct such that the number $-\alpha$ determines the decay rate of a functional $\left\langle u(t,\cdot), \hat{f} \right\rangle$, where $\hat{f} = X^* f \ne 0$. As we see later, the roles of the outputs $\left\langle v(t), \rho_k \right\rangle_{\mathscr{H}}$ and $\left\langle v(t), \left(\alpha - B^*\right) f \right\rangle_{\mathscr{H}}$ are, respectively, to determine the decay of $u(t,\cdot)$ and the decay of $\left\langle u(t,\cdot), \hat{f} \right\rangle$. In state stabilization problems only, the output $\left\langle v(t), \left(\alpha - B^*\right) f \right\rangle_{\mathscr{H}}$ does not appear. More precisely, let β be a number such that $\alpha > \beta > 0$. We seek a new feedback scheme such that the decay estimates

$$\|u(t)\| \leq \text{const}e^{-\beta t}, \text{ and } \left|\left\langle u(t), \hat{f}\right\rangle\right| \leq \text{const}e^{-\alpha t} \tag{12}$$

hold for every initial value and $t \geq 0$, such that the decay of $\|u(t)\|$ is no longer improved. To achieve the non-standard decay (12), introduce a new operator

$$L_f = L + \left\langle \cdot, X^*\left(\alpha - B^*\right)f\right\rangle \eta, \quad \mathscr{D}\left(L_f\right) = \mathscr{D}\left(L\right), \tag{13}$$

and assume conditions in terms of L_f. These conditions have never appeared in the literature. Note that conditions are posed on L for state stabilization.

The functional $\left\langle u(t), \hat{f}\right\rangle$ may be regarded as a kind of output of the system. It is worthwhile to refer to our previous results on output stabilization [10-13]: In [10], the decay of outputs is discussed with lack of observability conditions, but the relationship of the decay between $u(t)$ and the outputs is unclear. In [11-13], the problem is discussed, based on a different principle, i.e., a finitedimensional pole assignment theory with constraint. The controlled plants are, however, limited to those equipped with Riesz basis, and the actuators of the controlled plant, corresponding to our η, are restrictive, and must be subject to a strong constraint: the actuators have to be designed so that their spectral elements in each spectral subspace are orthogonal to the weights of the outputs at infinity. As we have seen, the controlled plants in the present paper do not necessarily allow a Riesz basis, although a somewhat stronger condition, i.e., complete observability, is assumed. An example of systems equipped with complete observability is illustrated in the end of Section 2.

The feedback law in (9) contains parameters: $\alpha > 0$, η, $c_k(\cdot)$, ξ_k, f, ρ_k, and ζ_k to achieve the nonstandard decays (12). They are designed in the following manner: 1) The actuator $\eta \neq 0$ is given in advance arbitrarily. 2) Given the parameters $c_k(\cdot)$ and ξ_k, the operator solution X is ensured. 3) Then, a vector f is chosen among a very wide range of sets in $\mathscr{H}$. 4) Finally, ζ_k, as well as ρ_k, are designed to satisfy a finite number of controllability conditions associated with the new operator L_f. It is generally desirable to pose less restrictive assumptions on the actuators of the controlled plant. In fact, $\eta \neq 0$ is arbitrary, and, as we see later that the conditions on ζ_k are much less restrictive than those in our preceding works above. In fact, ξ_k only have to be designed, belonging to an infinite dimensional subspace of H.

Our main results consist of Theorem 4 in Section 2 and a series of assertions in Section 3 (Theorem 7, Propositions 8, and Theorem 9): The former is on the control law ensuring the non-standard decays (12), and the latter on the relevance of the problem setting with Equation (8), which discusses a spectral property of the invariant subspaces in $H \times \mathscr{H}$ associated with the coefficient operator in (9), that is, unlikeliness of a vector of the form, $f = (f_1, 0)$ in these subspaces. In Section 3, the spectrum of the coefficient operator is characterized (Theorem 9). To the best of the author's knowledge, the latter has never been discussed so far, and clarifies a new property of internal structures of control systems.

2. DECAY ESTIMATES OF SOLUTIONS

To ensure well-posedness of our control system (8), let us begin with the operator Equation (10). Adjusting ω^+ and μ_i, $i \ge 1$, we may assume that

$$\begin{aligned}&\sigma(L) \cap \sigma(B) = \varnothing; \quad \arg \omega^+ > \theta_0;\\ &\text{and } \mu_i \le \text{const } i^\gamma, i \ge 1, \text{ for } 0 < \exists \gamma < 2\end{aligned} \tag{14}$$

(see (2) for θ_0). In (9), let us express the actuators ξ_k, $1 \le k \le N$ as Fourier series in terms of $\{\eta_{ij}^{+}\}$:

$$\xi_k = \sum_{i,j} \left(\xi_{ij}^k \eta_{ij}^+ + \overline{\xi_{ij}^k} \eta_{ij}^- \right), \quad \sum_{i,j} \left| \xi_{ij}^k \right|^2 < \infty,$$

where the overline denotes the complex conjugate. Before stating our first result, let us define the matrices Ξ_i, and W_i, $i \ge 1$ by

$$\Xi_i = \left(\xi_{ij}^k; \begin{matrix} k \to 1, \cdots, N \\ j \downarrow 1, \cdots, n_i \end{matrix} \right), \quad i \ge 1, \tag{15}$$

and

$$W_i = \begin{pmatrix} c_k(\varphi_{ij}); & k \downarrow 1,\cdots,N \\ & j \to 1,\cdots,m_i \end{pmatrix}, \quad i \geq 1, \tag{16}$$

respectively. Then our first result is stated as follows:

THEOREM 1. 1) By assuming (14), the operator equation (11) on $\mathscr{D}(L)$ admits a unique operator solution $X \in \mathscr{L}(H;\mathscr{H})$, which is expressed as

$$Xu = \sum_{k=1}^{N} \sum_{i,j} \Big(c_k\Big(\big(\mu_i\omega^+ - L\big)^{-1} u\Big)\xi_{ij}^k \eta_{ij}^+ \\ + c_k\Big(\big(\mu_i\omega^- - L\big)^{-1} u\Big)\overline{\xi_{ij}^k}\eta_{ij}^- \Big). \tag{17}$$

2) Assume further that

$$\operatorname{rank}\big(W_i\Lambda_i^l; l \downarrow 0,\cdots,m_i-1\big) = m_i, \\ \textit{and} \quad \operatorname{rank}\Xi_i = N, \quad i \geq 1. \tag{18}$$

Then we have $\overline{X^*\mathscr{H}} = H$. Here, $\overline{X^*\mathscr{H}}$ denotes the closure of $X^*\mathscr{H}$ in H.

Remark. 1) The first condition in (18), called the complete observability condition, is fulfilled with $N = 1$ in the case where $\alpha_{(j+1)j}^i \neq 0$, $1 \leq j < m_i$, $i \geq 1$. Actually, by choosing the w_1 with $c_1(\varphi_{i1}) \neq 0$, $i \geq 1$, the condition is fulfilled. In the case where $N_i = 0$, $i \geq 1$, however, the condition means that $\operatorname{rank} W_i = m_i$, $i \geq 1$, which requires that N be equal to or greater than $\sup_{i\geq 1} m_i$: this is the case, for example, where L is selfadjoint.

2) The condition: $\sup_{i\geq 1} m_i < \infty$ is the so called finite multiplicity condition. In the case of $m = 1$, we know that $m_i = 1$, $i \geq 1$. Thus, by choosing $N = 1$, the complete observability condition is automatically fulfilled. As another example, let L be a self-adjoint operator defined by $L = -\Delta$ in $\Omega = \{|x| < 1\} \subset \mathbb{R}^2$ equipped with the Dirichlet boundary. The eigenvalues of L consist of λ_{mk}, $m \geq 0$, $k \geq 1$, where $\lambda_{mk}^{1/2}$ are the zeros of the Bessel functions $J_m(\cdot)$ of m-th order. It is expected that $\sup m_i = 2$, if the well known Bourget's hypothesis (see pp. 484-485 of [14]) is proven. As long as the author knows, this conjecture has not been proven so far.

Proof. The result is a version of the results in [2-5], so that we give here only an outline of the proof. 1) Expression (17) and uniqueness of X are examined in a straightforward manner.

2) Relation $\overline{X^{*}\mathscr{H}} = H$ is equivalent to $\ker X = \{0\}$. Assuming that $Xu = 0$, we see by (17) that

$$\Xi_i \begin{pmatrix} c_1\left(\left(\mu_i\omega^{+} - L\right)^{-1} u\right) \\ \vdots \\ c_N\left(\left(\mu_i\omega^{+} - L\right)^{-1} u\right) \end{pmatrix} = \mathbf{0}, \quad i \ge 1.$$

Since $\operatorname{rank}\Xi_i = N$, we see that $c_k\left(\left(\mu_i\omega^{+} - L\right)^{-1} u\right) = 0$ for $i \ge 1$ and $1 \le k \le N$.

For a $c > 0$ such that $-c \in \rho(L)$ and each k, we introduce a series of meromorphic functions $f_k^l(\lambda)$, $l = 0,1,\cdots$ by the recursion formula:

$$f_k^0(\lambda) = c_k\left(\left(\lambda - L\right)^{-1} u\right), \quad f_k^{l+1}(\lambda) = \frac{f_k^l(\lambda)}{\lambda + c}. \tag{19}$$

Each function $f_k^l(\lambda)$ has the properties: 1) It has at least the zeros $\lambda = \mu_i\omega^{\pm}$, $i \ge 1$; and 2) the algebraic growth rate γ of the se zeros, $\mu_i\omega^{\pm}$ is smaller than 2 by (14). These properties combined with Carleman's theorem [15,16] imply that $f_k^l(\lambda) = 0$ for $\lambda \in \rho(L)$, $1 \le k \le N$, $l = 0,1,\cdots$. Following [5], we calculate the residue at each λ_i. Then, we see that

$$c_k\left(P_{\lambda_i} L_c^{-l} u\right) = 0, \quad 1 \le k \le N, \; l \ge 0. \tag{20}$$

Let $P_{\lambda_i} u = \sum_{1 \le j \le m_i} u_{ij}\varphi_{ij} \Leftrightarrow \boldsymbol{u}_i = \begin{pmatrix} u_{i1} & \cdots & u_{im_i} \end{pmatrix}^{\mathrm{T}}$. Note that the restriction of L_c^{-l} on $P_{\lambda_i} H$ is equivalent to the matrix $\left(\Lambda_i + c\right)^{-l}$, and thus, $P_{\lambda_i} L_c^{-l} u \Leftrightarrow \left(\Lambda_i + c\right)^{-l} \boldsymbol{u}_i$. The relation (20) is rewritten as

$$W_i\left(\Lambda_i + c\right)^{-l} \boldsymbol{u}_i = \mathbf{0}, \quad i \ge 1, l \ge 1. \tag{21}$$

The complete observability in (18) implies that $u_i = 0$ for $i \geq 1$. In view of (5), we see that $\langle u, \psi_{ij} \rangle = 0$ for every i and j. Since the set $\{\psi_{ij}\}$ spans the whole space H, we conclude that $u = 0$. Q.E.D.

Decay of solutions to Equation (9): In view of (11), it is easily seen that $(Xu - v)_t + B(Xu - v) = 0$, $t > 0$, or $Xu(t) - v(t) = e^{-tB}(Xu_0 - v_0)$. Thus,

$$\|Xu(t) - v(t)\|_{\mathscr{H}} \leq e^{-a\mu_1 t} \|Xu_0 - v_0\|_{\mathscr{H}} \tag{22}$$

for $t \geq 0$. In (9). let $\alpha < a\mu_1$, and $\alpha \notin \sigma(L)$ (see Proposition 2). Equation (9) contains various parameters: α, $c_k(\cdot)$, ξ_k, η, f, ς_k, and ρ_k, among which α, $c_k(\cdot)$ and ξ_k are already determined in Theorem 1. Let a non-trivial $\eta \in H$ be given arbitrarily. Then, $X\eta \neq 0$ by Theorem 1. We find a non-trivial vector $f = \sum_{i=1}^{R} \sum_{j=1}^{n_i} \left(f_{ij} \eta_{ij}^{+} + \overline{f_{ij}} \eta_{ij}^{-} \right) \in \mathscr{P}_R \mathscr{H}$ such that

$$\langle \xi_k, f \rangle_{\mathscr{H}} = 0, \quad 1 \leq k \leq N, \quad \langle X\eta, f \rangle_{\mathscr{H}} = 1. \tag{23}$$

There is a variety of choice of such an f. In fact, this is simply possible, e.g., by finding $i < R$ such that

$$\left| c_k \left(\left(\mu_R \omega^{\pm} - L \right)^{-1} \eta \right) \right| < \left| c_k \left(\left(\mu_i \omega^{\pm} - L \right)^{-1} \eta \right) \right|$$

for $1 \leq k \leq N$. Then $\mathscr{P}_R X\eta$ does not belong to the space spanned by $\mathscr{P}_R \xi_k$, $1 \leq k \leq N$. The integer R may be chosen arbitrarily large. The vectors ς_k will be determined in terms of the operator L_f.

The state $u(t, \cdot)$ in (9) satisfies the equation:

$$\frac{d}{dt} Xu + BXu + Cu$$
$$= -\langle v, (\alpha - B^*) f \rangle_{\mathscr{H}} X\eta + \sum_{k=1}^{M} \langle v, \rho_k \rangle_{\mathscr{H}} X\varsigma_k.$$

Actuators ζ_k are chosen so that $\langle X\zeta_k, f\rangle_{\mathscr{H}} = 0$, $1 \le k \le M$. An assumption on these ζ_k will be discussed later (see (28)). Then, it is immediately seen that

$$\frac{\mathrm{d}}{\mathrm{d}t}\langle Xu, f\rangle_{\mathscr{H}} + \alpha\langle Xu, f\rangle_{\mathscr{H}} = \langle Xu - v, (\alpha - B^*)f\rangle_{\mathscr{H}}$$
$$= \langle \mathrm{e}^{-tB}(Xu_0 - v_0), (\alpha - B^*)f\rangle_{\mathscr{H}}. \tag{24}$$

By the decay (22), we obtain the estimate:

$$\left|\langle u, \hat{f}\rangle\right| = \left|\langle Xu, f\rangle_{\mathscr{H}}\right| \le \mathrm{const}\,\mathrm{e}^{-\alpha t} \tag{25}$$

for $t \ge 0$. Here, $\hat{f} = X^* f \in H$ is non-trivial.

The operator L_f defined in (13) has a compact resolvent, and $\sigma(L_f)$ consists only of eigenvalues. Let

$$\sigma(L_f) = \{\gamma_i\}_{i \ge 1}, \tag{26}$$

where $\mathrm{Re}\,\gamma_1 \le \cdots \le \mathrm{Re}\,\gamma_i \le \cdots \to \infty$, and $\gamma_i \ne \gamma_j$ for i ≠ j. Each γ_i may admit generalized eigenfunctions. Let P_i^f be the projector corresponding to the eigenvalue γ_i which is calculated as $P_i^f = \frac{1}{2\pi\sqrt{-1}}\int_{C_i}(\lambda - L_f)^{-1}\,\mathrm{d}\lambda$, C_i being the small counterclockwise circle with center γ_i. Then

PROPOSITION 2. 1) The number $\alpha \notin \sigma(L)$ belongs to $\sigma(L_f)$. 2) Any generalized eigenfunction of L_f in $P_i^f H$ $(\gamma_i \ne \alpha)$ and $X^* f$ are orthogonal to each other.

Proof. 1) Suppose that there is a $\chi \ne 0$ such that $(\alpha - L_f)\chi = 0$. Then, $(\alpha - L)\chi = \langle \chi, X^*(\alpha - B^*)f\rangle \eta$. Thus we have, as a necessary condition,

$$\chi = \langle \chi, X^*(\alpha - B^*)f\rangle (\alpha - L)^{-1}\eta.$$

Now set $\chi=(\alpha-L)^{-1}\eta\,(\neq 0)$, and calculate as

$$\begin{aligned}\langle \chi, X^*(\alpha-B^*)f\rangle &= \langle X\chi,(\alpha-B^*)f\rangle_{\mathscr{H}}\\ &=\alpha\langle X\chi,f\rangle_{\mathscr{H}}-\langle XL\chi,f\rangle_{\mathscr{H}}\\ &=\langle X(\alpha-L)\chi,f\rangle_{\mathscr{H}}\\ &=\langle X\eta,f\rangle_{\mathscr{H}}=1.\end{aligned}$$

Thus we see that

$$\alpha\chi = L\chi+\eta = L\chi+\langle \chi, X^*(\alpha-B^*)f\rangle\eta = L_f\chi.$$

2) Let $\gamma\in\sigma(L_f)$, $\gamma\neq\alpha$, and $W_\gamma^{(k)}=\left\{\chi;(\gamma-L_f)^k\chi=0\right\}$, $k\geq 1$, possible generalized eigenspaces of γ. For a $\chi\in W_\gamma^{(1)}$, we calculate by (23) as

$$\begin{aligned}\gamma\langle \chi, X^*f\rangle &= \langle L\chi, X^*f\rangle+\langle \chi, X^*(\alpha-B^*)f\rangle\\ &=\langle (XL-BX)\chi,f\rangle_{\mathscr{H}}+\alpha\langle \chi,X^*f\rangle\\ &=\langle C\chi,f\rangle_{\mathscr{H}}+\alpha\langle \chi,X^*f\rangle=\alpha\langle \chi,X^*f\rangle.\end{aligned}$$

This implies that $\langle \chi, X^*f\rangle=0$, or $W_\gamma^{(1)}\perp X^*f$. Suppose then that $W_\gamma^{(k)}\perp X^*f$,$i\geq 1$. For a $\chi\in W_\gamma^{(k+1)}$, the function $(\gamma-L_f)\chi$ is in $W_\gamma^{(k)}$, and $\langle(\gamma-L_f)\chi, X^*f\rangle=0$. The same calculation as above immediately shows that $\langle \chi, X^*f\rangle=0$. Thus,

$$X^*f\perp\bigcup_{k\geq 1}W_\gamma^{(k)}. \tag{27}$$

The integer k varies over a finite set of positive integers depending on γ. Q.E.D.

We now choose the actuators ζ_k in (9) such that

$$\zeta_k \in \overline{\bigcup_{\gamma_i \neq \alpha}\left(\bigcup_{i \geq 1} W_{\gamma_i}^{(k)}\right)}, \quad 1 \leq k \leq M. \tag{28}$$

Then, $\zeta_k \perp X^* f$ by the above proposition. All parameters except for ρ_k in (9) are determined.

Rewrite the equation for $u(t)$ in (9) as

$$\frac{\mathrm{d}u}{\mathrm{d}t} + L_f u - \sum_{k=1}^{M} \langle u, X^* \rho_k \rangle \zeta_k$$
$$= \left\langle Xu - v, \left(\alpha - B^*\right) f \right\rangle_{\mathscr{H}} \eta - \sum_{k=1}^{M} \langle Xu - v, \rho_k \rangle_{\mathscr{H}} \zeta_k. \tag{29}$$

Let us introduce an operator L_{ff} as

$$L_{ff} = L + \left\langle \cdot, X^* \left(\alpha - B^*\right) f \right\rangle \eta - \sum_{k=1}^{M} \langle \cdot, c_k \rangle \zeta_k.$$

Let the integer l be such that $l = \min\{i; \alpha < \operatorname{Re}\gamma_i\}$. Then, $\alpha = \gamma_{l-1}, l \geq 2$. The following proposition is just a simple version of the result in [17].

PROPOSITION 3. Let $\hat{P}$ be the projector defined by $\hat{P} = \sum_{\substack{1 \leq i \leq l, \\ i \neq l-1}} P_i^f$. Choose a β such that $0 < \beta < \alpha$. Suppose that the pair $\left(L_f\big|_{\hat{P}H}, \left\{\hat{P}\zeta_1, \cdots, \hat{P}\zeta_M\right\}\right)$ is a controllable one. Then we find $c_k \in \hat{P}^* H$ such that

$$\min \operatorname{Re}\sigma\left(L_{ff}\right) = \beta; \quad \beta \in \sigma\left(L_{ff}\right).$$

Thus, $\left\| e^{-tL_{ff}} \right\| \leq C e^{-\beta t}$, $t \geq 0$. We are ready to state the non-standard deay of solutions to Equation (9).

THEOREM 4. Let $0 < \beta < \alpha < a\mu_1$ such that $\alpha \notin \sigma(L)$. Suppose that

1) ξ_k and $c_k(\cdot)$ satisfy the rank conditions (18);

2) $\eta \neq 0$ is arbitrarily given;

3) $f \in \mathscr{P}_R\mathscr{H}$ is chosen to satisfy (23); and 4) $\zeta_k \in \overline{\bigcup_{\gamma_i \neq \alpha}\left(\bigcup_{k \geq 1} W_{\gamma_i}^{(k)}\right)}$ satisfiy the controllability condition in Proposition 3.

Then we find a large integer n; vectors $\rho_k \in \mathscr{P}_n\mathscr{H}$; and a postive $\tilde{\beta} < \alpha$ close to β, and subsequently the control system in the product space $H \times \mathscr{P}_n\mathscr{H}$;

$$
\begin{aligned}
&\frac{du}{dt} + Lu = -\left\langle v_1, \left(\alpha - B^*\right) f\right\rangle_{\mathscr{P}_n\mathscr{H}} \eta \\
&\qquad + \sum_{k=1}^{M} \left\langle v_1, \rho_k\right\rangle_{\mathscr{P}_n\mathscr{H}} \zeta_k, \\
&\frac{dv_1}{dt} + B_1 v = \sum_{k=1}^{N} c_k(u) \mathscr{P}_n\xi_k + \sum_{k=1}^{M} \left\langle v_1, \rho_k\right\rangle_{\mathscr{P}_n\mathscr{H}} \mathscr{P}_n X \zeta_k \\
&\qquad - \left\langle v_1, \left(\alpha - B^*\right) f\right\rangle_{\mathscr{P}_n\mathscr{H}} \mathscr{P}_n X \eta,
\end{aligned}
\tag{30}
$$

where $B_1 = B|_{\mathscr{P}_n\mathscr{H}}$. Every solution $\left(u(t,\cdot), v_1(t)\right) \in H \times \mathscr{P}_n\mathscr{H}$ satisfies the decay estimate:

$$
\begin{aligned}
&\|u(t,\cdot)\| + \|v_1(t)\|_{\mathscr{H}} \leq \text{const}\, e^{-\tilde{\beta}t}\left(\|u_0\| + \|v_{10}\|_{\mathscr{H}}\right), \\
&\left|\left\langle u(t,\cdot), \hat{f}\right\rangle\right| \leq \text{const}\, e^{-\alpha t}\left(\|u_0\| + \|v_{10}\|_{\mathscr{H}}\right)
\end{aligned}
\tag{31}
$$

for $t \geq 0$. The estimates for $\|u(t,\cdot)\|$ and $\|v_1(t)\|_{\mathscr{H}}$ can be no longer improved.

Proof. Choose the functions $c_k \in \hat{P}^* H$, $1 \leq k \leq M$ stated in Proposition 3. In view of Theorem 1, we find $\rho_k \in \mathscr{H}$ such that $X^*\rho_k$ arbitrarily approximate c_k in the topology of H. Since $\mathscr{H}$ is separable, we may assume with no loss of generality that these ρ_k are constructed in $\mathscr{P}_n\mathscr{H}$ for an enough large $n > R$.

Let us consider the operator L_{ff} and the perturbed $\tilde{L}_{ff} = L_f - \sum_{k=1}^{M}\left\langle \cdot, X^*\rho_k\right\rangle \zeta_k$. The right-hand side of (29) is dominated by the decay estimate (22). Since $X^*\rho_k$ are chosen close to c_k, the semigroup $e^{-t\tilde{L}_{ff}}$ is stable, and satisfies the standard estimate:

$$\left\| e^{-t\tilde{L}_{ff}} \right\| \le Ce^{-(\beta - C\delta)t}, \quad t \ge 0, \tag{32}$$

where $\delta = \sum_k \left\| X^* \rho_k - c_k \right\| \left\| \zeta_k \right\|$. Thus every solution $(u(t,\cdot), v(t))$ to Equation (9) satisfies the estimate:

$$\left\| u(t,\cdot) \right\| + \left\| v(t) \right\|_{\mathscr{H}} \le \text{const} e^{-\hat{\beta}t} \left(\left\| u(0) \right\| + \left\| v_0 \right\| \right), \tag{33}$$

where $\hat{\beta} = \beta - C\delta > 0$. The decay estimate for the functional $\left\langle u(t,\cdot), \hat{f} \right\rangle$ is already obtained in (25).

The control system (30) is derived in the following manner: Set $v_1(t) = \mathscr{P}_n v(t)$, and apply the projector $\mathscr{P}_n$ to the equation for v in (9). By noting that $\left(\alpha - B^* \right) f \in \mathscr{P}_R \mathscr{H}$, then, (30) is immediately obtained. Equation (30) is clearly well posed in $H \times \mathscr{P}_n \mathscr{H}$: Thus every solution to (30) is derived from the solution $(u(t,\cdot), v(t))$ to (9) with initial value (u_0, v_0) such that $\mathscr{P}_n v_0 = v_{10}$, by setting $v_1(t) = \mathscr{P}_n v(t)$. Thus the first estimate of (31) is derived from (33). The second estimate of (31) is clear by (25).

Finally we show that the first estimate of (31) for $(u(t,\cdot), v_1(t))$ is no more improved. The spectrum $\sigma\left(\tilde{L}_{ff}\right)$ of the perturbed operator $\tilde{L}_{ff}$ consists only of eigenvalues. It is expected that the value of $\min \operatorname{Re} \sigma\left(\tilde{L}_{ff}\right)$ would be close to $\beta = \min \operatorname{Re} \sigma\left(L_{ff}\right)$ as long as $X^* \rho_k$ are close to c_k: When both L_{ff} and $\tilde{L}_{ff}$ are selfadjoint, it is well known—via the min-max principle (see [18])—that each eigenvalue of L_{ff} is continuous relative to the coefficient parameters. In our problem, the following result holds:

PROPOSITION 5. The minimum of $\operatorname{Re} \sigma\left(\tilde{L}_{ff}\right)$ is continuous relative to $\left\| X^* \rho_k - c_k \right\|$, $1 \le k \le M$.

Proof. Set $L_{ff} + D_\varepsilon$ $\tilde{L}_{ff} = L_{ff} + \sum_{1 \le k \le M} \left\langle \cdot, c_k - X^* \rho_k \right\rangle \zeta_k =$. In view of (32), the left half-plane: $\operatorname{Re} \lambda < \beta - C\delta$ is contained in $\rho\left(\tilde{L}_{ff}\right)$. Thus we see that $\beta - C\delta \le \min \operatorname{Re} \sigma\left(\tilde{L}_{ff}\right)$. Choose an $r > 0$ enough small so that $\{\lambda; |\lambda-\beta|<r\}\cap\sigma(L_{ff})=\{\beta\}$. Let Γ be the counterclockwise circle: $\Gamma=\{\lambda; |\lambda-\beta|=r\}$, and suppose that $\left\|(\lambda-L_{ff})^{-1}\right\|\le d=d(r)$ for $\lambda\in\Gamma$. Choose $X^*\rho_k$ such that $\|D_\varepsilon\|\le\delta<(2d)^{-1}$. Then, Γ belongs to $\rho(\tilde{L}_{ff})$. In fact, we have the relation:

$$(\lambda-\tilde{L}_{ff})^{-1}=(\lambda-L_{ff})^{-1}\left(1-D_\varepsilon(\lambda-L_{ff})^{-1}\right)^{-1},$$
$$\left\|(\lambda-\tilde{L}_{ff})^{-1}\right\|\le 2d,\quad \lambda\in\Gamma. \tag{34}$$

Recall that, for $\lambda\in\rho(L_{ff})\cap\rho(\tilde{L}_{ff})$, the (second) resolvent equation:

$$(\lambda-L_{ff})^{-1}-(\lambda-\tilde{L}_{ff})^{-1}=-(\lambda-L_{ff})^{-1}D_\varepsilon(\lambda-\tilde{L}_{ff})^{-1}$$

holds. Then we see that

$$\frac{1}{2\pi i}\int_\Gamma(\lambda-L_{ff})^{-1}u\,d\lambda-\frac{1}{2\pi i}\int_\Gamma(\lambda-\tilde{L}_{ff})^{-1}u\,d\lambda$$
$$=\frac{-1}{2\pi i}\int_\Gamma(\lambda-L_{ff})^{-1}D_\varepsilon(\lambda-\tilde{L}_{ff})^{-1}u\,d\lambda,\quad \forall u\in H. \tag{35}$$

The first term of the above left-hand side of (35) is the projector, corresponding to the eigenvalue β of L_{ff}. Choose $X^*\rho_k$ closer to c_k, if necessary, so that

$$\delta=\sum_{k=1}^{M}\left\|X^*\rho_k-c_k\right\|\left\|\varsigma_k\right\|<\min\left(\frac{1}{2d},\frac{1}{2d^2r},r\right).$$

Supposing that $\sigma(\tilde{L}_{ff})$ is contained in the half-plane: $\{\lambda;\, \mathrm{Re}\,\lambda>\beta+r\}$, we then derive a contradiction. If so, the resolvent $(\lambda-\tilde{L}_{ff})^{-1}u$ is analytic inside and on Γ. Thus the second term of the left-hand side of (35) must be equal to 0. Let u be an eigenfunction of L_{ff}, corresponding to the eigenvalue β. Then,

$$u = \frac{-1}{2\pi i}\int_{\Gamma}\left(\lambda - L_{ff}\right)^{-1} D_{\varepsilon}\left(\lambda - \tilde{L}_{ff}\right)^{-1} u\,\mathrm{d}\lambda \neq 0.$$

The right-hand side is, however, estimated as follows:

$$\begin{aligned}&\left\|\frac{-1}{2\pi i}\int_{\Gamma}\left(\lambda - L_{ff}\right)^{-1} D_{\varepsilon}\left(\lambda - \tilde{L}_{ff}\right)^{-1} u\,\mathrm{d}\lambda\right\| \\ &\leq \frac{1}{2\pi}\int_{\Gamma}\left\|\left(\lambda - L_{ff}\right)^{-1} D_{\varepsilon}\left(\lambda - \tilde{L}_{ff}\right)^{-1} u\right\|\left|\mathrm{d}\lambda\right| \\ &\leq 2d^2 r\left\|D_{\varepsilon}\right\|\left\|u\right\| < \left\|u\right\|,\end{aligned}$$

which is a contradiction. Therefore, the spectrum $\sigma\left(\tilde{L}_{ff}\right)$ also lies in the left-half plane: $\mathrm{Re}\,\lambda \leq \beta + r\}\ \{\lambda \in \mathrm{C};$. As a conclusion, the minimum of $\mathrm{Re}\,\sigma\left(\tilde{L}_{ff}\right)$ satisfies the estimate:

$$\beta - Cr \leq \min \mathrm{Re}\,\sigma\left(\bar{M}\right) \leq \beta + r,$$

as long as $\left\|X^* \rho_k - c_k\right\|$, $1 \leq k \leq M$ are small. Q.E.D.

Let us turn to the proof of Theorem 4. Choose an r in Proposition 5 such that $\beta + r < \alpha$. Let ν_1 be the eigenvalue of $\tilde{L}_{ff}$ such that $\mathrm{Re}\,\nu_1 = \min \mathrm{Re}\,\sigma\left(\tilde{L}_{ff}\right)$, and u_0 a corresponding eigenfunction:

$$\left(\nu_1 - \tilde{L}_{ff}\right)u_0 = 0, \quad \mathrm{Re}\,\nu_1 \leq \beta + r < \alpha.$$

Set $v_0 = Xu_0$. As easily seen from Equation (29), the function $\left(u(t,\cdot), v(t)\right)$ given by

$$u(t,\cdot) = \mathrm{e}^{-\nu_1 t} u_0, \quad v(t) = Xu(t,\cdot) = \mathrm{e}^{-\nu_1 t} v_0$$

is the solution to Equation (9) with the initial value (u_0, v_0). In view of the reduction process to Equation (30), the function $\left(\mathrm{e}^{-\nu_1 t} u_0, \mathrm{e}^{-\nu_1 t} \mathscr{P}_n v_0\right)$ is thus a non-trivial solution to (30). This shows that the decay (31) for

$\|u(t,\cdot)\|+\|v_1(t)\|_{\mathscr{H}}$ is no longer improved. This finishes the proof of Theorem 4. Q.E.D.

EXAMPLE. In (1), let us consider the case where Ω is a bounded interval $I=(0,1)\subset R^1$. The pair of differential operators $(\mathscr{L},\tau)$ is then rewritten as

$$\mathscr{L}u=-(a(x)u_x)_x+b(x)u_x+c(x)u,$$
$$\tau u=\alpha(\xi)u+(1-\alpha(\xi))u_\nu,\quad \xi=0,1, \tag{36}$$

where $\alpha(0),\alpha(1)\in[0,1]$, and $a(x)\geq\delta(>0)$. Let $T\in\mathscr{L}(H)$ be an operator defined, for $u\in H$, by

$$Tu=\kappa(x)u,\quad \kappa(x)=\exp\left(-\int_0^x\frac{b(s)}{2a(s)}ds\right).$$

Clearly T defines an isomorphism in H. Let us consider the case where τ is of the third kind, i.e., $\alpha(1)<1\ \alpha(0)$. Then, T transforms $(\mathscr{L},\tau)$ into another pair $(\tilde{\mathscr{L}},\tilde{\tau})$, which defines a self-adjoint operator $\tilde{L}$ with dense domain $\mathscr{D}(\tilde{L})$. In $(\tilde{\mathscr{L}},\tilde{\tau})$, a is unchanged; b and c are changed, respectively, to 0 and $c-(\kappa b)'/(2\kappa)$; and $\tilde{\tau}$ of the third kind. The idea is a slightly modified version of the well known result (see page 292 of [18]). Based on this, we have

PROPOSITION 6. 1) The spectrum $\sigma(L)$ consists of real and simple eigenvalues: $\sigma(L)=\{\lambda_i\}_{i=1}^{\infty}$, $\cdots<\lambda_i<\cdots\to\infty\ \lambda_1<\lambda_2<$.

2) The eigenfunctions $\varphi_i(x)$ of λ_i forms a Riesz basis. Any $u\in H$ is uniquely expressed as $u=\sum_{i=1}^{\infty}c_i\varphi_i$.

In our problem, we know that $m_i=1$, $i\geq1$. Thus we choose $N=1$, so that the output of the system is a single observation at the end point $x=0$ as

$$u(t,0)=\langle u(t,\cdot),w\rangle_{\Gamma}, \tag{37}$$

that is, $w(0)=1$, $w(1)=0$. Let us examine some assumptions in Theorem 4 in this example. Most important is the complete observability (18). The matrices W_i in (16) are now $W_i = c_1(\varphi_i) = \varphi_i(0) \neq 0$, $i \geq 1$. Thus, we see that the complete observability is satisfied. Since L_f in (13) is a one dimensional operator, the multiplicities of the eigenvalues γ_i are equal to 1. This enables us to choose $M=1$ in (9). In Proposition 3, the controllability condition on the actuator $\zeta_1 \in \overline{\bigcup_{\gamma_i \neq a}\left(W_{\gamma_i}^{(1)}\right)}$ is stated as follows: let χ_i be eigenfunctions of γ_i. By setting $\hat{P}\zeta_1 = \sum_{i \neq l-1}^{1 \leq i \leq l} c_i \chi_i$, the controllability condition is simply that $c_i \neq 0$, $1 \leq i \leq l$, $i \neq l-1$.

In the case where τ is of the first kind, the output is a single observation at the end point $x=0$ as $u_x(t,0) = \langle u_x(t,\cdot), w \rangle_\Gamma$. Proposition 6 also holds in this case. Since $W_i = c_1(\varphi_i) = -\varphi_i'(0) \neq 0$, $i \geq 1$, the complete observability is similarly satisfied.

3. SPECTRAL PROPERTY OF THE COEFFICIENT OPERATOR

We go back to the problem raised in Section 1: Unlikeliness of a vector of the form $(f_1 \quad 0)^{\mathrm{T}}$. The basic control system is Equation (30) in the product space $H \times \mathscr{P}_n\mathscr{H}$. To avoid any unnecessary technical complexity, we limit ourselves to the simple case of one dimensioanl equations raised in (36), τ being of the third kind. In the setting of the space $\mathscr{H}$ as well as B in (6), we can choose $n_i = 1$, $i \geq 1$. Thus, $M = N = 1$. Equation (30) is simply rewritten as

$$\frac{\mathrm{d}}{\mathrm{d}t}\begin{pmatrix} u \\ v \end{pmatrix} + A\begin{pmatrix} u \\ v \end{pmatrix} = 0, \quad \left.\begin{pmatrix} u \\ v \end{pmatrix}\right|_{t=0} \in H \times \mathscr{P}_n\mathscr{H}, \tag{38}$$

where A is defined as

$$A\begin{pmatrix} u \\ v \end{pmatrix} = \begin{pmatrix} Lu - \langle v,\rho\rangle_{\mathscr{P}_n\mathscr{H}}\zeta + \langle v,(\alpha - B^*)f\rangle_{\mathscr{P}_n\mathscr{H}}\eta \\ \hline B_1 v - \langle u,w\rangle_\Gamma \mathscr{P}_n\xi - \langle v,\rho\rangle_{\mathscr{P}_n\mathscr{H}}\mathscr{P}_n X\zeta \\ + \langle v,(\alpha - B^*)f\rangle_{\mathscr{P}_n\mathscr{H}}\mathscr{P}_n X\eta \end{pmatrix}, \tag{39}$$

and $\mathscr{D}(A) = \mathscr{D}(L)\times\mathscr{P}_n\mathscr{H}$. Here, $\langle u,w\rangle_\Gamma = u(0)$ (see (37)). The operator A is sectorial, and every solution to (30) or (38) is expressed as $(u(t)\,v(t))^{\mathrm{T}} = \mathrm{e}^{-tA}(u_0 v_{10})^{\mathrm{T}}$, $t\ge 0$. Let $\boldsymbol{P}$ be the projector corresponding to a $\lambda\in\sigma(A)$ with $\operatorname{Re}\lambda \ge \alpha$. In view of the relation:

$$\left\langle \boldsymbol{P}\mathrm{e}^{-tA}\begin{pmatrix} u_0 \\ v_{10}\end{pmatrix}, \begin{pmatrix}\varphi \\ \psi\end{pmatrix}\right\rangle_{H\times\mathscr{P}_n\mathscr{H}} = \left\langle \mathrm{e}^{-tA}\begin{pmatrix} u_0 \\ v_{10}\end{pmatrix}, \begin{pmatrix} f_1 \\ f_2\end{pmatrix}\right\rangle_{H\times\mathscr{P}_n\mathscr{H}}$$
$$= \langle u(t,\cdot), f_1\rangle + \langle v(t), f_2\rangle_{\mathscr{P}_n\mathscr{H}}, \quad \begin{pmatrix} f_1 \\ f_2\end{pmatrix} = \boldsymbol{P}^*\begin{pmatrix}\varphi \\ \psi\end{pmatrix},$$

the right-hand side of which decays as $t\to\infty$ with decay rate $-\alpha$ for every initial state. Now we ask: Does the range of the $\boldsymbol{P}^*$ contain a vector of the form $(f_1 \;\; 0)^{\mathrm{T}}$? This problem immediately leads to the structure of the eigenspaces of the operator A^*. In the operator A, the vectors $\rho\in\mathscr{P}_n\mathscr{H}$ and $f\in\mathscr{P}_R\mathscr{H}$ of the compensator are the parameters to be designed $(n>R)$. In designing these parameters, they are generally influenced by small perturbations. It is thus implausible to assume that some Fourier coefficients of these parameters would be designed to be 0: such conditions are very easily broken. Thus we may henceforth assume that

$$\begin{aligned} &\rho_i^\pm \ne 0, \quad 1\le i\le n, \\ &\tilde{f}_i^\pm = (\alpha - \mu_i\omega^\mp) f_i^\pm \ne 0, \quad 1\le i\le R, \end{aligned} \tag{40}$$

where $\rho_i^\pm$ and f_i^+ denote, respectively, $\langle \rho,\eta_i^\pm\rangle_{\mathscr{P}_n\mathscr{H}}$ and $\langle f,\eta_i^\pm\rangle_{\mathscr{P}_n\mathscr{H}}$. The actuators ζ and η of the controlled plant are the given parameters in advance. It is also implausible to assume that some Fourier coefficients of ζ and η relative to L^* might be equal to 0. Thus we may also assume that

$$\langle \zeta, \psi_i \rangle \neq 0, \quad \langle \eta, \psi_i \rangle \neq 0, \quad i \geq 1. \tag{41}$$

The main results in this section are Theorem 7, Proposition 8, and Theorem 9 stated just below. The proof of these results will be given later.

THEOREM 7. Let $1 \leq K \leq \min(R, n-R+1)$. Suppose that $\{\alpha_k\}_{k=1}^K \subset \sigma(A^*)$ and that

$$A^* \begin{pmatrix} p_k \\ q_k \end{pmatrix} = \alpha_k \begin{pmatrix} p_k \\ q_k \end{pmatrix}, \quad \begin{pmatrix} p_k \\ q_k \end{pmatrix} \neq \begin{pmatrix} 0 \\ 0 \end{pmatrix}, \quad 1 \leq k \leq K, \tag{42}$$

where $\alpha_i \neq \alpha_j$ for $i \neq j$. Then any linear combination of these eigenvectors $(p_k \ q_k)^{\mathrm{T}}$ of A^* cannot generate a vector of the form, $(f_1 \ 0)^{\mathrm{T}}$, $f_1 \neq 0$.

Remark. The adjoint operator A^* will be characterized later in (49). Theorem 7 also asserts that there is no eigenvector of the form, $(f_1 \ 0)^{\mathrm{T}}$, $f_1 \neq 0$. The restriction on K is derived from our setting of the operator A in (39): The setting is made for constructing a finitedimensional compensator. In the original equation (9), however, the parameters are constructed in a more general setting. The operator A is then replaced by

$$A\begin{pmatrix} u \\ v \end{pmatrix} = \begin{pmatrix} Lu - \langle v, \rho \rangle_{\mathscr{H}} \zeta + \langle v, (\alpha - B^*) f \rangle_{\mathscr{H}} \eta \\ \hline Bv - \langle u, w \rangle_{\Gamma} \xi - \langle v, \rho \rangle_{\mathscr{H}} X\zeta \\ + \langle v, (\alpha - B^*) f \rangle_{\mathscr{H}} X\eta \end{pmatrix},$$
$$\mathscr{D}(A) = \mathscr{D}(L) \times \mathscr{D}(B). \tag{39'}$$

Then, the above restriction on the K is removed: In fact, the integer K may be chosen arbitrarily large.

We hope to know more on $\sigma(A^*)$. The following proposition partly gives concrete informations on what $\sigma(A^*)$ consists of. It shows that $\sigma(B_1)$ is contained in $\sigma(A^*)$, regardless of the assumptions (40) and (41).

PROPOSITION 8. The numbers $\mu_i\omega^\pm$, $1 \le i \le n$ belong to $\sigma(A^*)$. Actually we have the relations:

$$A^*\begin{pmatrix} -X^*\eta_i^\mp \\ \eta_i^\mp \end{pmatrix} = \mu_i\omega^\pm \begin{pmatrix} -X^*\eta_i^\mp \\ \eta_i^\mp \end{pmatrix} \tag{43}$$

for $1 \le i \le n$. Since the set $\{\eta_i^\mp\}$ forms an orthonormal system for $\mathscr{H}$, any linear combinations of these eigenvectors cannot generate a vector of the form, $(f_1 \quad 0)^{\mathrm{T}}$, $f_1 \ne 0$.

To seek eigenvalues of A^* other than $\{\mu_i\omega^\pm\}_{i=1}^n$, let us recall the operator $\tilde{L}_{ff}$ which appeared in (32), where

$$\tilde{L}_{ff} = L - \langle \cdot, X^*\rho \rangle \zeta + \langle \cdot, X^*(\alpha - B^*) f \rangle \eta \text{ with } \mathscr{D}(\tilde{L}_{ff}) =$$

$\mathscr{D}(L)$. The adjoint operator $\tilde{L}_{ff}^*$ is clearly given by $\tilde{L}_{ff}^* =$

$$L^* - \langle \cdot, \zeta \rangle X^*\rho + \langle \cdot, \eta \rangle X^*(\alpha - B^*) f \text{ with } \mathscr{D}(\tilde{L}_{ff}^*) = \mathscr{D}(L^*).$$

In the following result, we characterize $\sigma(A^*)$ by introducing an operator L_g^*, a slightly perturbed operator of $\tilde{L}_{ff}^*$:

THEOREM 9. Let L_g^* be an operator defined as

$$L_g^* = L^* + \langle \cdot, \eta \rangle X_1^*(\alpha - B^*) f - \langle \cdot, \zeta \rangle X_1^*\rho,$$
$$\mathscr{D}(L_g^*) = \mathscr{D}(L^*), \tag{44}$$

where $X_1 = \mathscr{P}_n X$. Then, we have the relation

$$\sigma(A^*) = \sigma(L_g^*) \cup \sigma(B_1^*). \tag{45}$$

Let $\{\hat{\lambda}, \hat{\psi}\}$ be an arbitrary eigenpair of L_g^* such that $\hat{\lambda}$ is not contained in $\sigma(B_1^*)$. Then, the corresponding eigenvector $(p \quad q)^{\mathrm{T}}$ of A^* is given by

$$\begin{pmatrix} \hat{\psi} - X_1^* \left(\hat{\lambda} - B_1^*\right)^{-1} \left(-\langle \hat{\psi}, \zeta \rangle \rho + \langle \hat{\psi}, \eta \rangle \left(\alpha - B^*\right) f\right) \\ \left(\hat{\lambda} - B_1^*\right)^{-1} \left(-\langle \hat{\psi}, \zeta \rangle \rho + \langle \hat{\psi}, \eta \rangle \left(\alpha - B^*\right) f\right) \end{pmatrix}, \tag{46}$$

where $\left(\hat{\lambda} - B_1^*\right)^{-1} \left(-\langle \hat{\psi}, \zeta \rangle \rho + \langle \hat{\psi}, \eta \rangle \left(\alpha - B^*\right) f\right) \neq 0$.

In the above assertions, we need to characterize the adjoint operator A^*, which will be described later by (51). To seek the structure of A^*, let us begin with the operator equation:

$$X_1 L - B_1 X_1 = C_1, \; C_1 = -\langle \cdot, w \rangle_\Gamma \, \xi_1, \; \xi_1 = \mathscr{P}_n \xi. \tag{47}$$

It is clear that (47) admits a unique solution $X_1 \in \mathscr{L}(H; \mathscr{P}_n \mathscr{H})$, and that the solution is expressed as $X_1 = \mathscr{P}_n X$ (see (17)). Let $g \in H^2(I)$ be a unique solution to the boundary value problem: $(\mathscr{L}^* + c) g = 0$ in I, $\tau^* g = w$ on Γ. We note that $\|g\|$ remains bounded when $c \to \infty$ (this fact will be used in Lemma 10 below). For any $\lambda \in \rho(L)$, note that

$$\begin{aligned} (\lambda - L)^{-1} u \Big|_{x=0} &= \left\langle (\lambda - L)^{-1} u, w \right\rangle_\Gamma \\ &= \left\langle u, L_c^* \left(\bar{\lambda} - L^*\right)^{-1} g \right\rangle, \; L_c^* = L^* + c. \end{aligned}$$

Then the adjoint $X_1^* \in \mathscr{L}(\mathscr{P}_n \mathscr{H}; H)$ is expressed as

$$\begin{aligned} X_1^* v &= \sum_{i \le n} \left(v_i^+ \overline{\xi_i} L_c^* \left(\mu_i \omega^- - L^* \right)^{-1} g + v_i^- \xi_i L_c^* \left(\mu_i \omega^+ - L^* \right)^{-1} g \right) \\ &= -\langle v, \xi_1 \rangle_{\mathcal{R}_n \mathcal{H}} g + \sum_{i \le n} v_i^+ \overline{\xi_i} \left(\mu_i \omega^- + c \right) \left(\mu_i \omega^- - L^* \right)^{-1} g \\ &\quad + \sum_{i \le n} v_i^- \xi_i \left(\mu_i \omega^+ + c \right) \left(\mu_i \omega^+ - L^* \right)^{-1} g, \end{aligned}$$

where $v = \sum_{i \le n} v_i^\pm \eta_i^\pm$, $v_i^\pm = \langle v, \eta_i^\pm \rangle_{\mathcal{R}_n \mathcal{H}}$. Thus,

$$X_1^* v \in H^2(I), \ \tau^* \left(X_1^* v \right) = -\langle v, \xi_1 \rangle_{\mathcal{R}_n \mathcal{H}} w. \tag{48}$$

Let us find the equation for X_1^*. For $u \in \mathcal{D}(L)$ and $v \in \mathcal{R}_n \mathcal{H}$, we calculate through Green's formula, (47), and the boundary condition (48) as

$$\begin{aligned} -\langle u, w \rangle_\Gamma \langle \xi_1, v \rangle_{\mathcal{R}_n \mathcal{H}} &= \langle X_1 L u, v \rangle_{\mathcal{R}_n \mathcal{H}} - \langle B_1 X_1 u, v \rangle_{\mathcal{R}_n \mathcal{H}} \\ &= \langle L u, X_1^* v \rangle - \langle X_1 u, B_1^* v \rangle_{\mathcal{R}_n \mathcal{H}} \\ &= \langle u, \mathcal{L}^* X_1^* v \rangle - \langle \xi_1, v \rangle_{\mathcal{R}_n \mathcal{H}} \langle u, w \rangle_\Gamma \\ &\quad - \langle u, X_1^* B_1^* v \rangle, \end{aligned}$$

and thus $\langle u, \mathcal{L}^* X_1^* v \rangle - \langle u, X_1^* B_1^* v \rangle = 0$. Since $\mathcal{D}(L)$ is dense in H, we see that

$$\mathcal{L}^* X_1^* v = X_1^* B_1^* v, \quad \forall v \in \mathcal{R}_n \mathcal{H}. \tag{49}$$

Let us calculate the adjoint A^*. By assuming that p is in $H^2(I)$ and satisfies the boundary condition: $\tau^* p = \langle q, \xi_1 \rangle_{\mathcal{R}_n \mathcal{H}} w$, $\left\langle A \begin{pmatrix} u \\ v \end{pmatrix}, \begin{pmatrix} p \\ q \end{pmatrix} \right\rangle_{H \times \mathcal{R}_n \mathcal{H}}$ is calculated as

$$\begin{aligned}
&\Big\langle Lu - \langle v,\rho\rangle_{\mathscr{T}_n\mathscr{H}}\,\zeta + \big\langle v,(\alpha - B^*)f\big\rangle_{\mathscr{T}_n\mathscr{H}}\,\eta, p\Big\rangle \\
&+\Big\langle B_1 v - \langle u,w\rangle_\Gamma\,\xi_1 - \langle v,\rho\rangle_{\mathscr{T}_n\mathscr{H}}\,X_1\zeta \\
&\qquad + \big\langle v,(\alpha - B^*)f\big\rangle_{\mathscr{T}_n\mathscr{H}}\,X_1\eta, q\Big\rangle_{\mathscr{T}_n\mathscr{H}} \\
&=\big\langle u,\mathscr{L}^* p\big\rangle + \big\langle u,\tau^* p\big\rangle_\Gamma - \langle v,\rho\rangle_{\mathscr{T}_n\mathscr{H}}\langle \zeta,p\rangle \\
&\quad+\big\langle v,(\alpha - B^*)f\big\rangle_{\mathscr{T}_n\mathscr{H}}\langle \eta,p\rangle \\
&\quad+\big\langle v,B_1^* q\big\rangle_{\mathscr{T}_n\mathscr{H}} - \langle u,w\rangle_\Gamma\langle \xi_1,q\rangle_{\mathscr{T}_n\mathscr{H}} \\
&\quad-\langle v,\rho\rangle_{\mathscr{T}_n\mathscr{H}}\langle X_1\zeta,q\rangle_{\mathscr{T}_n\mathscr{H}} \\
&\quad+\big\langle v,(\alpha - B^*)f\big\rangle_{\mathscr{T}_n\mathscr{H}}\langle X_1\eta,q\rangle_{\mathscr{T}_n\mathscr{H}} \\
&=\big\langle u,\mathscr{L}^* p\big\rangle + \Big\langle v, B_1^* q - \big\langle p + X_1^* q,\zeta\big\rangle\rho \\
&\qquad\qquad + \big\langle p + X_1^* q,\eta\big\rangle(\alpha - B^*)f\Big\rangle_{\mathscr{T}_n\mathscr{H}} \\
&=\left\langle \begin{pmatrix} u \\ v\end{pmatrix}, A^\dagger\begin{pmatrix} p \\ q\end{pmatrix}\right\rangle_{H\times\mathscr{T}_n\mathscr{H}},
\end{aligned}$$

where $A^\dagger\begin{pmatrix} p \\ q\end{pmatrix}$ is given by

$$\begin{pmatrix} \mathscr{L}^* p \\ B_1^* q - \big\langle p + X_1^* q,\zeta\big\rangle\rho + \big\langle p + X_1^* q,\eta\big\rangle(\alpha - B^*)f\end{pmatrix}, \tag{50}$$

and $\mathscr{D}(A^\dagger) = \Big\{(p \quad q)^T;\, p\in H^2(I),\, q\in\mathscr{T}_n\mathscr{H},\ \tau^* p - \langle q,\xi_1\rangle_{\mathscr{T}_n\mathscr{H}}\, w = 0\Big\}.$

We see that $\mathscr{D}(A^\dagger)\subset\mathscr{D}(A^*)$, and thus $A^\dagger\subset A^*$. In order to show that $A^\dagger = A^*$, we need the following elementary result:

LEMMA 10. The operator $A^\dagger$ is densely defined, and the bounded inverse, $A_c^{\dagger-1} = (A^\dagger + c)^{-1}$ exists for a sufficiently large $c > 0$.

Proof. Given a $(u \quad v)^{\mathrm{T}} \in H \times \mathscr{P}_n\mathscr{H}$, we solve the equation: $A_c^{\dagger}(p \quad q)^{\mathrm{T}} = (u \quad v)^{\mathrm{T}}$, where $A_c^{\dagger} = A^{\dagger} + c$. Set $\tilde{f} = (\alpha - B^*) f$, and define an operator $\hat{B}$ as

$$\hat{B} = B_1^* + c - \left(\langle \cdot, \xi_1 \rangle_{\mathscr{P}_n\mathscr{H}} \langle g, \zeta \rangle + \langle \cdot, X_1 \zeta \rangle_{\mathscr{P}_n\mathscr{H}}\right) \rho$$
$$+ \left(\langle \cdot, \xi_1 \rangle_{\mathscr{P}_n\mathscr{H}} \langle g, \eta \rangle + \langle \cdot, X_1 \eta \rangle_{\mathscr{P}_n\mathscr{H}}\right) \tilde{f}.$$

The function g depends on c. However, since g remains bounded as $c \to \infty$, there is a bounded inverse: $\hat{B}^{-1} \in \mathscr{L}(\mathscr{P}_n\mathscr{H})$ for a sufficiently large $c > 0$. A straightforward calculation shows that $(p \quad q)^{\mathrm{T}}$ defined by $p = \langle q, \xi_1 \rangle_{\mathscr{P}_n\mathscr{H}} g + L_c^{*-1} u$ and $q = \hat{B}^{-1}\left(v + \langle L_c^{*-1} u, \zeta \rangle \rho - \langle L_c^{*-1} u, \eta \rangle \tilde{f}\right)$ uniquely solves the above equation. Thus the bounded inverse $A_c^{\dagger}$ exists.

Denseness of $\mathscr{D}(A^{\dagger})$: it is enough to show that

$$\left\langle A_c^{\dagger -1} \begin{pmatrix} u \\ v \end{pmatrix}, \begin{pmatrix} \varphi \\ \psi \end{pmatrix} \right\rangle_{H \times \mathscr{P}_n\mathscr{H}} = 0 \quad \text{for } \forall \begin{pmatrix} u \\ v \end{pmatrix}$$

implies $(\varphi \quad \psi)^{\mathrm{T}} = (0 \quad 0)$. The above left-hand side is calculated for every $(u \; v)^{\mathrm{T}} \in H \times \mathscr{P}_n\mathscr{H}$ as

$$0 = \langle L_c^{*-1} u, \zeta \rangle \langle \hat{B}^{-1} \rho, \xi_1 \rangle_{\mathscr{P}_n\mathscr{H}} \langle g, \varphi \rangle$$
$$- \langle L_c^{*-1} u, \eta \rangle \langle \hat{B}^{-1} \tilde{f}, \xi_1 \rangle_{\mathscr{P}_n\mathscr{H}} \langle g, \varphi \rangle$$
$$+ \langle L_c^{*-1} u, \varphi \rangle + \langle L_c^{*-1} u, \zeta \rangle \langle \hat{B}^{-1} \rho, \psi \rangle_{\mathscr{P}_n\mathscr{H}}$$
$$- \langle L_c^{*-1} u, \eta \rangle \langle \hat{B}^{-1} \tilde{f}, \psi \rangle_{\mathscr{P}_n\mathscr{H}}$$
$$+ \langle \hat{B}^{-1} v, \xi_1 \rangle_{\mathscr{P}_n\mathscr{H}} \langle g, \varphi \rangle + \langle \hat{B}^{-1} v, \psi \rangle_{\mathscr{P}_n\mathscr{H}}.$$

Since $\mathscr{D}(L_c^*)$ is dense, this means that

$$0=\left\langle \xi_1, \hat{B}^{-1}\rho \right\rangle_{\mathcal{R}_n\mathcal{H}} \langle \varphi, g\rangle \zeta - \left\langle \xi_1, \hat{B}^{-1}\tilde{f} \right\rangle_{\mathcal{R}_n\mathcal{H}} \langle \varphi, g\rangle \eta$$
$$+\varphi + \left\langle \psi, \hat{B}^{-1}\rho \right\rangle_{\mathcal{R}_n\mathcal{H}} \zeta - \left\langle \psi, \hat{B}^{-1}\tilde{f} \right\rangle_{\mathcal{R}_n\mathcal{H}} \eta,$$

and $0=\langle \varphi, g\rangle \xi_1 + \psi$, from which we conclude that $\varphi = 0$ and $\psi = 0$. Q.E.D.

In view of the fact that both $A_c^{*-1} = \left(A_c^{-1}\right)^*$ and $A_c^{\dagger-1}$ exist as a bounded inverse, it is immediate that $\mathcal{D}\left(A^*\right)$ is contained in $\mathcal{D}\left(A^\dagger\right)$. We have proven that

$$= A^\dagger . A^* \tag{51}$$

Proof of Proposition 8. By setting $p = -X_1^* q$ and $q = \eta_i^{\mp}$, $1 \le i \le n$, $(p \quad q)^{\mathrm{T}}$ belongs to $\mathcal{D}\left(A^*\right)$. By (49), we see that $B_1^* q = \mu_i \omega^{\pm} q$, and $\mathcal{L}^* p = -X_1^* B_1^* q = -\mu_i \omega^{\pm} X_1^* q = \mu_i \omega^{\pm} p$, which shows (43) Q.E.D.

Proof of Theorem 7. Assuming that $\sum_{k=1}^K q_k = 0$ in (42), we derive a contradiction. In (42), adding the equations of q_k over 1 through K, we see that

$$-c_1\rho + c_2\tilde{f} = \sum_{k=1}^K \alpha_k q_k = \sum_{k=1}^{K-1} (\alpha_k - \alpha_K) q_k, \tag{52}$$

where $c_1 = \sum_{k=1}^K \langle p_k, \zeta\rangle$, $c_2 = \sum_{k=1}^K \langle p_k, \eta\rangle$, and $\tilde{f} = \left(\alpha - B^*\right) f$. The Fourier coefficients of these vectors relative to the orthonormal system $\left\{\eta_i^{\pm}\right\}$ satisfy

$$-c_1\rho_i^{\pm} + c_2\tilde{f}_i^{\pm} = \sum_{k=1}^{K-1} (\alpha_k - \alpha_K) q_{ki}^{\pm}, \quad i \ge 1, \tag{53}$$

where $q_{ki}^{\pm} = \left\langle q_k, \eta_i^{\pm} \right\rangle_{\mathcal{R}_n\mathcal{H}}$. Note that $\tilde{f}_i^{\pm} = \left(\alpha - \mu_i \omega^{\mp}\right) f_i^{\pm} = 0$ for $i > R$. We show that $c_1 = 0$. Supposing the contrary, we must have

$$\rho_i^{\pm} = -\sum_{k=1}^{K-1} \frac{\alpha_k - \alpha_K}{c_1} q_{ki}^{\pm}, \quad R < i \le n. \tag{54}$$

Set $d_k = \langle p_k + X_1^* q_k, \zeta \rangle$, $1 \le k \le K$ for simplicity. Then, $\sum_{k=1}^{K} d_k = c_1$. In the equation for q_k, $K-1$ $1 \le k \le$, we see that

$$\mu_i \omega^{\mp} q_{ki}^{\pm} + d_k \sum_{j=1}^{K-1} \frac{\alpha_j - \alpha_K}{c_1} q_{ji}^{\pm} = \alpha_k q_{ki}^{\pm}, \ R < i \le n.$$

The number of these $\mu_i \omega^{\top}$ is $2(n-R)$. Consider the algebraic equation in τ :

$$\begin{vmatrix} \tau + d_1 \frac{\alpha_1 - \alpha_K}{c_1} - \alpha_1 & d_1 \frac{\alpha_2 - \alpha_K}{c_1} & \cdots & d_1 \frac{\alpha_{K-1} - \alpha_K}{c_1} \\ d_2 \frac{\alpha_1 - \alpha_K}{c_1} & \tau + d_2 \frac{\alpha_2 - \alpha_K}{c_1} - \alpha_2 & \cdots & d_2 \frac{\alpha_{K-1} - \alpha_K}{c_1} \\ \vdots & \vdots & \ddots & \vdots \\ d_{K-1} \frac{\alpha_1 - \alpha_K}{c_1} & d_{K-1} \frac{\alpha_2 - \alpha_K}{c_1} & \cdots & \tau + d_{K-1} \frac{\alpha_{K-1} - \alpha_K}{c_1} - \alpha_{K-1} \end{vmatrix} = 0.$$

The equation admits $(K-1)$ solutions. The number of the solutions τ which agree with one of the $\mu_i \omega^{\top}$ is at most $(K-1)$. In other words, the determinant is not equal to 0 for the other $\mu_i \omega^{\top}$, the number of which is atleast $2(n-R)-(K-1) > 0$. Thus for these $\mu_i \omega^{\top}$, we must have $q_{1i}^{\pm} = q_{2i}^{\pm} = \cdots = q_{(K-1)i}^{\pm} = 0$. By (53), this implies that $\rho_i^{\pm} = 0$, which contradicts our assumption (40). We have shown that $c_1 = 0$. Thus we have, for $i > 1$,

$$c_2 \tilde{f}_i^{\pm} = \sum_{k=1}^{K-1} (\alpha_k - \alpha_K) q_{ki}^{\pm}. \tag{55}$$

Comparing the Fourier coefficients in the equations to q_k in (42), we see that

$$\mu_i \omega^{\mp} q_{ki}^{\pm} - d_k \rho_i^{\pm} = \alpha_k q_{ki}^{\pm}, \text{ or}$$
$$\left(\mu_i \omega^{\mp} - \alpha_k\right) q_{ki}^{\pm} = d_k \rho_i^{\pm}, \ R < i \le n, \ 1 \le k \le K-1.$$

The number of the eigenvalues α_k which agree with one of the $\mu_i \omega^{\mp}$ is at most $(K-1)$. In other words, the number of the $\mu_i \omega^{\mp}$ which does not agree with any of the α_k is at least $2(n-R)-(K-1)>0$. For these $\mu_i \omega^{\mp} \ne \alpha_k$, we see from (55) that

$$0 = \sum_{k=1}^{K-1} \left(\alpha_k - \alpha_K\right) q_{ki}^{\pm} = \sum_{k=1}^{K-1} \frac{d_k \left(\alpha_k - \alpha_K\right)}{\mu_i \omega^{\mp} - \alpha_k} \rho_i^{\pm}.$$

Since $\rho_i^{\pm} \ne 0$, this means that the relation in σ:

$$0 = \sum_{k=1}^{K-1} \frac{d_k \left(\alpha_k - \alpha_K\right)}{\sigma - \alpha_k} \tag{56}$$

holds for the above $\sigma = \mu_i \omega^{\mp}$, the distinct number of which is at least $2(n-R)-(K-1) \ge K-1$. This implies that the relation (56) holds for any $\sigma (\in \mathrm{C}) \ne \alpha_1, \cdots, \alpha_{K-1}$. calculating the residue at each α_k, we find that $d_1 = d_2 = \cdots = d_{K-1} = 0$, and thus $d_K = c_1 - \sum_{k=1}^{K-1} d_k = 0$, too.

We go back to the equations to q_k in (42) again. Since $d_k = 0$, $1 \le k \le K$, we have

$$B_1^* q_k + \left\langle p_k + X_1^* q_k, \eta \right\rangle \tilde{f} = \alpha_k q_k, \quad 1 \le k \le K.$$

Set $e_k = \left\langle p_k + X_1^* q_k, \eta \right\rangle$. Then, $\sum_{k=1}^{K} e_k = c_2$. Calculating the Fourier coefficients, we have

$$\left(\mu_i \omega^{\mp} - \alpha_k\right) q_{ki}^{\pm} = -e_k \tilde{f}_i^{\pm}, \quad 1 \le i \le R, \ 1 \le k \le K.$$

The numbers of these $\mu_i\omega^{\top}$ and α_k are $2R$ and $K-1$, respectively. Thus, the number of $\mu_i\omega^{\top}$ which does not agree with any of α_k is at least $2R-K \geq K$. For these $\mu_i\omega^{\mp} \neq \alpha_k$, we see by the relation (54) that

$$c_2\tilde{f}_i^{\pm} = \sum_{k=1}^{K-1}\left(\alpha_k-\alpha_K\right)q_{ki}^{\pm}$$
$$= -\sum_{k=1}^{K-1}\frac{e_k\left(\alpha_k-\alpha_K\right)}{\mu_i\omega^{\mp}-\alpha_k}\tilde{f}_i^{\pm}.$$

But, since $\tilde{f}_i^{\pm} \neq 0$ by (40), this means that the relation in σ :

$$c_2 = -\sum_{k=1}^{K-1}\frac{e_k(\alpha_k-\alpha_K)}{\sigma-\alpha_k},$$
$$\text{or } c_2\prod_{k=1}^{K-1}(\sigma-\alpha_k)$$
$$+\sum_{k=1}^{K-1}e_k\left(\alpha_k-\alpha_K\right)\prod_{j=1,\,j\neq k}^{K-1}\left(\sigma-\alpha_j\right)=0 \tag{57}$$

holds for the above $\sigma=\mu_i\omega^{\mp}$, the distinct number of which is at least $2R-(K-1)\geq K-1$. The situation is the same as in (56). Thus the relation (57) holds for any $\sigma \neq \alpha_1,\cdots,\alpha_{K-1}$. Calculating the residue at each α_k, we similarly find that $e_1=\cdots=e_{K-1}=0$, $c_2=0$, and thus $e_K=c_2-\sum_{k=1}^{K-1}e_k=0$, too. Since $d_k=e_k=0$, $1\leq k\leq K$, we have finally obtained from (42) and (55) that $B_1^{*}q_k=\alpha_k q_k$, $1\leq k\leq K$, and $c_2\tilde{f}=0$ $\sum_{k=1}^{K-1}\left(\alpha_k-\alpha_K\right)q_k=$. Applying B_1^{*l} to the both sides of the above second equation, we see that, for $l=1,2,\cdots$,

$$B_1^{*l}\sum_{k=1}^{K-1}\left(\alpha_k-\alpha_K\right)q_k=\sum_{k=1}^{K-1}\left(\alpha_k-\alpha_K\right)\alpha_k^l q_k=0.$$

In other words, we have the relation:

$$\begin{pmatrix} 1 & 1 & \cdots & 1 \\ \alpha_1 & \alpha_2 & \cdots & \alpha_{K-1} \\ \vdots & \vdots & \ddots & \vdots \\ \alpha_1^{K-2} & \alpha_2^{K-2} & \cdots & \alpha_{K-1}^{K-2} \end{pmatrix} \begin{pmatrix} (\alpha_1 - \alpha_K) q_1 \\ (\alpha_2 - \alpha_K) q_2 \\ \vdots \\ (\alpha_{K-1} - \alpha_K) q_{K-1} \end{pmatrix} = \mathbf{0}.$$

But $\alpha_i \neq \alpha_j$ for $i \neq j$. Thus, we immediately find that $(\alpha_k - \alpha_K) q_k = 0$, $1 \le k \le K-1$, i.e., $\cdots = q_{K-1} = 0$ $q_1 = q_2 =$, and that $q_K = -\sum_{k=1}^{K-1} q_k = 0$.

Recall that $\tau^* p_k = \langle q_k, \xi_1 \rangle$ $w = 0$, $1 \le k \le K$. Thus, $p_k (\neq 0)$ belongs to $\mathscr{D}(L^*)$, and $L^* p_k = \alpha_k p_k$ by (42). Each α_k is found an eigenvalue of L^*, and $p_k (\neq 0)$ must be an eigenfunction. In addition,

$$\begin{aligned} d_k &= \langle p_k + X_1^* q_k, \zeta \rangle = \langle p_k, \zeta \rangle = 0, \quad 1 \le k \le K, \\ e_k &= \langle p_k + X_1^* q_k, \eta \rangle = \langle p_k, \eta \rangle = 0, \quad 1 \le k \le K. \end{aligned}$$

But, this contradicts our assumption (41). Q.E.D.

Proof of Theorem 9. We already know that $\sigma(B_1^*) \subset$

$\sigma(A^*)$ by Proposition 8. Let $\hat{\lambda}$ be in $\sigma(A^*)$ $\sigma(B_1^*)$.

In view of (51), the relation: $A^* (p\ q)^{\mathrm{T}} = \hat{\lambda} (p\ q)^{\mathrm{T}}$, $(p\ q)^{\mathrm{T}} \neq (0\ 0)^{\mathrm{T}}$ means that

$$\begin{aligned} &\mathscr{L}^* p = \hat{\lambda} p, \\ &B_1^* q - \langle p + X_1^* q, \zeta \rangle \rho + \langle p + X_1^* q, \eta \rangle \tilde{f} = \hat{\lambda} q, \end{aligned} \tag{58}$$

where $\tilde{f} = (\alpha - B^*) f$. The calculation of: (the first equation) + X_1^* × (the second equation) yields that

$$\begin{aligned} &\mathscr{L}^* (p + X_1^* q) - \langle p + X_1^* q, \zeta \rangle X_1^* \rho \\ &+ \langle p + X_1^* q, \eta \rangle X_1^* \tilde{f} = \hat{\lambda} (p + X_1^* q). \end{aligned}$$

By noting that $\tau^*\left(p+X_1^*q\right)=\tau^*p-\langle q,\xi_1\rangle_{\mathbb{R}^n} w=0$ (see (48)), the function $p+X_1^*q$ belongs to $\mathscr{D}\left(L^*\right)=\mathscr{D}\left(L_g^*\right)$. Thus we see that $L_g^*\left(p+X_1^*q\right)=\hat{\lambda}\left(p+X_1^*q\right)$. Supposing that $\hat{\psi}=p+X_1^*q=0$, we show a contradiction. In fact, if so, the second equation of (58) becomes $B_1^*q=\hat{\lambda}q$. Since $\hat{\lambda}\notin\sigma\left(B_1^*\right)$, however, we see that $q=0$, and $p=-X_1^*q=0$, or $(p\ q)^{\mathrm{T}}=(0\ 0)^{\mathrm{T}}$. Thus, $\hat{\lambda}$ belongs to $\sigma\left(L_g^*\right)$, and the corresponding eigenvector of A^* is given by the form $(p\ q)^{\mathrm{T}}=\left(\hat{\psi}-X_1^*q\ q\right)^{\mathrm{T}}$, where $q\neq 0$. We have also shown that $\sigma\left(A^*\right)\subset\sigma\left(L_g^*\right)\cup\sigma\left(B_1^*\right)$.

Conversely, let $\{\hat{\lambda},\hat{\psi}\}$ be an arbitrary eigenpair of L_g^* such that $\hat{\lambda}\in\sigma\left(L_g^*\right)\ \sigma\left(B_1^*\right)$. Then we solve the equation: $B_1^*q-\langle\hat{\psi},\zeta\rangle\rho+\langle\hat{\psi},\eta\rangle\tilde{f}=\hat{\lambda}q$, the unique solution of which is given by $q=\left(\hat{\lambda}-B_1^*\right)^{-1}\left(-\langle\hat{\psi},\zeta\rangle\rho+\langle\hat{\psi},\eta\rangle\tilde{f}\right)$.

By setting $p=\hat{\psi}-X_1^*q$, the vector $(p\ q)^{\mathrm{T}}$ means (46), and clearly satisfies the relation: $\hat{\lambda}(p\ q)^{\mathrm{T}}\ A^*(p\ q)^{\mathrm{T}}=$. To show that $q\neq 0$, we suppose the contrary: $q=0$, or $-\langle\hat{\psi},\zeta\rangle\rho+\langle\hat{\psi},\eta\rangle\tilde{f}=0$. Then, $p=\hat{\psi}\neq 0$, and $\tau^*p=0$. Thus, $\{\hat{\lambda},p\}$ must be an eigenpair of L^*. But, this contradicts the assumptions (40) and (41). We have shown that $(p\ q)^{\mathrm{T}}$ given by (46) is an eigenvector of A^*. Q.E.D.

Takao Nambu (45), it is not certain if $\sigma\left(L_g^*\right)\cap\sigma\left(B_1^*\right)\neq\varnothing$. This problem seems a pathological one. If $(\mu_i\omega^{\pm}-L_g^*)\hat{\psi}=0$, $\hat{\psi}\neq 0$ for some i, $1\leq i\leq n$, and, in addition, $\langle\hat{\psi},\zeta\rangle\rho_i^{\pm}=\langle\hat{\psi},\eta\rangle\tilde{f}_i^{\pm}$, then the equation $\left(\mu_i\omega^{\pm}-B_1^*\right)q=-\langle\hat{\psi},\zeta\rangle\rho+\langle\hat{\psi},\eta\rangle\tilde{f}$ admits a (non-unique) solution q (see the second equation of (58)). By setting $p=\hat{\psi}-X_1^*q$, the vector $(p\ q)^{\mathrm{T}}$ belongs to the eigenspace of A^* for $\mu_i\omega^{\pm}$.

REFERENCES

1. R. F. Curtain, "Finite Dimensional Compensators for Parabolic Distributed Systems with Unbounded Control and Observation," SIAM Journal on Control and Optimization, Vol. 22, 1984, pp. 255-276.
2. T. Nambu, "On Stabilization of Partial Differential Equations of Parabolic Type: Boundary Observation and Feedback," Funkcialaj Ekvacioj, Vol. 28, No. 3, 1985, pp. 267- 298.

3. T. Nambu, "An $L^2(\Omega)$-Based Algebraic Approach to Boundary Stabilization for Linear Parabolic Systems," Quarterly of Applied Mathematics, Vol. 62, No. 4, 2004, pp. 711-748.
4. T. Nambu, "A New Algebraic Approach to Stabilization for Boundary Control Systems of Parabolic Type," Journal of Differential Equations, Vol. 218, No. 1, 2005, pp. 136-158.
5. T. Nambu, "Alternative Algebraic Approach to Stabilization for Linear Parabolic Boundary Control Systems," Mathematics of Control, Signals, and Systems, 2013, in Press.
6. Y. Sakawa, "Feedback Stabilization of Linear Diffusion Systems," SIAM Journal on Control and Optimization, Vol. 21, No. 5, 1983, pp. 667-676.
7. S. Itô, "Diffusion Equations," American Mathematical Society, Providence, 1992.
8. S. Agmon, "Lectures on Elliptic Boundary Value Problems," Van Nostrand, Princeton, 1965.
9. E. Taylor, "Introduction to Functional Analysis," John Wiley & Sons, New York, 1958.
10. T. Nambu, "Stability Enhancement of Output for a Class of Linear Parabolic Systems," Proceedings of Royal Society Edinburgh, Section A, Vol. 133A, No. 1, 2003, pp. 157-175.
11. T. Nambu, "Stabilization and Decay of Functionals for Linear Parabolic Control Systems," Proceedings of the Japan Academy, Series A, Vol. 84, No. 2, 2008, pp. 19- 24.
12. T. Nambu, "On State and Output Stabilization of Linear Parabolic Systems," Funkcialaj Ekvacioj, Vol. 52, No. 3, 2009, pp. 321-341. http://dx.doi.org/10.1619/fesi.52.321
13. T. Nambu, "Stabilization and a Class of Functionals for Linear Parabolic Control Systems," Proceedings of Royal Society Edinburgh, Section A, Vol. 140, No. 1, 2010, pp. 153-174.
14. G. N. Watson, "A Treatise on the Theory of Bessel Functions," Cambridge University Press, Cambridge, 1922.
15. N. Levinson, "Gap and Density Theorems," American Mathematical Society Colloquium Publications, New York, 1940.
16. E. C. Titchmarsh, "The Theory of Functions," The Clarendon Press, Oxford, 1939.
17. Y. Sakawa and T. Matsushita, "Feedback Stabilization of a Class of Distributed Systems and Construction of a State Estimator," IEEE Transactions on Automatic Control, Vol. 20, No. 6, 1975, pp. 748-753.
18. R. Courant and D. Hilbert, "Methods of Mathematical Physics, I," Wiley Interscience, New York, 1953.

NOTES

[1]For an $f = (f_1, f_2)$ with $f_2 \neq 0$ in a spectral subspace of $H \times \mathscr{H}$, a faster decay of $\langle \boldsymbol{x}, \boldsymbol{f} \rangle_{H \times \mathscr{H}} = \langle u, f_1 \rangle + \langle v, f_2 \rangle_{\mathscr{H}}$ does not ensure the same decay of $\langle u, f_1 \rangle$.

CHAPTER 5

Differential Transform Method for Some Delay Differential Equations

Baoqing Liu[1], Xiaojian Zhou[2], Qikui Du[3]

[1]*School of Applied Mathematics, Jiangsu Provincial Key Laboratory for NSLSCS, Nanjing University of Finance and Economics, Nanjing, China*

[2]*School of Science, Jiangsu Provincial Key Laboratory for NSLSCS, Nantong University, Nantong, China*

[3]*School of Mathematical Sciences, Jiangsu Key Laboratory for NSLSCS, Nanjing Normal University, Nanjing, China*

ABSTRACT

This paper concentrates on the differential transform method (DTM) to solve some delay differential equations (DDEs). Based on the method of steps for DDEs and using the computer algebra system Mathematica, we successfully apply DTM to find the analytic solution to some DDEs, including a neural delay differential equation. The results confirm the feasibility and efficiency of DTM.

Keywords: Differential Transform Method, Delay Differential Equation, Method of Steps, Analytic Solution, Approximate Solution

1. INTRODUCTION

The differential transform method (DTM) is a semi analytical-numerical technique depending on Taylor series for solving integral-differential equations (IDEs). The method was first introduced by Pukhov [1] for solving linear and nonlinear initial value problems in physical processes. Zhou, at the same time, had also introduced DTM to study electrical circuits [2] . Since the main advantage of this method is that it can be applied directly to nonlinear ordinary and partial differential equations without requiring linearization, discretization or perturba- tion, it has been studied and applied during the last two decades widely. DTM has been used to obtain numerical and analytical solutions of ordinary differential equations [3] , partial differential equations [4] , eigenvalue pro- blems [5] , differential algebraic equations [6] [7] , integral equations [8] and so on.

Delay differential equations (DDEs) arise in many applied fields, such as control technology, communication networks, and biological population management, and hence they have attracted considerable attention. There are many papers devoted to the problem of approximate solution of DDEs [9] - [15] . Recently, F. Karako and H. Bereketoğlu [13] extend the method of differential transformation for solving the following two types of DDEs:

$$\begin{cases} y'(t) = f(t, y(t), y(at)), & 0 \le t \le T, \\ y(0) = y_0, \end{cases} \tag{1}$$

and

$$\begin{cases} y'(t) = f(t, y(t), y(t-\tau)), & 0 \le t \le T, \\ y(0) = y_0, \end{cases} \tag{2}$$

with $0 < a < 1$ and $\tau > 0$ and the constant $T > 0$.

It should be pointed out that the solution to DDEs (2) maybe be non-unique (see Section 2 in [16]). So usually, researchers pay more attention to the following DDEs, instead of (2)

$$\begin{cases} y'(t) = f(t, y(t), y(t-\tau)), & 0 \le t \le T, \\ y(t) = g(t), & t \le 0, \end{cases} \tag{3}$$

where $g(t)$ is a given function, called initial function.

In this paper, we will apply DTM to find the analytic solution to DDEs (3) with the help of the computer algebra system Mathematica. Thus, in some sense, our work can be viewed as a supplement to [13] .

2. DIFFERENTIAL TRANSFORM

The basic theory of differential transform can be found in [1] [2] , in this section we will state it in brief.

Consider a function $y(t)$ be analytic in the time domain I , and let $t_0 \in I$. The function $y(t)$ is then

represented by one series whose center is located at t_0. The differential transform of the function $y(t)$ is the form

$$Y(k) = \frac{1}{k!} \left. \frac{\mathrm{d}^k y(t)}{\mathrm{d}t^k} \right|_{t=t_0} \tag{4}$$

where $Y(k)$ is the transformed function of the original function $y(t)$.

Differential inverse transformation of $Y(k)$ is defined as follows:

$$y(t) = \sum_{k=0}^{\infty} Y(k)(t-t_0)^k \tag{5}$$

From (4) and (5), it is easy to see that the concept of the differential transformation is derived from the Taylor series expansion. By our assumption, t_0 is taken as zero, then the function $y(t)$ is expressed by a finite series and (5) can be written as

$$y(t) = \sum_{k=0}^{\infty} Y(k) t^k$$

Table 1. Estimates of τ_0, λ_0, and λ_1 from simulated data.

λ_0	λ_1	Sample Size	MLE			LSE		
			τ_0	λ_0	λ_1	τ_0	λ_0	λ_1
0.3	0.1	100	2.8	0.33	0.150	3.925	0.239	0.159
		200	2.701	0.315	0.156	5.117	0.233	0.157
		300	2.979	0.312	0.147	5.917	0.222	0.155
0.25	0.15	100	2.809	0.271	0.173	3.860	0.234	0.188
		200	2.93	0.263	0.176	3.808	0.254	0.184
		300	3.146	0.262	0.171	4.232	0.251	0.182
0.2	0.15	100	3.44	0.208	0.161	4.136	0.212	0.169
		200	3.403	0.208	0.159	4.72	0.225	0.166
		300	3.261	0.208	0.158	5.111	0.242	0.164

In this study, we use the lower case letters to represent the original functions and upper case letters to stand for the transformed functions (T-functions). The fundamental mathematical operations performed by differential transform method are listed in Table 1.

3. DTM FOR DDES (3)

There are many methods to deal with the delay differential Equation (3). For example, linear multistep (LM) methods, Runge-Kutta (RK) methods, waveform relaxation (WR) methods, etc. However, the basic idea to solve the DDE (3) is to solve the following system of ODEs step by step:

$$y_i'(t) = f\left(t, y_i(t), y_{i-1}(t-\tau)\right), \quad i = 1,2,\cdots, \quad t \in \left[(i-1)\tau, i\tau\right] \tag{6}$$

with $y_0(t) = g(t)$. In brief, this idea is to shift the interval from $\left[(i-1)\tau, i\tau\right]$ to $\left[i\tau, (i+1)\tau\right]$ and extend the solution from $\left[0, i\tau\right]$ to $\left[0, (i+1)\tau\right]$ by using the component in the current interval. This procedure can, in

principle, be continued as far as desired. It is called, quite naturally, the method of steps [16] .

Using the basic idea of the method of steps, first, we apply the DTM to find the solution to the following ODEs:

$$y_1'(t) = f\left(t, y_1(t), g(t-\tau)\right)$$

with $t \in [0, \tau]$.

Suppose the approximate solution is given by

$$y_1(t) = \sum_{k=0}^{\infty} a_{1k} t^k, \quad t \in [0, \tau]$$

If $\tau > T$, $y_1(t)$ is the solution to (3). Otherwise, we should continue to find the solution in the interval $[\tau, 2\tau]$. At this time, we should solve the following ODEs

$$y_2'(t) = f(t, y_2(t), y_1(t-\tau)), \quad t \in [\tau, 2\tau]$$

Applying the DTM to the differential equation above again, we will obtain the following solution

$$y_2(t) = \sum_{k=0}^{\infty} a_{2k}(t-\tau)^k, \quad t \in [\tau, 2\tau]$$

Of course, we should go on if $2\tau < T$ holds also. In generally, applying the DTM to ODEs (6), we can obtain the analytic solution

$$y_{n+1}(t) = \sum_{k=0}^{\infty} a_{nk}(t-n\tau)^k, \quad t \in [n\tau, (n+1)\tau]$$

until for some n, $n\tau < T \le (n+1)\tau$. In fact, after necessary steps, we have the following solution to (3)

$$y(t) = \begin{cases} y_1(t), & t \in [0, \tau], \\ y_2(t), & t \in [\tau, 2\tau], \\ \vdots & \\ y_{n+1}(t), & t \in [n\tau, T]. \end{cases}$$

Remark 1 If $y_1(t) = g(t)$, we can conclude that $y = g(t)$ is the analytic solution to (3) directly.

Remark 2 If $y_{i+1}(t) = y_i(t)$ for some integer i, we can conclude that the analytic solution (3) is

$$y(t) = \begin{cases} y_1(t), & t \in [0, \tau], \\ \vdots & \\ y_{i-1}(t), & t \in [(i-2)\tau, (i-1)\tau], \\ y_i(t), & t \in [(i-1)\tau, T]. \end{cases}$$

Remark 3 If we want to improve the accuracy of the approximate solution in each interval, we can combine the above method with the multi-step method given by [17] .

Remark 4 In fact, the DTM based on the method of steps can also be applied to solve the following neutral delay differential equations

$$\begin{cases} y'(t) = f\left(t, y(t), y(t-\tau), y'(t-\tau)\right), & 0 \le t \le T, \\ y(t) = g(t), & t \le 0. \end{cases}$$

4. NUMERICAL EXPERIMENTS

In this section, four examples are given to show the performance of the DTM based on the method of steps. First, we want to solve the following simple but classical DDE to further illustrate the process of DTM.

Example 4.1 Consider the DDE [18]

$$\begin{cases} y'(t) = -y(t-1), & t \ge 0, \\ y(t) = 1, & -1 \le t \le 0. \end{cases} \tag{7}$$

First, since $\tau = 1$, we apply the DTM to obtain the solution in the interval $[0,1]$. In this interval, (7) can be written as $y'(t) = -1$, and the initial condition is $y(0) = 1$. Taking the differential transform, we have

$$(k+1)Y(k+1) = -\delta(k), \quad Y(0) = 1$$

It is easy to get

$$Y(k) = \begin{cases} 1, & k = 0, \\ -1, & k = 1, \\ 0, & k \ge 2. \end{cases}$$

Thus we have the analytic solution $y(t) = 1 - t$ of (7) defined on $[0,1]$.

Second, we should continue to solve the following DDE:

$$\begin{cases} y'(t) = -y(t-1), & 1 \le t \le 2, \\ y(t) = 1-t, & 0 \le t \le 1, \end{cases}$$

or equivalently,

$$y'(t) = t-2 = -1+(t-1), \quad 1 \le t \le 2 \tag{8}$$

with initial condition $y(1)=0$.

From (8), we have the following differential transform

It is easy to get

$$Y(k) = \begin{cases} 0, & k = 0, \\ -1, & k = 1, \\ \frac{1}{2}, & k = 2, \\ 0, & k \ge 3. \end{cases}$$

Thus we have the analytic solution

$$y(t) = -(t-1) + \frac{(t-1)^2}{2} = 1-t+\frac{(t-1)^2}{2}, \quad 1 \le t \le 2$$

Now, if we want to obtain the solution in the interval [2, 3], we should deal with the following DDE:

$$\begin{cases} y'(t) = -y(t-1), & 2 \le t \le 3, \\ y(t) = 1-t+\frac{(t-1)^2}{2}, & 1 \le t \le 2. \end{cases}$$

or equivalently,

$$y'(t) = (t-2) - \frac{(t-2)^2}{2}, \quad 2 \le t \le 3$$

with the initial condition $y(2) = -\frac{1}{2}$.

Then we have the following differential transform

and get

$$Y(k) = \begin{cases} -\frac{1}{2}, & k = 0, \\ 0, & k = 1, \\ \frac{1}{2}, & k = 2, \\ -\frac{1}{6}, & k = 3, \\ 0, & k \geq 4. \end{cases}$$

Thus the analytic solution defined on [2, 3] is given by

$$y(t) = -\frac{1}{2} + \frac{(t-2)^2}{2} - \frac{(t-2)^3}{6} = 1 - t + \frac{(t-1)^2}{2} - \frac{(t-2)^3}{6}, \quad 2 \leq t \leq 3$$

The DTM can be proceed till the desire solution is obtained.

Example 4.2 Consider the nonlinear DDE of third-order [11] [13]

$$\begin{aligned} &y'''(x) = -y(x) - y(x-0.3) + e^{-x+0.3}, && 0 \leq x \leq 1, \\ &y(0) = 1,\ f'(0) = -1,\ y''(0) = 1,\ y(x) = e^{-x}, && x \leq 0. \end{aligned} \tag{9}$$

Since $\tau = 0.3$, according to the foregoing, we have the following ODE, defined in the interval [0, 0.3]

$$y_1'''(x) = -y_1(x)$$

Thus, applying DTM to the equation above, we obtain

$$(k+1)(k+2)(k+3)Y(k+3) = -Y(k)$$

The initial conditions lead to $Y(0)=1$, $Y(1)=-1$, and $Y(2)=\frac{1}{2}$. It is easy to have $Y(k)=\frac{(-1)^k}{k!}\ (k\geq 0)$. Thus we have the solution

$$y_1(x)=\sum_{k=0}^{\infty}\frac{(-1)^k}{k!}x^k=\mathrm{e}^{-x},\quad x\in[0,0.3]$$

Noting that $y_1(x)=g(x)=\mathrm{e}^{-x}$, Remark 1 tells us $y(x)=\mathrm{e}^{-x}$ is the analytic solution to (9) in the whole interval $[0,1]$.

Remark 5 In [13] , F. Karako and H. Bereketoğlu also apply DTM to DDE (9) where the initial function $y(x)=\mathrm{e}^{-x}$, $x\leq 0$ is omitted, i.e.

$$\begin{aligned}&y'''(x)=-y(x)-y(x-0.3)+\mathrm{e}^{-x+0.3}, && 0\leq x\leq 1,\\ &y(0)=1,\ f'(0)=-1,\ y''(0)=1, && x\leq 0.\end{aligned} \tag{10}$$

It's worth pointing out that, using the method given in [13] , only approximate solution can be obtained. On the other hand, for DDEs (3), the initial function has the vital role. Without it, the DDEs may have un-unique solution. In fact, Example 4.2 shows that $y(x)=\mathrm{e}^{-x}$ is a solution to (10). Let's suppose $y(x)=\mathrm{e}^{-x}+f(x)$ is also a solution to (10), then $f(x)$ should satisfy the following DDEs

$$\begin{aligned}&f'''(x)=-f(x)-f(x-0.3), && 0\leq x\leq 1,\\ &f(0)=f'(0)=f''(0)=0, && x\leq 0.\end{aligned} \tag{11}$$

Figure 1 shows the solution to (11), together with it's first and second derivative value on $[0,1]$, with the

initial function, which satisfies the initial conditions in (10) obviously. It can be seen $f(x)\neq 0$, then $y(x)=\mathrm{e}^{-x}+f(x)$ is also the solution to (11). So, (11) has infinite solutions. Maybe, the authors "happen to" get the approximate solution of $y(x)=\mathrm{e}^{-x}$.

Example 4.3 Consider a single delay equation with a stiffness parameter [10]

$$y'(t) = Ay(t) + y\left(t - \frac{3\pi}{2}\right) - A\sin(t) \tag{12}$$

with $y(0) = 1$, $y(t) = e^{pt} + \sin(t)$, $t \in \left[-\frac{3\pi}{2}, 0\right]$.

Similarly, we should solve the following DDE limited in the interval $\left[0, \frac{3\pi}{2}\right]$:

$$y'(t) = Ay(t) + e^{-\frac{3p\pi}{2}} e^{pt} + \cos(t) - A\sin(t)$$

Thus applying DTM to this equation, we obtain

$$(k+1)Y(k+1) = AY(k) + e^{-\frac{3p\pi}{2}} \frac{p^k}{k!} + \frac{1}{k!}\cos\left(\frac{\pi k}{2}\right) - A\frac{1}{k!}\sin\left(\frac{\pi k}{2}\right)$$

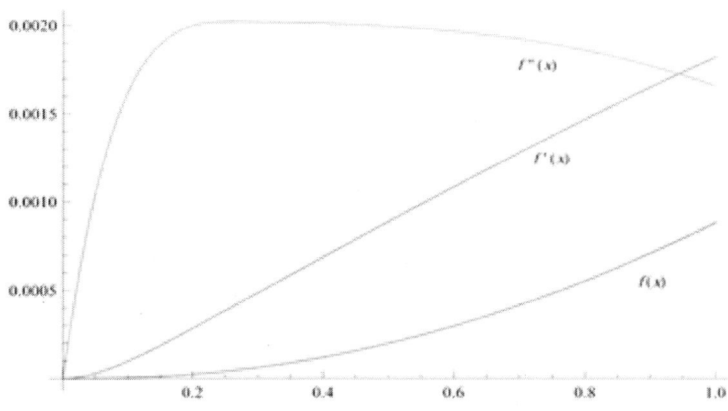

Figure 1. $f(x), f'(x)$ and $f''(x)$ of (11) with $g(x) = x^3$.

The initial values lead to $Y(0)=1$. With the help of Mathematica, we have

$$
\begin{aligned}
Y(1) &= 1 + A + e^{-\frac{3p\pi}{2}} = \frac{1}{1!} + \frac{A}{1!} + \frac{1}{1!} e^{-\frac{3p\pi}{2}} \frac{A-p}{A-p}, \\
Y(2) &= \frac{1}{2} e^{-\frac{3p\pi}{2}} \left(A + A^2 e^{\frac{3p\pi}{2}} + p \right) = \frac{1}{2!} A^2 + \frac{1}{2!} e^{-\frac{3p\pi}{2}} (A+p) \\
&= 0 + \frac{A^2}{2!} + \frac{1}{2!} e^{-\frac{3p\pi}{2}} \frac{A^2 - p^2}{A-p}, \\
Y(3) &= \frac{1}{6} e^{-\frac{3p\pi}{2}} \left(A^2 - e^{\frac{3p\pi}{2}} + A^3 e^{\frac{3p\pi}{2}} + Ap + p^2 \right) \\
&= -\frac{1}{3!} + \frac{A^3}{3!} + \frac{1}{3!} e^{-\frac{3p\pi}{2}} \frac{A^3 - p^3}{A-p}, \\
Y(4) &= \frac{1}{24} e^{-\frac{3p\pi}{2}} \left(A^3 + A^4 e^{\frac{3p\pi}{2}} + A^2 p + Ap^2 + p^3 \right) \\
&= 0 + \frac{A^4}{4!} + \frac{1}{4!} e^{-\frac{3p\pi}{2}} \frac{A^4 - p^4}{A-p}, \\
Y(5) &= \frac{1}{120} e^{-\frac{3p\pi}{2}} \left(A^4 + e^{\frac{3p\pi}{2}} + A^5 e^{\frac{3p\pi}{2}} + A^3 p + A^2 p^2 + Ap^3 + p^4 \right) \\
&= \frac{1}{5!} + \frac{A^5}{5!} + \frac{1}{5!} e^{-\frac{3p\pi}{2}} \frac{A^5 - p^5}{A-p}, \\
Y(6) &= \frac{1}{720} e^{-\frac{3p\pi}{2}} \left(A^5 + A^6 e^{\frac{3p\pi}{2}} + A^4 p + A^3 p^2 + A^2 p^3 + Ap^4 + p^5 \right) \\
&= 0 + \frac{A^6}{6!} + \frac{1}{6!} e^{-\frac{3p\pi}{2}} \frac{A^6 - p^6}{A-p}, \\
&\vdots
\end{aligned}
$$

Then, we obtain the solution to (12):

$$
\begin{aligned}
y(t) &= \left(\frac{t}{1!} - \frac{t^3}{3!} + \frac{t^5}{5!} + \cdots \right) + \sum_{k=0}^{\infty} \frac{A^k}{k!} t^k + e^{-\frac{3p\pi}{2}} \sum_{k=0}^{\infty} \frac{1}{k!} \frac{A^k - p^k}{A-p} t^k \\
&= \sin(t) + e^{At} + \frac{e^{-\frac{3p\pi}{2}}}{A-p} \left(\sum_{k=0}^{\infty} \frac{A^k}{k!} t^k - \sum_{k=0}^{\infty} \frac{p^k}{k!} t^k \right) \\
&= \sin(t) + e^{At} + \frac{e^{-\frac{3p\pi}{2}}}{A-p} \left(e^{At} - e^{pt} \right).
\end{aligned}
$$

This is the analytic solution to (12). Particularly, if $A = p - e^{\frac{3p\pi}{2}}$, then the above solution can be simplified to $y(t) = e^{pt} + \sin(t)$, which coincides with the definition on $\left[-\frac{3\pi}{2}, 0\right]$. So in this case, $y(t) = e^{pt} + \sin(t)$ is the analytic solution to (12) in the whole interval $[0, \infty]$.

As the last example, we apply the DTM based on the method of steps to solve a neutral delay differential equations.

Example 4.4 Consider the neutral delay differential equation

$$y'(x) = y'(x-1) - y(x), \quad 0 \le x \le 1 \tag{13}$$

with the initial function $y(x) = \sin(x)$, $x \le 0$

According to the idea of the method of steps, DDE (13) becomes

$$y'(x) = \cos(x-1) - y(x), \quad 0 \le x \le 1$$

Applying DTM to this equation, we have

$$(k+1)Y(k+1) = \frac{1}{k!}\cos\left(\frac{\pi k}{2} - 1\right) - Y(k)$$

From the initial function, we get $y(0) = 0$, so $Y(0) = 0$. With the help of Mathematica, we have

$$\begin{aligned}
&Y(1) = \cos(1), \quad Y(2) = \frac{1}{2!}(-\cos(1) + \sin(1)), \quad Y(3) = -\frac{\sin(1)}{3!}, \quad Y(4) = 0, \\
&Y(5) = \frac{\cos(1)}{5!}, \quad Y(6) = \frac{1}{6!}(-\cos(1) + \sin(1)), \quad Y(7) = -\frac{\sin(1)}{7!}, \quad Y(8) = 0, \\
&Y(9) = \frac{\cos(1)}{9!}, \quad Y(10) = \frac{-\cos(1) + \sin(1)}{10!}, \quad Y(11) = -\frac{\sin(1)}{11!}, \quad Y(12) = 0, \\
&\ldots
\end{aligned}$$

Then, the solution to (13) is

$$y(x)=\cos(1)\sum_{k=0}^{+\infty}\frac{x^{4k+1}}{(4k+1)!}+(\sin(1)-\cos(1))\sum_{k=0}^{+\infty}\frac{x^{4k+2}}{(4k+2)!}-\sin(1)\sum_{k=0}^{+\infty}\frac{x^{4k+3}}{(4k+3)!}$$
$$=\cos(1)\frac{\sin(x)+\sinh(x)}{2}+(\sin(1)-\cos(1))\frac{-\cos(x)+\cosh(x)}{2}-\sin(1)\frac{-\sin(x)+\sinh(x)}{2}$$
$$=\frac{1}{2}e^{-x}(\sin(1)-\cos(1))+\frac{1}{2}(\cos(1-x)-\sin(1-x)).$$

5. CONCLUSION

Although the theory of differential transform method is not complete yet, it has been successfully applied to solve ordinary differential equations, partial differential equations, integral-differential equations, differential-algebraic equations and etc. In this paper, we apply DTM based on the method of steps to solve some delay differential equations, including neutral delay differential equations, successfully. Numerical experiments show that DTM is feasible and efficient for them. We believe that the operations of DTM presented in this paper also can be used to solve some partial delay differential equations (PDDEs), which is worth while studying in the future work.

ACKNOWLEDGEMENTS

Baoqing Liu, Xiaojian Zhou, Qikui Du This work is supported by the National Natural Science Foundation of China, contract/grant number 11371198 and 11401296, Jiangsu Provincial Natural Science Foundation of China, contact/grant no. BK20141008, Natural Science Fund for colleges and universities in Jiangsu Province contact/grant no. 14KJB110007, Jiangsu Provin- cial Key Laboratory for Numerical Simulation of Large Scale Complex Systems contract/grant number 201305 and 201401.

REFERENCES

1. Pukhov, G.E. (1986) Differential Transformations and Mathematical Modelling of Physical Processes. Naukova Dumka, Kiev.
2. Zhou, J.K. (1986) Differential Transformation and Its Application for Electrical Circuits. Huazhong University Press, Wuhan. (In Chinese)
3. Ayaz, F. (2004) Solutions of the System of Differential Equations by Differential Transform Method. Applied Mathematics and Computation, 147, 547-567.
4. Kurnaz, A., Oturnaz, G. and Kiris, M.E. (2005) n-Dimensional Differential Transformation Method for Solving PDEs. International Journal of Computer Mathematics, 82, 369-380.

5. Ayaz, F. (2004) Applications of Differential Transform Method to Differential-Algebraic Equations. Applied Mathematics and Computation, 152, 649-657.
6. Liu, H. and Song, Y.Z. (2007) Differential Transform Method Applied to High Index Differential-Algebraic Equations. Applied Mathematics and Computation, 184, 748-753.
7. Abdel-Halim Hassan, I.H. (2002) On Solving Same Eigenvalue Problems by Using a Differential Transformation. Applied Mathematics and Computation, 127, 1-22.
8. Biazar, J., Eslami, M. and Islam, M.R. (2012) Differential Transform Method for Special Systems of Integral Equations. Journal of King Saud University (Science), 24, 211-214.
9. Shadia, M. (1992) Numerical Solution of Delay Differential and Neutral Differential Equations Using Spline Methods. Ph.D. Thesis, Assuit University, Assuit.
10. El-Hawary, H.M. and Mahmoud, S.M. (2003) Spline Collocation Methods for Solving Delay-Differential Equations. Applied Mathematics and Computation, 146, 359-372.
11. Evans, D.J. and Raslan, K.R. (2005) The Adomian Decomposition Method for Solving Delay Differential Equation. International Journal of Computer Mathematics, 82, 49-54.
12. Vanani, S.K. and Aminataei, A. (2008) On the Numerical Solution of Neutral Delay Differential Equations Using Multiquadric Approximation Scheme. Bulletin of the Korean Mathematical Society, 45, 663-670.
13. Karako, F. and Bereketoğlu, H. (2009) Solutions of Delay Differential Equations by Using Differential Transform Method. International Journal of Computer Mathematics, 86, 914-923.
14. Lainscsek, C. and Sejnowski, T.J. (2015) Delay Differential Analysis of Time Series. Neural Computation, 27, 594- 614.
15. Abazari, R. and Kilicman, A. (2014) Application of Differential Transform Method on Nonlinear Integro-Differential Equations with Proportional Delay. Neural Computing and Applications, 24, 391-397.
16. Driver, R.D. (1977) Ordinary and Delay Differential Equations. Applied Mathematical Sciences, 20.
17. Odibat, Z.M., et al. (2010) A Multi-Step Differential Transform Method and Application to Non-Chaotic or Chaotic Systems. Computers & Mathematics with Applications, 59, 1462-1472.
18. Hairer, E., et al. (1993) Solving Ordinary Differential Equations I: Nonstiff Problems. 2nd Edition, Springer-Verlag, Berlin.

CHAPTER 6

Classification of Single Traveling Wave Solutions to the Generalized Kadomtsev-Petviashvili Equation without Dissipation Terms in p = 2

Xinghua Du, Hua Xin

Department of Mathematics, Northeast Petroleum University, Daqing, China

ABSTRACT

By using the complete discrimination system for the polynomial method, the classification of single traveling wave solutions to the generalized Kadomtsev-Petviashvili equation without dissipation terms in $p=2$ is obtained.

Keywords: complete discrimination system for polynomial; traveling wave solution; generalized Kadomtsev-Petviashvili equation without dissipation terms

1. INTRODUCTION

In mathematics and physics, the Kadomtsev-Petviashvili (KP) equation is a partial differential equation to describe nonlinear wave motion. It can be used to model water waves of long wavelength with weakly nonlinear restoring forces and frequency dispersion [1]. A number of modified forms of the KP equation have been studied [2-6]. In [1,7], the generalized Kadomtsev-Petviashvili equation without dissipation terms was given by

$$\frac{\partial}{\partial x}\left(u_t + du^p u_x + bu^{2p} u_x + \delta u_{xxx}\right) + 3k^2 u_{yy} = 0, \tag{1}$$

where d,b,δ are constants, $\delta \neq 0$, $p > 0$. Some of modified form of the KP equation can be written in the form of Equation (1).

Many reliable methods are used in the literature to examine the completely integrable nonlinear evolution equations. The Hirota bilinear method, the Bäcklund transformation method, the inverse scattering method, the Painlevé analysis, the simplified Hirotas method established by Hereman et al. [8], and others were effectively used in [1-13]. Liu proposed a complete discrimination system for polynomial method [10-13]. That is, by using of elementary integral method and complete discrimination system for polynomial, the single wave solutions can be classified for some nonlinear differential equations which can be directly reduced to integral forms.

In this paper, we consider the following generalized Kadomtsev-Petviashvili equation without dissipation terms in $p = 2$:

$$\frac{\partial}{\partial x}\left(u_t + du^2 u_x + bu^4 u_x + \delta u_{xxx}\right) + 3k^2 u_{yy} = 0, \tag{2}$$

where d,b,δ are constants, $\delta \neq 0$. By using Liu's complete discrimination system for polynomial method, the classification of single traveling wave solutions to Equation (2) is obtained.

2. CLASSIFICATION OF SOLUTIONS TO EQUATION (2)

Take wave transformation

$$u(x,y,t)=u(\xi) \text{ and } \xi = lx+my-\omega t$$

into Equation (1), the following nonlinear ordinary difference equation is given:

$$l\left(-\omega u'+dlu^2u'+blu^4u'+l^3\delta u'''\right)'+3k^2m^2u''=0, \tag{3}$$

Integrating Equation (3) once with respect to ξ, and setting the integral constant to zero yields:

$$l\left(-\omega u'+dlu^2u'+blu^4u'+l^3\delta u'''\right)+3k^2m^2u'=0. \tag{4}$$

Integrating Equation (4) twice yields

$$(u')^2=-\frac{b}{15l^2\delta}u^6-\frac{d}{6l^2\delta}u^4-\frac{3k^2m^2-\omega l}{l^4\delta}u^2+a_1u+a_0, \tag{5}$$

where a_1, a_0 are arbitrary constants.

Case 2.1. $a_1=a_0=0$, we substitute the transformation

$$u(\xi)=v^{\frac{1}{2}}(\xi)$$

into Equation (5) yields

$$\int\frac{dv}{v\sqrt{F(v)}}=\pm 2(\xi-\xi_0), \tag{6}$$

where

$$F(v)=-\frac{b}{15l^2\delta}v^2-\frac{d}{6l^2\delta}v-\frac{3k^2m^2-\omega l}{l^4\delta} \tag{7}$$

Let

$\Delta=\left(-\frac{d}{6l^2\delta}\right)^2-\frac{4b\left(3k^2m^2-\omega l\right)}{15l^6\delta^2}$ and Δ is the discriminant of the polynomial $F(v)$. According to the classification of the roots of $F(v)$, there are three cases to be discussed.

Case 2.1.1. $\Delta=0$, when $\frac{b}{\delta}<0$ from Equation (6), we have

$$v=\frac{5d}{b\left(1\mp e^{\pm\frac{5d}{2b}\sqrt{\frac{b}{15l^2\delta}}(\xi-\xi_0)}\right)}, \tag{8}$$

Case 2.1.2. $\Delta>0$, when

$\frac{b}{\delta}<0$ from Equation (6), we have

$$\pm 2\sqrt{-\frac{b}{15l^2\delta}}\left(\xi-\xi_0\right)=\frac{1}{\sqrt{\beta\gamma}}\ln\frac{\left[\sqrt{(-\gamma)(v-\beta)}-\sqrt{(-\beta)(v-\gamma)}\right]^2}{|v|}, \tag{9}$$

$$\pm 2\sqrt{-\frac{b}{15l^2\delta}}\left(\xi-\xi_0\right)=\frac{1}{\sqrt{\beta\gamma}}\ln\frac{\left[\sqrt{(\gamma)(v-\beta)}-\sqrt{(\beta)(v-\gamma)}\right]^2}{|v|}, \tag{10}$$

$$\pm 2\sqrt{-\frac{b}{15l^2\delta}}\left(\xi-\xi_0\right)=\frac{1}{\sqrt{-\beta\gamma}}\arcsin\frac{(-\gamma)(v-\beta)+(-\beta)(v-\gamma)}{|v||\beta-\gamma|}. \tag{11}$$

When

$\frac{b}{\delta}>0$ from Equation (6), we have

$$\pm 2\sqrt{\frac{b}{15l^2\delta}}\left(\xi-\xi_0\right)=\frac{1}{\sqrt{-\beta\gamma}}\ln\frac{\left[\sqrt{(-\gamma)(v+\beta)}-\sqrt{(\beta)(v-\gamma)}\right]^2}{|v|}, \tag{12}$$

$$\pm 2\sqrt{\frac{b}{15l^2\delta}}\left(\xi-\xi_0\right)=\frac{1}{\sqrt{-\beta\gamma}}\ln\frac{\left[\sqrt{(-\gamma)(-v+\beta)}-\sqrt{(-\beta)(v-\gamma)}\right]^2}{|v|}, \tag{13}$$

$$\pm 2\sqrt{\frac{b}{15l^2\delta}}\left(\xi-\xi_0\right)=\frac{1}{\sqrt{\beta\gamma}}\arcsin\frac{(-\gamma)(v+\beta)+\beta(v-\gamma)}{|v||\beta-\gamma|}. \tag{14}$$

where

$$\beta=\frac{5d+30l^2\delta\sqrt{\Delta}}{-b},\quad \gamma=\frac{5d-30l^2\delta\sqrt{\Delta}}{-b}.$$

Case 2.1.3. $\Delta<0$. From Equation (6), we have

$$\mp\frac{6k^2m^2-2\omega l}{l^4\delta}\left(\xi-\xi_0\right)=\ln\left|\frac{\dfrac{3l^2\delta}{d\sqrt{-\dfrac{3k^2m^2-\omega l}{l^4\delta}}}v+\sqrt{\dfrac{3k^2m^2-\omega l}{l^4\delta}}-\sqrt{-\dfrac{b}{15l^2\delta}v^2-\dfrac{d}{6l^2\delta}v-\dfrac{3k^2m^2-\omega l}{l^4\delta}}}{v}\right|, \tag{15}$$

where

$$\frac{3k^2m^2-\omega l}{l^4\delta}<0.$$

Case 2.2. $a_1=0, a_0\neq 0$. Substituting the transformation

$$u(\xi)=w^{\frac{1}{2}}(\xi)$$

into Equation (5) yields

$$\pm 2\sqrt{-\frac{\grave{o}b}{15l^2\delta}}\left(\xi-\xi_0\right)=\int\frac{\mathrm{d}w}{\sqrt{\grave{o}wF(w)}}, \tag{16}$$

where

$$F(w)=w^3+\frac{d}{2b}w^2+\frac{15\left(3k^2m^2-\omega l\right)}{bl^2}w-\frac{15l^2\delta a_0}{b} \tag{17}$$

If

$\frac{b}{\delta}>0$ we take

ò $=1$; if $\frac{b}{\delta}<0$ we take ò $=-1$. The complete discrimination system for the third order polynomial $F(w)$ is given as follows:

$$\Delta=\frac{4860l^4\delta a_0 b^3-3d^2l^2+4d^2l^2b-\left(24300l^2b\delta a_0+720b^2\right)\left(3k^2m^2-\omega l\right)}{12b^3l^2},$$

$$D_1=\frac{180bK^2m^2-60b\omega l-d^2l^2}{4b^2l^2}, \tag{18}$$

According to the classification of the roots of $F(w)$, there are four cases to be discussed.

Case 2.2.1. $\Delta=0, D_1<0$. Then

$F(w)=(w-\alpha)^2(w-\beta)$ where α,β are real constants, $\alpha\neq\beta$, and $\beta>0$. If ò $=1$, when $\alpha>\beta$ and $w>\beta$, or when $\alpha<0$ and $w<0$, from Equation (16), we have

$$\pm 2\sqrt{-\frac{b}{15l^2\delta}}\alpha(\alpha-\beta)(\xi-\xi_0)=\ln\frac{\left[\sqrt{\alpha(w-\beta)}-\sqrt{w(\alpha-\beta)}\right]^2}{|w-\alpha|}, \tag{19}$$

when $\alpha>\beta$, and $w<0$, or $\alpha<w<\beta$, we have

$$\pm 2\sqrt{-\frac{b}{15l^2\delta}}\alpha(\alpha-\beta)(\xi-\xi_0)=\ln\frac{\left[\sqrt{-\alpha(w-\alpha)}-\sqrt{w(\beta-\alpha)}\right]^2}{|w-\alpha|}, \tag{20}$$

when $\beta>\alpha>0$, we have

$$\pm 2\sqrt{-\frac{b}{15l^2\delta}}\alpha(\beta-\alpha)(\xi-\xi_0)=\arcsin\frac{\alpha(w-\beta)+w(\alpha-\beta)}{|\beta(w-\alpha)|}. \tag{21}$$

If ò $= -1$, when $\alpha > \beta$ and $w > \beta$, or when $\alpha < 0$ and $w < 0$, from Equation (16), we have

$$\pm 2\sqrt{-\frac{b}{15l^2\delta}}\alpha(\beta-\alpha)(\xi-\xi_0)=\ln\frac{\left[\sqrt{\alpha(-w+\beta)}-\sqrt{w(\beta-\alpha)}\right]^2}{|w-\beta|}, \tag{22}$$

when $\alpha > \beta$, and $w < 0$, or $\alpha < 0$, and $w < \beta$, we have

$$\pm 2\sqrt{-\frac{b}{15l^2\delta}}\alpha(\beta-\alpha)(\xi-\xi_0)=\ln\frac{\left[\sqrt{-\alpha(-w+\beta)}-\sqrt{w(\alpha-\beta)}\right]^2}{|w-\beta|}, \tag{23}$$

when $\beta > \alpha > 0$, we have

$$\pm 2\sqrt{-\frac{b}{15l^2\delta}}\alpha(\alpha-\beta)(\xi-\xi_0)=\arcsin\frac{\alpha(-w+\beta)+w(\beta-\alpha)}{|\beta(w-\alpha)|}. \tag{24}$$

Case 2.2.2. $\Delta = 0, D_1 = 0$. Then

$F(w)=(w-\alpha)^3$ where α is a real constant. If ò $= 1$, when $w > \alpha$, and $w > 0$, or $w < \alpha$, and $w < 0$, we have

$$w=\frac{15\alpha l^2\delta}{-b\alpha^2(\xi-\xi_0)^2-15l^2\delta}+\alpha, \tag{25}$$

If ò $= -1$, when $w > \alpha$, and $w < 0$, or $w < \alpha$, and $w > 0$, we have

$$w=\frac{15\alpha l^2\delta}{b\alpha^2(\xi-\xi_0)^2-15l^2\delta}+\alpha, \tag{26}$$

Case 2.2.3. $\Delta>0, D_1<0$. Then

$F(w)=(w-\alpha)(w-\beta)(w-\gamma)$ where α,β,γ are different real constants. If ò$=1$, when $w>0$, or $w<\gamma$, we have

$$w=\frac{-\gamma\alpha\,\mathrm{sn}^2\left(\sqrt{\frac{b}{15l^2\delta}\beta(\alpha-\beta)}(\xi-\xi_0),n\right)}{-\gamma\,\mathrm{sn}^2\left(\sqrt{\frac{b}{15l^2\delta}\beta(\alpha-\gamma)}(\xi-\xi_0),n\right)-(\alpha-\gamma)} \tag{27}$$

$$w=\frac{-\gamma(\alpha-\beta)\,\mathrm{sn}^2\left(\sqrt{\frac{b}{15l^2\delta}\beta(\alpha-\beta)}(\xi-\xi_0),n\right)-\beta(\alpha-\gamma)}{-\gamma\,\mathrm{sn}^2\left(\sqrt{\frac{b}{15l^2\delta}\beta(\alpha-\gamma)}(\xi-\xi_0),n\right)-(\alpha-\gamma)} \tag{28}$$

where

$$n^2=\frac{\gamma(\alpha-\beta)}{\beta(\alpha-\gamma)}.$$

If ò$=-1$, when $0>w>\alpha$, and $\gamma<w<\beta$, we have

$$w=\frac{-\beta\alpha\,\mathrm{sn}^2\left(\sqrt{\frac{b}{15l^2\delta}\beta(\alpha-\beta)}(\xi-\xi_0),n\right)+\alpha\beta}{-\alpha\,\mathrm{sn}^2\left(\sqrt{\frac{b}{15l^2\delta}\beta(\alpha-\gamma)}(\xi-\xi_0),n\right)+\beta} \tag{29}$$

$$w=\frac{-\gamma\beta}{(\beta-\gamma)\,\mathrm{sn}^2\left(\sqrt{\frac{b}{15l^2\delta}\beta(\alpha-\gamma)}(\xi-\xi_0),n\right)-\beta} \tag{30}$$

where

$$n^2=\frac{\alpha(\beta-\gamma)}{\beta(\alpha-\gamma)}.$$

Case 2.2.4. $\Delta < 0,$

$F(w)=(w-\alpha)\left[(w-\beta)^2+\gamma^2\right]$ where α,β,γ are all real constants, and $\alpha>0,\beta>0$, and $\gamma>0$. we have

$$w=\frac{a\,\mathrm{cn}\left(\frac{\sqrt{\pm\frac{2b\gamma m_2\alpha}{15l^2\delta}}}{m_1m_2}(\xi-\xi_0),m_1\right)+b_1}{c\,\mathrm{cn}\left(\frac{\sqrt{\pm\frac{2b\gamma m_2\alpha}{15l^2\delta}}}{m_1m_2}(\xi-\xi_0),m_1\right)+d} \tag{31}$$

where

$$a=\frac{1}{2}\alpha(c-d),b_1=\frac{1}{2}\alpha(d-c),c=\alpha-\beta-\frac{\gamma}{m_2},$$

$$d=-\beta-\gamma m_2,$$

$$E=\frac{\gamma^2-\beta(\alpha-\beta)}{\gamma\alpha},$$

$$m_2=E\pm\sqrt{E^2+1},m_1^2=\frac{1}{1+m_2^2}.$$

Case 2.3. $a_1\neq 0,a_0=0$. The Equation (5) becomes

$$\pm(\xi-\xi_0)=\int\frac{\mathrm{d}u}{\sqrt{-\frac{b}{15l^2\delta}uF(u)}}, \tag{32}$$

where $F(u)=u^5+qu^3+ru+s$, and

$$r=\frac{15(3k^2m^2-\omega l)}{bl^2},\quad q=\frac{d}{2b},\quad s=-\frac{15l^2\delta a_1}{b}.$$

The complete discrimination system for the fifth order polynomial $F(u)$ is given as follows:

$$D_2 = -q, D_3 = 40rq - 12q^3,$$

$$D_4 = 12q^4 r - 88r^2 q^2 + 125qs^2 + 160r^3,$$

$$E_2 = 160r^2 q^3 - 48rq^5 + 625s^2 q^2, \quad D_5 = 2000qs^2 r^2 - 900rs^2 q^3 + 16q^4 r^3 + 108q^5 s^2 - 128r^4 q^2 + 256r^5 + 3125s^4,$$

$$F_2 = -8rq. \tag{33}$$

According to the classification of the roots of $F(u)$, there are seven cases to be discussed.

Case 2.3.1. $D_5 = 0, D_4 = 0, D_3 > 0, E_2 \neq 0$, then

$$F(u) = (u-\alpha)^2 (u-\beta)^2 (u-\gamma),$$

$\alpha, \beta,$ and γ are real numbers, $\alpha \neq \beta \neq \gamma \neq 0.$ From Equation (32), we have

$$\pm(\alpha-\beta)(\xi-\xi_0) = \frac{1}{\sqrt{-\frac{b}{15l^2\delta}\alpha(\alpha-\gamma)}} \ln \frac{\left(\sqrt{-\frac{\varepsilon_1 b}{15l^2\delta}u(\alpha-\gamma)} - \sqrt{-\frac{\varepsilon_1 b}{15l^2\delta}\alpha(u-\gamma)}\right)^2}{|u-\alpha|}$$
$$-\frac{1}{\sqrt{-\frac{b}{15l^2\delta}\beta(\beta-\gamma)}} \arcsin \frac{-\frac{b}{15l^2\delta}u(\beta-\gamma) - \frac{b}{15l^2\delta}\beta(u-\gamma)}{\left|-\frac{b}{15l^2\delta}\gamma(u-\beta)\right|}, \tag{34}$$

$$\pm(\alpha-\beta)(\xi-\xi_0)=\frac{1}{\sqrt{-\frac{b}{15l^2\delta}\alpha(\alpha-\gamma)}}\ln\frac{\left(\sqrt{-\frac{\varepsilon_1 b}{15l^2\delta}u(\alpha-\gamma)}-\sqrt{-\frac{\varepsilon_1 b}{15l^2\delta}\alpha(u-\gamma)}\right)^2}{|u-\alpha|}$$
$$-\frac{1}{\sqrt{-\frac{b}{15l^2\delta}\beta(\beta-\gamma)}}\ln\frac{\left(\sqrt{-\frac{\varepsilon_2 b}{15l^2\delta}u(\beta-\gamma)}-\sqrt{-\frac{\varepsilon_2 b}{15l^2\delta}\beta(u-\gamma)}\right)^2}{|u-\alpha|},\tag{35}$$

$$\pm(\alpha-\beta)(\xi-\xi_0)=\frac{1}{\sqrt{-\frac{b}{15l^2\delta}\alpha(\alpha-\gamma)}}\arcsin\frac{-\frac{b}{15l^2\delta}u(\alpha-\gamma)-\frac{b}{15l^2\delta}\alpha(u-\gamma)}{\left|-\frac{b}{15l^2\delta}\gamma(u-\alpha)\right|}$$
$$-\frac{1}{\sqrt{-\frac{b}{15l^2\delta}\beta(\beta-\gamma)}}\ln\frac{\left(\sqrt{-\frac{\varepsilon_2 b}{15l^2\delta}u(\beta-\gamma)}-\sqrt{-\frac{\varepsilon_2 b}{15l^2\delta}\beta(u-\gamma)}\right)^2}{|u-\beta|},\tag{36}$$

$$\pm(\alpha-\beta)(\xi-\xi_0)=\frac{1}{\sqrt{-\frac{b}{15l^2\delta}\alpha(\alpha-\gamma)}}\arcsin\frac{-\frac{b}{15l^2\delta}u(\alpha-\gamma)-\frac{b}{15l^2\delta}\alpha(u-\gamma)}{\left|-\frac{b}{15l^2\delta}\gamma(u-\alpha)\right|}$$
$$-\frac{1}{\sqrt{-\frac{b}{15l^2\delta}\beta(\beta-\gamma)}}\arcsin\frac{-\frac{b}{15l^2\delta}u(\beta-\gamma)-\frac{b}{15l^2\delta}\beta(u-\gamma)}{\left|-\frac{b}{15l^2\delta}\gamma(u-\beta)\right|},\tag{37}$$

where $\varepsilon_1=\pm1,\varepsilon_2=\pm1$.

Case 2.3.2.

$$D_5=0,D_4=0,D_3=0,D_2\neq0,F_2\neq0,$$

$$F(u)=(u-\alpha)^3(u-\beta)^2,$$

α and β are real numbers, $\alpha\neq\beta$. From Equation (32), we have

$$\pm(\alpha-\beta)(\xi-\xi_0)=\frac{2}{\frac{b}{15l^2\delta}\alpha}\sqrt{\frac{-\frac{b}{15l^2\delta}u}{u-\alpha}}-\frac{1}{\sqrt{-\frac{b}{15l^2\delta}\beta(\beta-\gamma)}}\arcsin\frac{-\frac{b}{15l^2\delta}u(\beta-\alpha)-\frac{b}{15l^2\delta}\beta(u-\alpha)}{\left|-\frac{b}{15l^2\delta}\alpha(u-\beta)\right|},\tag{38}$$

$$\pm(\alpha-\beta)(\xi-\xi_0)=\frac{2}{\frac{b}{15l^2\delta}\alpha}\sqrt{\frac{-\frac{b}{15l^2\delta}u}{u-\alpha}}-\frac{1}{\sqrt{-\frac{b}{15l^2\delta}\beta(\beta-\alpha)}}\ln\frac{\left(\sqrt{-\frac{\varepsilon_2 b}{15l^2\delta}u(\beta-\alpha)}-\sqrt{-\frac{\varepsilon_2 b}{15l^2\delta}\beta(u-\alpha)}\right)^2}{|u-\beta|},\tag{39}$$

where $\varepsilon_2=\pm1$.

Case 2.3.3.

$$D_5=0, D_4=0, D_3=0, D_2\neq 0, F_2=0,$$

$$F(u)=(u-\alpha)^4(u-\beta),$$

α,β are real numbers, $\alpha\neq\beta\neq 0$. From Equation (32), we have

$$\mp\frac{b}{15l^2\delta}(\alpha-\beta)(\xi-\xi_0)=-\frac{\sqrt{-\frac{b}{15l^2\delta}u(u-\beta)}}{u-\alpha}+\frac{\frac{b}{15l^2\delta}\left(\alpha-\frac{1}{2}\beta\right)}{\sqrt{-\frac{b}{15l^2\delta}\alpha(\alpha-\beta)}}\ln\frac{\left(\sqrt{-\frac{\varepsilon_2 b}{15l^2\delta}u(\alpha-\beta)}-\sqrt{-\frac{\varepsilon_2 b}{15l^2\delta}\alpha(u-\beta)}\right)^2}{|u-\alpha|},\tag{40}$$

$$\mp\frac{b}{15l^2\delta}(\alpha-\beta)(\xi-\xi_0)=-\frac{\sqrt{-\frac{b}{15l^2\delta}u(u-\beta)}}{u-\alpha}+\frac{\frac{b}{15l^2\delta}\left(\alpha-\frac{1}{2}\beta\right)}{\sqrt{-\frac{b}{15l^2\delta}\alpha(\alpha-\beta)}}\arcsin\frac{-\frac{b}{15l^2\delta}u(\alpha-\beta)-\frac{b}{15l^2\delta}\alpha(u-\beta)}{\left|-\frac{b}{15l^2\delta}\beta(u-\alpha)\right|},\tag{41}$$

where $\varepsilon_2=\pm1$.

Case 2.3.4.

$$D_5=0, D_4=0, D_3<0, E_2\neq 0,$$

$s > 0$ $F(u) = \left((u-\beta)^2 + \gamma^2\right)^2 (u-\alpha),$.

Respectively, from Equation (32), we have

$$\pm(\xi - \xi_0) = \frac{\beta_1}{\alpha\rho^2}\left(\arctan\frac{\gamma}{u-\beta} + \arctan\frac{\gamma_2(u-\beta)+\delta_2}{\gamma_1(u-\beta)+\delta_1 - \sqrt{-\frac{b}{15l^2\delta}u(u-\alpha)}}\right)$$
$$+\frac{\beta_2}{2\alpha\rho^2}\ln\frac{(u-\beta)^2+\gamma^2}{\left(\gamma_1(u-\beta)+\delta_1\sqrt{-\frac{b}{15l^2\delta}u(u-\alpha)}\right)^2 + \left(\gamma_2(u-\beta)+\delta_2\right)^2}, \quad (42)$$

where

$$\rho^2 = \sqrt{\frac{b^2\left(\gamma^2(2\beta-\alpha)^2 + \left(\beta(\beta-\alpha)-\gamma^2\right)\right)^2}{225l^4\delta^2}}, \beta_1 = \pm\sqrt{\frac{\rho^2}{2} - \frac{b\left(\beta(\beta-\alpha)-\gamma^2\right)}{30l^2\delta}},$$
$$\beta_2 = \pm\sqrt{\frac{\rho^2}{2} + \frac{b\left(\beta(\beta-\alpha)-\gamma^2\right)}{30l^2\delta}}, \gamma_1 = \frac{-b\gamma\beta_2 + b\beta_1\left(\beta - \frac{\gamma}{2}\right)}{15\beta^2\rho^2\delta}, \gamma_2 = \frac{-b\gamma\beta_2 - b\beta_1\left(\beta-\frac{\gamma}{2}\right)}{15\beta^2\rho^2\delta},$$
$$\delta_1 = \frac{-b\beta(\beta-\alpha)\beta_1 - b\left(\beta-\frac{\gamma}{2}\right)\gamma\beta_2}{15l^2\delta\rho^2}, \delta_2 = \frac{b\beta(\beta-\alpha)\beta_2 - b\left(\beta-\frac{\gamma}{2}\right)\gamma\beta_1}{15l^2\delta\rho^2}, \quad (43)$$

where the signs of β_1 and β_2 must satisfy

$$\beta_1\beta_2 = \gamma - \frac{b}{15l^2\delta}\left[\beta - \frac{\alpha}{2}\right].$$

Case 2.3.5. $D_5 = 0, D_4 > 0$,

$$F(u) = (u-\alpha)^2(u-\alpha_1)(u-\alpha_2)(u-\alpha_3), \alpha_1 > \alpha > \alpha_3.$$

Respectively, from Equation (32), we have

$$\pm(\xi-\xi_0)=\frac{2\delta_1}{(a-\alpha c)\gamma_1}\left[cF(\psi,k_1)+\frac{\delta_1}{b_1-\alpha d}\Pi\left(\psi,\frac{a-\alpha c}{b_1-\alpha d},k_1\right)\right], \tag{44}$$

where we renew to queue the orders of $\alpha_1,\alpha_2,\alpha_3$, and α_4, denote $\alpha_1>\alpha_2>\alpha_3>\alpha_4$.When

$$\frac{b}{\delta}>0,u\geq\alpha_1 \text{ or } u\leq\alpha_4$$

(other cases can be written similarly, they are omitted), the meaning of every parameter in Equation (44) are given as follows:

$$a=\alpha_2(\alpha_1-\alpha_4),b_1=-\alpha_1(\alpha_2-\alpha_4),c=\alpha_1-\alpha_4,d=\alpha_2-\alpha_4,\delta_1=(\alpha_1-\alpha_2)(\alpha_1-\alpha_4)(\alpha_2-\alpha_4),$$
$$\frac{\delta_1}{\gamma_1}=\frac{1}{\sqrt{\left|\frac{b}{\delta}(\alpha_1-\alpha_3)(\alpha_2-\alpha_4)\right|}},k_1^2=\frac{(\alpha_1-\alpha_4)(\alpha_2-\alpha_3)}{(\alpha_1-\alpha_3)(\alpha_2-\alpha_4)},F(\varphi,k_1)=\int_0^{\varphi}\frac{\mathrm{d}\varphi}{\sqrt{1-k_1^2\sin^2\varphi}},$$
$$\Pi(\varphi,n,k_1)=\int_0^{\varphi}\frac{\mathrm{d}\varphi}{(1+n\sin^2\varphi)\sqrt{1-k_1^2\sin^2\varphi}} \tag{45}$$

Case 2.3.6.

$$D_5=0,D_4=0,D_3>0,$$
$$E_2=0,F(u)=(u-\alpha)^3(u-\beta)(u-\rho).$$

where we renew to queue the orders of $\alpha,\beta,\rho,$ and 0, and denote $\alpha_1>\alpha_2>\alpha_3>\alpha_4$. When

$$\frac{b}{\delta}>0,u\geq\alpha_1 \text{, or } u\leq\alpha_4$$

(other cases can be written similarly, they are omitted), we have

$$\pm(\xi-\xi_0)=\frac{-2c}{k_1^2\gamma}\left[(c+k_1^2d)F(\varphi,k_1)-cE(\varphi,k_1)\right],\alpha_1=\alpha \tag{46}$$

$$\pm(\xi-\xi_0)=\frac{2d}{\gamma}\left[(c+d)F(\varphi,k_1)-dE(\varphi,k_1)-d\coth\left(\phi\sqrt{1-k_1^2\sin^2\phi}\right)\right],\alpha_2=\alpha. \tag{47}$$

The signs are the same as the ones in Equation (45), furthermore,

$$E(\varphi,k_1)=\int_0^{\varphi_1}\sqrt{1-k_1^2\sin^2\psi}\,\mathrm{d}\psi. \tag{48}$$

Case 2.3.7.

$$D_5=0,D_4=0,D_3<0,$$
$$E_2=0,F(u)=(u-\alpha)^3\left((u-\beta)^2+\gamma^2\right)$$

Now we renew to queue the orders of α and 0, and denote $\alpha_1>\alpha_2$, we have

$$\pm(\xi-\xi_0)$$
$$=\frac{cd}{\gamma}\left[F(\varphi,k_1)+\frac{c^2}{k_1\gamma}\arcsin(k_1\sin\varphi)\right],\alpha_1=\alpha. \tag{49}$$

$$\pm(\xi-\xi_0)=\frac{-d^2}{\gamma\sqrt{1-k_1^2}}\ln\frac{\sqrt{1-k_1^2\sin^2\varphi}+\sqrt{1-k_1^2}\sin\varphi}{\cos\varphi}$$
$$-\frac{cd}{\gamma}F(\varphi,k_1),\alpha_2=\alpha. \tag{50}$$

where

$$c=\alpha_1-\beta-\frac{\gamma}{m_1},m_1=E\pm\sqrt{E^2+1},$$
$$d=\alpha_1-\beta-\gamma m_1,E=\frac{\gamma^2+(\alpha_1-\beta)(\alpha_2-\beta)}{\gamma(\alpha_1-\alpha_2)}, \tag{51}$$

where the positive sign and negative sign for m_1 must satisfy

$\frac{bm_1}{\delta}<0$ other signs are the same with the former.

From the description above, using elementary integral method and complete discrimination system for polynomial, we have obtained the solutions of equations (6), (16) and (32) that can be expressed by elementary functions and elliptic functions. What's more, some solutions are explicit, but some solutions are implicit functions. So we can write concretely the exact traveling wave solutions of Equation (5) in some special cases. They are omitted for simplicity.

3. CONCLUSION

Using the complete discrimination system for polynomial method, we have obtained the classification of single traveling wave solutions to the generalized KadomtsevPetviashvili equation without dissipation terms in $p=2$. With the same method, some of other evolution equations can be dealt with.

ACKNOWLEDGEMENTS

Xinghua Du, Hua Xin The project is supported by Scientific Research Fund of Education Department of Heilongjiang Province of China under Grant No. 12521049.

REFERENCES

1. M. J. Ablowits, "Lectures on the Inverse Scattering Transform," Studies in Applied Mathematics, Vol. 58, No. 11, 1978, pp. 17-94.
2. G. C. Das and J. Sarma, "Evolution of Solitary Waves in Multicomponent Plasmas," Chaos, Solitons and Fractals, Vol. 9, No. 6, 1998, pp. 901-911.
3. X. Zhao, W. Xu, H. Jia and H. Zhou, "Solitary Wave Solutions for the Modified Kadomtsev-Petvisahvili Equation," Chaos, Solitons, and Fractals, Vol. 74, 2007, pp. 465-475.
4. W. X. Ma, A. Abdeljabbar and M. G. Assad, "Wronskian and Grammian Solutions to a (3 + 1)-Dimensional Generalized KP Equation," Applied Mathematics and Computation, Vol. 217, No. 24, 2011, pp. 10016-10023.

A. M. Wazwaz, "Multiple-Soliton Solutions for a (3 + 1)-Dimensional Generalized KP Equation," Communications in Nonlinear Science and Numerical Simulation, Vol. 17, No. 2, 2012, pp. 491-495.

5. W. X. Ma and A. Abdeljabbar, "A Bilinear Bäcklund Transformation of a (3 + 1)-Dimensional Generalized KP Equation," Applied Mathematics Letters, Vol. 25, No. 10, 2012, pp. 1500-1504.
6. G. B. Whitham, "Linear and Nonlinaer Wave," John Wiley, New York, 1974.
7. W. Hereman and A. Nuseir, "Symbolic Methods to Construct Exact Solutions of Nonlinear Partial Differential Equations," Mathematics and Computers in Simulation, Vol. 43, No. 1, 1997, pp. 13-27.

8. W. X. Ma and A. Pekcan, "Uniqueness of the Kadomtsev-Petviashvili and Boussinesq Equations," Zeitschrift für Naturforschung A, Vol. 66, 2011, pp. 377-382.
9. C. S. Liu, "Applications of complete Discrimination System for Polynomial for Classifications of Traveling Wave Solutions to Nolinear Differential Equations," Computer Physics Communications, Vol. 181, No. 2, 2010, pp. 317- 324.
10. C. S. Liu, "Classification of All Single Traveling Wave Solutions to Calogero-Focas Equation," Communications in Theoretical Physics, Vol. 48, 2007, pp. 601-604.
11. C. S. Liu, "Exact Traveling Wave Solutions for (1 + 1)-Dimentional Dispersive Long Wave Equation," Chinese Physical Society, Vol. 14, No. 9, 2005, pp. 1710- 1716.
12. C. S. Liu, "Direct Integral Method, Complete Discrimination System for Polynomial and Applications to Classifications of All Single Traveling Wave Solutions to Nonlinear Differential Equations: A Survey," arXiv:nlin/ 0609058v1, 2006.

CHAPTER 7

Identification of Rotor Unbalance as Inverse Problem of Measurement

Yuri Menshikov

Faculty of Mechanics & Mathematics, Dnepropetrovsk University, Dnepropetrovsk, Ukraine

ABSTRACT

In this paper, the problem of identification of the characteristics of the rotor unbalance on two supports is investigated as the inverse problem of measurement. The vibration of rotor supports in two mutually perpendicular directions used as the initial information. The inverse problem is considered, taking into account the error of the mathematical description of rotor-bearings system. To obtain estimates of real unbalance characteristics, the hypothesis as to the exact solutions is applied. The method of Tikhonov regularization is used to obtain stable results. Test calculations are given to illustrate the proposed approach.

Keywords: mathematical model; unbalance identification; solution estimation; regularization; numerical test

1. INTRODUCTION

The constructional differences of rotors and their domains of exploitation led to the creation of special methods of balancing. In many cases, the only criterion of rotor balancing is the absence (or the permissible value) of dynamical responses of supports. The compensation of deflections on length of rotor is considered only as the means to arrive of minimum of main criterion. In other cases, the reach of minimum of its deflections or its bending moment is taken as the criterion of rotor balancing. Such difference in choice of criterion can explain that for each case the parameters that are the main for given type of rotors are chosen. The reactions of supports or corresponding vibrations are taken for criteria of rotors balancing in particular to turbine-generator-building and the rotor deflection axis in jet engine building [1,2].

The basis of the most existing methods of flexible rotors balancing is the measuring of rotor vibrations and its supports, namely measuring of deflection and phases of rotating rotor followed by the choice and putting of trial plummets according to the shape of normal mode of vibrations.

The motion of flexible rotor relative to the rotating together with its coordinate system (one of axis coincides with the geometric axis of rotor) is described by Fredholm equation of the first kind. This equation is substituted for the matrix equation of form:

$$\boldsymbol{y} = \omega^2 A(\boldsymbol{y} + \boldsymbol{e}) = \omega^2 A\boldsymbol{y} + \omega^2 A\boldsymbol{e},$$

where $\boldsymbol{y}$ and $\boldsymbol{e}$ are vectors of dimension n; $A = \{a_{ik}\}$ is the quadratic matrix of dimension $n \times n$; ω is the frequency of rotating. The coefficients of influence $A = \{a_{ik}\}$ are defined experimentally for each plane of correction by trial starts.

Balancing plummets are calculated with the help of the initial values of vibrations y_i from condition of remaining amplitudes minimum. The definition of balancing plummets is made by least squares method. The information about experimental complex values of influence coefficients that are obtained by different balancing is neutralized as a rule.

The main tendencies of balancing methods development are connected with ways of development of coefficient influence definition [2,3].

Besides, current passive methods do not give the complete information about the position of unbalance if the rotor has a large length along the axis of rotation.

The suggested algorithms of unbalance evaluation use the experimental data about vibrations (accelerations) of two rotors' support in two mutually perpendicular directions during the work and a few rotor rotations as the initial information. These algorithms do not demand special conditions of work or the installation of trial plummets.

2. PROBLEM DEFINITION

Let us consider a deformable rotor rotating on two non-rigid supports [4,5]. We introduce rectangular righthand coordinates system $O\xi\eta\varsigma$. The axis $O\varsigma$ coincides with axis of the rotating shaft of rotor. The axis $O\varsigma$ belongs to the plane of rotor in horizontal position. The axis $O\eta$ has vertical direction. We obtain equations of rotor motion in the following [4,5]:

1) Weight's centers and stiffness's centers of crosssection of rotor coincide;
2) Eccentricity of rotor's disk is one-order infinitesimal with a displacement under vibrations.

The motion of rotor on two non-region supports is described by system of ordinary differential equations of 18th order [4,5]. Unbalance of rotor is modeled by some external load (EL). The value of this EL and the place of its action is it necessary to find. It is assumed that the vibrations of rotor supports in two mutual perpendicular directions are obtained from experiment. Let us suppose that the functions $z_1(t), z_2(t), z_3(t)$ characterize the unbalance of rotor (EL)

$$z_1(t) = m_r r\dot{\varphi}^2 \sin(\vartheta + \varphi), z_2(t) = m_r r\dot{\varphi}^2 \cos(\vartheta + \varphi),$$
$$z_3(t) = hm_r r\dot{\varphi}^2 \sin(\vartheta + \varphi);$$

where r is the radius of rotor, m_r is the mass of unbalance reducing to a surface of rotor, $\dot{\varphi}$ is the angular velocity of rotation, h is unbalance arm, ϑ is angular deviation of the factor of EL with respect to correction plane. If the unbalance is absent then the functions $z_1(t), z_2(t), z_3(t)$ will be equal to zero. We suppose that with the help of acceleration transducers the function have been recorded ($\ddot{\xi}_A(t), \ddot{\xi}_B(t)$ are the acceleration of supports in horizontal direction, $\ddot{\eta}_A(t), \ddot{\eta}_B(t)$ are the acceleration of supports in vertical direction). As an example, we consider the equation for the unknown function $z_1(t)$ only.

Then the problem of unbalance measurement is reduced to the solution of integral equations of Volltera first kind

$$\int_0^t (t-\tau) z_1(t) \mathrm{d}\tau = u_1(t), t \in [0, T];$$

or

$$A_1 z_1 = B_{1,p} x_\delta = u_{1,\delta}, \tag{1}$$

where A_1 is a linear integral operator $(A_1 : Z \to U)$, z_1 is the searched characteristic of EL, $B_{1,p}$ is a linear irreversible operator $(B_{1,p} : X \to U)$ depending on vector parameters of mathematical model (MM) of “rotorsupports” system $p = (p_1, p_2, \cdots, p_n)^{\mathrm{T}}$ $(()^{\mathrm{T}}$ is the sign of transposition); x_δ is the vector-function of initial data. Subjective factors influence on the definition of parameters of system “rotor-supports” MM and therefore the parameters are supposed to have their values within certain limits: $p_i^l \leq p_i \leq p_i^l, 1 \leq i \leq n$. In this way the vector p can be changed inside the known closed region $p \in D \subset R^n$.

The equations for required functions $z_2(t), z_3(t)$ will be similar to the Equation (1).

For a rotor on two supports for function $z_1(t)$, $z_2(t), z_3(t)$ the vector parameters $p \in D \subset R^{11}$ of MM has a kind $p = (E, J, m, m_A, m_B, c_A^\xi, c_B^\xi, b_A^\xi, b_B^\xi, a, b)^{\mathrm{T}}$ where E is module of Jung of rotor material, m is the mass of rotor, m_A is the mass of the A support, m_B is the mass of the B support; c_A^ξ, c_B^ξ are the stiffness of supports A and B with respect to the horizontal and vertical direction; b_A^ξ, b_B^ξ are the coefficients of external friction; a is the distance of gravity centre of rotor to the A support, a + b = l is the shaft length of rotor.

The vector function x_δ is obtained using the experimental data (vibrations of supports) where the noise is present. Therefore it is convenient to think that each component of vector function x_δ and function u_δ belongs to L_2 [0,

T]. Under this conditions the problem of equation (1) solution belongs to ill-posed problems if the searched functions $z_1(t)$ belong to $C[0,T]$ as the operator A in (1) is completely continuous [6].

The value of function deviation $u_{1,\delta}$ from the exact function u_{ex} is given (if exact operator $B_{1,ex}$ is linear):

$$\left\|u_{1,\delta}-u_{1,ex}\right\|_U=\left\|B_{1,p}\boldsymbol{x}_\delta-B_{1,ex}\boldsymbol{x}_{ex}\right\|_U\le\delta_0,\left\|A_1-A_{ex}\right\|\le h,$$

where $\delta_0=\delta b_0+d\left\|\boldsymbol{x}_\delta\right\|_X$;

$$b_0=\sup_{p\in D}\left\|B_{1,p}\right\|_{X\to U},\left\|B_{1,p}-B_{1,ex}\right\|_{X\to U}\le d,\left\|\boldsymbol{x}_\delta-\boldsymbol{x}_{ex}\right\|_X\le\delta;$$

$\boldsymbol{x}_{ex}$ is the exact vector function of initial data; $A_{1,ex}$ $B_{1,ex}$, are the exact operators; d, b_0, d, h are given values.

3. STATEMENTS OF IDENTIFICATION PROBLEM AS INVERSE PROBLEM OF MEASUREMENT

Let us consider the set of possible solutions of Equation (1) with account of whole error of initial data

$$Q_{h,\delta_0}=\left\{z:z\in Z,\left\|A_1z-B_{1,p}\boldsymbol{x}_\delta\right\|_U\le\delta_0+h\left\|z\right\|_Z\right\}.$$

The set Q_{h,δ_0} is unbounded for any $\delta_0>0$ as this problem is ill-posed [6].

For definition of stable approximate solution is used the regularization method of Tikhonov [6]. This way is based on the search of following extreme problem solution:

$$\Omega[z_0]=\inf_{z\in Q_{h,\delta_0}}\Omega[z], \qquad (2)$$

where $\Omega[z]$ is the stabilizing functional which is defined on Z_1 ($Z_1$is the everywhere dense set into Z).

The functional $\Omega[z]$ is chosen as follows

$$\Omega[z]=\|z\|^2_{W_1^2[0,T]}=\int_0^T\left(q_0z^2+q_1\dot{z}^2\right)dt$$

Where

$$q_0\geq 0, q_1>0.$$

The choice such functional is explained by the following reasons:

- The solution z_0 of an extreme problem (2) least will deviate zero and to have least first derivative in root-mean-square sense;
- The solution z_0 will give an estimation from below of exact solution z_T of equation (1) $\Omega[z_0]\leq\Omega[z_T]$.

From the practical point of view the function z_0 gives a guaranteed estimation from below sizes of real of a rotor in sense of functional $\Omega[z]$. If $\Omega^0\leq\Omega[z_0]$ (Ω^0 there is known limiting an allowable size for the given type of rotor machine) then the rotor is working in emergency operation with guarantee.

If the inequality $\Omega[z_0]<\Omega^0$ is carried out, then no objective conclusions can be made. We will be named the solution of extreme problem (2) as estimation from below of real unbalance.

In work [7] regularizing algorithm was suggested for equation (1) with approximate linear operator $\tilde{A}$ for Banach spaces Z, U, which based on the regularization method [6].

The solution of problem (1) is reduced to the solution of following extreme problem:

$$\inf_{z\in Z_1} M^\alpha\left[z,u_{1,\delta},A_1\right]=\inf_{z\in Z_1}\left\{\left\|A_1z-u_{1,\delta}\right\|^2_U+\alpha\Omega[z]\right\}$$
$$=M^\alpha\left[z_{1,\alpha},u_{1,\delta},A_1\right], \qquad (3)$$

where $\Omega[z]$ is stabilizing functional for equation (1) which defined on Z_1, the set Z_1 is everywhere dense into Z.

Regularization parameter α can be obtained from equation of general discrepancy:

$$\left\|A_1 z_{1,\alpha} - u_{1,\delta}\right\|_U^2 = \left(\delta_0 + h\left\|z_{1,\alpha}\right\|_{Z_1}\right)^2 + \mu^2\left(u_{1,\delta}, A_1\right), \tag{4}$$

where $\mu\left(u_{1,\delta}, A_1\right)$ is measure of discrepancy.

However at realization of such approach there are large difficulties at definition of size d, h as the absolute exact operators $B_{1,ex}, A_{1,ex}$ are unknown and basically they cannot be constructed. However at realization of such approach there are large difficulties at definition of size d, h as the absolute exact operators $B_{1,ex}$ is unknown and basically it cannot be constructed. Therefore size d is determined with the large overestimate and in set $Q_{d\delta}$ the "extraneous" functions get, that considerably reduces accuracy of the regularized solution.

So in this paper the estimation of inverse problem solution instead of solution of equation (1) is suggested. The following hypothesis is assumed for this purpose [8,9]: the such inequality is valid

$$\Omega\left[z_{1,ex}\right] \geq \Omega\left[z_{1,\alpha}\right] \tag{5}$$

for any approximate operators $A_1, B_{1,p}$ in equation (1) which corresponding to adequate mathematical description of vibration process [10]; $z_{1,ex}$ is the solution of equation (1) with exact right part $u_{1,ex}$ and exact operator $A_{1,ex}$. The exact operators $A_{1,ex}, B_{1,ex}$ can are nonlinear.

The inequality (5) is evident if the exact operators are linear one.

In the given work it is supposed, that all approximate operators $B_{1,p}$ in (1) have the same structures which depend from some vector parameters $p \in D$. In this case extreme problem (2) can be replaced by the following extreme problem [11]:

$$\Omega\left[z^*\right] = \inf_{z \in Q_\delta^*} \Omega[z] = \inf_{p \in D} \inf_{z \in Q_{p,\delta}} \Omega[z], \tag{6}$$

where $Q^* = \bigcup_{p \in D} Q_{p,\delta}$ ($\bigcup$ is the union).

Now in set of the possible solutions "the extraneous functions" have not got.

It is evident that $Q^* \subset Q_{h,\delta}$ for any d > 0, h > 0, d > 0. Therefore the use of the set Q^* instead of $Q_{h,\delta}$ allows obtaining the most "thin" solution.

From the practical point of view the solution z^* of an extreme problem (6) has the same meaning, as well as z_0, but gives an exacter estimation from below. We will be named the solution of extreme problem (6) as estimation from below of real unbalance also. But the inequality $\Omega[z_0] \le \Omega[z^*]$ is valid.

For the solution of an extreme problem (6) it is offered to use a method of a choice of the minimal special mathematical model of system "rotor-supports" [11,12]. It allows getting more exact estimation of exact solution.

For the realization of such approach it is necessary to choose within the vectors $p \in D$ some vector $p_0 \in D$ such that

$$\Omega\left[A_1^{-1}B_{1,p_0}\boldsymbol{x}_\delta\right] \le \Omega\left[A_1^{-1}B_{1,p}\boldsymbol{x}_\delta\right]$$

for all possible $x_\delta \in X$ and all $p \in D$. The operator $B_{1,p}$ with parameter $p_0 \in D$ will be called the minimal operator. Appropriate to this operator the model is named as the minimal MM [12,13].

If the minimal MM exists, then the extreme problem (6) can be replaced by an equivalent simpler extreme problem:

$$\Omega\left[z^*\right] = \inf_{z \in Q_{p_0,\delta}} \Omega[z] \quad (7)$$

Let's consider the problem on existence of the minimal MM in a problem of unbalance identification.

From physical sense of a problem follows, that the vibrations of support of a rotor $\ddot{\xi}_A(t), \ddot{\xi}_B(t), \dot{\xi}_A(t), \dot{\xi}_B(t)$, $\xi_A(t), \xi_B(t)$ at constant speed of rotation of a rotor w and at constant size of unbalance in time $t \in [0,T]$ are periodic functions with zero average for the minimal period $T_1 = 2\pi/\omega$.

The function $z_1(t)$ on physical sense represents also periodic function of a type $A_1 \sin(\omega t + \varphi_1)$. At ω = const and at constant size of unbalance between functions $z_1(t), \xi_A(t)$ and $\xi_B(t)$, also there is a connection $\xi_A(t) = \mu^{-1}\xi_B(t)$, where m-Const, m > 0.

Then the identity is valid

$$\alpha_1 \dddot{\xi}_A(t) + \alpha_2 \ddot{\xi}_A(t) + \alpha_3 \ddot{\xi}_A(t) + \alpha_4 \dot{\xi}_A(t) + \alpha_5 \xi_A(t) = z_1(t), \quad (8)$$

where $\alpha_1, \alpha_2, \alpha_3, \alpha_4, \alpha_5$ are the function from T and p.

In this case function $\Omega[z(p)] = \Omega[p]$ at the fixed functions $\xi_A(t), \xi_B(t)$ is continuous on components of a vector p. Under the well-known theorem of Weierstrass the function $\Omega[p]$ reaches on the convex closed set D of the greatest lower bound.

Let's calculate partial derivatives of function $\Omega[p]$ on parameters (in the dimensionless form):

$$\frac{\partial \Omega}{\partial m} = (\varphi_1\varphi_2 + \varphi_1 k_{11})k_{12} + k_{13}(\varphi_1\varphi_2 + \varphi_3 k_{14}),$$

$$\frac{\partial \Omega}{\partial (EJ)} = -\left[\varphi_1\varphi_4 + \varphi_3 k_{12} + k_{22}(\varphi_3 k_{23} + \varphi_1\varphi_4)\right], \quad (9)$$

$$\frac{\partial \Omega}{\partial m_A} = \varphi_1\varphi_5 k_{31}, \frac{\partial \Omega}{\partial m_B} = \varphi_1\varphi_6 k_{41}, \frac{\partial \Omega}{\partial a} = \varphi_1\varphi_7 k_{51} + \varphi_3 k_{52},$$

$$\frac{\partial \Omega}{\partial b} = \varphi_1\varphi_8 k_{61} + k_{62}\varphi_3 + (\varphi_3 k_{63} + \varphi_1 k_{64}), \frac{\partial \Omega}{\partial b_A^{\xi}} = \varphi_3\varphi_9 k_{71},$$

$$\frac{\partial \Omega}{\partial b_B^{\xi}} = \varphi_3\varphi_{10} k_{81}, \frac{\partial \Omega}{\partial c_A^{\xi}} = \varphi_1\varphi_{11} k_{91}, \frac{\partial \Omega}{\partial c_B^{\xi}} = \varphi_1\varphi_{12} k_{92};$$

where k_{ik} are positive constants.

The signs of the partial derivatives are determined by signs and sizes of functions $\varphi_1, \varphi_2, \dots, \varphi_{12}$, which depend on angular speed of rotation of a rotor w, parameters p of MM and size of T: $\varphi_k = \varphi_k(\omega, p, T), k = \overline{1,12}$.

The functions $\varphi_k\left(k = \overline{1,12}\right)$ from variable $z = \omega^2 T^2$ are either square-law or linear. Therefore their signs are enough easily determined.

At rather small T the minimal operator for (1) exists and associated the corner point of D.

In a problem of modeling of fluctuations in ventilator of the furnace the parameters of MM were the following [14]:

$$m^l = 0.84 \times 10^4 \text{ kg}, \; m^t = 0.9 \times 10^4 \text{ kg},$$

$$E^l = E^t = 2 \times 10^{11} \text{ H/m}^2, \; J^l = 3 \times 10^{-3} \text{ m}^4,$$

$$J^t = 3.1 \times 10^{-3} \text{ m}^4,$$

$$m_A^l = m_B^l = 0.34 \times 10^5 \text{kg}, \; m_A^t = m_B^t = 0.35 \times 10^5 \text{kg},$$

$$c_A^l = c_B^l = 0.65 \times 10^9 \text{ H/m}, \; c_A^t = c_B^t = 0.7 \times 10^9 \text{ H/m},$$

$$b_A^l = b_B^l = 0, \; b_A^t = b_B^t = 100 \text{ Hs/m}^2,$$

$$a^l = 5 \text{ m}, \; a^t = 5.1 \text{ m}, \; b^l = 3 \text{ m}, \; b^t = 3.1 \text{ m}.$$

At $T = 0.1$ s and $\omega = 311 \text{ s}^{-1}$ the value $\omega^2 T^2$ has appeared less least roots of the equations $\varphi_k = 0, \; k = \overline{1.12}$ (in view of allowable disorder of parameters). Therefore, in this example of function j_k have signs:

$$\varphi_1 > 0, \varphi_3 < 0, \varphi_4 < 0, \varphi_5 < 0, \varphi_6 < 0, \varphi_7 < 0, \varphi_8 < 0,$$

$$\varphi_9 < 0, \varphi_{10} < 0, \varphi_{11} > 0, \varphi_{12} > 0.$$

Then, taking into account expressions (6), it is possible to conclude, that

$$\frac{\partial \Omega}{\partial m} < 0, \frac{\partial \Omega}{\partial (EJ)} > 0, \frac{\partial \Omega}{\partial m_A} < 0, \frac{\partial \Omega}{\partial m_B} < 0, \frac{\partial \Omega}{\partial a} < 0,$$

$$\frac{\partial \Omega}{\partial b} < 0, \frac{\partial \Omega}{\partial b_A^{\xi}} > 0, \frac{\partial \Omega}{\partial b_B^{\xi}} > 0, \frac{\partial \Omega}{\partial c_A^{\xi}} > 0, \frac{\partial \Omega}{\partial c_A^{\xi}} > 0.$$

Hence, in a considered example the minimal model exists and corresponds to a vector

$$p_0 = \left(E^l, J^l, m^l, m^l{}_A, m^l{}_B, c^{l\xi}_A, c^{l\xi}_B, b^{l\xi}_A, b^{l\xi}_A, a^l, b^l\right)^{\mathrm{T}}.$$

It is possibly that the size of $\Omega\left[z_{\delta,p}\right]$ for function $z_{\delta,p}$ from the some set $Q_{p,\delta}$ which satisfies to equality

$$\Omega\left[z_{\delta,p}\right] = \inf_{z \in Q_{p,\delta}} \Omega[z],$$

exceeds admissible size of Ω^0. But the vector parameters p of MM is given with error. To make this conclusion about a similar situation with a guarantee it is necessary to consider all possible sets $Q_{p,\delta}$, then in them to find all functions $\hat{z}_p$ which minimize $\Omega[z]$ on $Q_{p,\delta}$, further among them to find greatest in sense of functional value $\Omega\left[z_{\delta,p}\right]$. Thus we are getting to necessity of statement of the following problem:

$$\Omega\left[\tilde{z}_p\right] = \sup_{p \in D} \inf_{z \in Q_{p,\delta}} \Omega[z] \quad . \tag{10}$$

It is obvious that $\Omega\left[z^*\right] \le \Omega\left[\tilde{z}_p\right]$. If $\Omega^0 \le \Omega\left[\tilde{z}_p\right]$, then the machine probably works in emergency operation. If $\Omega\left[\tilde{z}_p\right] \le \Omega^0$, then no certain conclusions can be made.

For the solution of an extreme problem (10) it is offered to use a method of a choice of the special maximal MM of system "rotor-supports".

It is possible to show that the solution of an extreme problem (10) under some conditions always exists.

For the realization of such approach it is necessary to choose within the vectors $p \in D$ some vector $p^1 \in D$ such that

$$\Omega\left[A_1^{-1}B_{1,p^1}\boldsymbol{x}_\delta\right] \geq \Omega\left[A_1^{-1}B_{1,p}\boldsymbol{x}_\delta\right]$$

for all possible $\boldsymbol{x}_\delta \in X$ and all $p \in D$. The operator B_{1,p^1} with parameter $p^1 \in D$ will be called the special maximal operator. Appropriate to this operator the model is named as the special maximal MM [9,10].

If the special maximal MM exists then the solution of an extreme problem (10) will coincide with the solution of the following extreme problem:

$$\Omega\left[\tilde{z}_{1,p}\right] = \inf_{z \in Q_{p^1,\delta}} \Omega[z] \quad . \tag{11}$$

For considered before an example the special maximal MM exists, unique and corresponds to a vector

$$p^1 = \left(E^l, J^l, m^l, m^l_A, m^l_B, c^{l\xi}_A, c^{l\xi}_B, b^{l\xi}_A, b^{l\xi}_A, a^l, b^l\right)^{\mathrm{T}}.$$

If at a vector parameters p is inexact the part of parameters is given only then a situation essentially to not change but only the vector parameters p will have only smaller dimension. In a number of cases it is possible to carry out a choice of the special minimal or special maximal MM only in part of parameters of a vector p. And in this case it is possible to receive some prize in accuracy of the approximate estimation.

Let us consider the following statement of a problem of rotor unbalance identification: to find the function z^{pl} among set of the possible solutions of the equation (1) which would give the least maximal deviation from the experimentally measured vibrations of support of a rotor for all operators $B_{1,p}$. Such statement is reduced to the solution of the following extreme problem:

$$\left\|A_1 z^{pl} - B_{1,p}\boldsymbol{x}_\delta\right\| = \inf_{z_p} \sup_{B_{1,p}} \left\|A_1 z_{p,\delta} - B_{1,p}\boldsymbol{x}_\delta\right\|_U \quad , \tag{12}$$

where $z_{p,\delta}$ is the solution of extreme problem (11) on set $Q_{p,\delta}$.

As all operators $B_{1,p}$ it is possible to consider equivalent within the limits of the specified accuracy, it is possible to consider function z^{pl} as the most probable solution of a problem of unbalance identification. The function z^{pl} will be call the most plausible estimation of unbalance. The most probable estimation z^{pl} will coincide with classical regularizing the solution of an inverse problem of unbalance identification of a rotor if there is the one operator $B_{1,p}$ only. The function z^{pl} is the best approximation of real unbalance characteristics and also is steady to small deviations of the initial data.

Suggested algorithm can be used in case if the exact operator $B_{1,ex}$ does not belong to set of operators $B_{1,p}$ and the operator A does not coincide with exact operator A_{ex}.

4. TEST CALCULATION

For suggested algorithm examination of unbalance characteristics' evaluation, there was a calculated case when functions $\ddot{\xi}_A(t), \ddot{\xi}_B(t), \ddot{\eta}_A(t), \ddot{\eta}_A(t)$ are the results of mathematical simulation of rotor vibrations with given unbalance. The parameters of rotor unbalance were chosen as:

$$m_r = 0.5 \text{ kg} \text{ by } r = 0.25 \text{ m, h} = 0.25 \text{ m}, \vartheta = 0.5 \text{ rad.}$$

The values of initial data inaccuracy were chosen after filtering as the following:

$$\left\| \ddot{\xi}_A(t) - \ddot{\xi}_{Aex}(t) \right\|_C \le \delta_1 = 0.08 \text{ m/s}^2,$$

$$\left\| \ddot{\xi}_B(t) - \ddot{\xi}_{Bex}(t) \right\|_C \le \delta_2 = 0.1 \text{ m/s}^2,$$

$$\left\| \ddot{\eta}_A(t) - \ddot{\eta}_{Aex}(t) \right\|_C \le \delta_3 = 0.1 \text{ m/s}^2,$$

$$\left\|\ddot{\eta}_B(t)-\ddot{\eta}_{Bex}(t)\right\|_C \le \delta_4 = 0.1\ \text{m/s}^2.$$

The whole inaccuracy of function $u_d(t)$ in equation (1) is $\delta = 0.266\times10^4$ by chosen inaccuracy of initial data. The discrepancy method defined the parameter of regularization a [7].

The functions $z_1(t), z_2(t), z_3(t)$ with parameters $m_r = 0.02\ \text{kg}, h = 0.01\ \text{m},\ \vartheta = 0.0\ \text{rad}$ are the results of identification as solution of extreme problem (2).

The results of identification with using the special minimal MM are followings:

$$m_r = 0.23\ \text{kg}, h = 0.12\ \text{m},\ \vartheta = 0.45\ \text{rad}.$$

The results of identification with using the special maximal MM are followings:

$$m_r = 0.37\ \text{kg}, h = 0.23\ \text{m},\ \vartheta = 0.45\ \text{rad}.$$

The solution of extreme problem (12) gives the results:

$$m_r = 0.41\ \text{kg}, h = 0.23\ \text{m},\ \vartheta = 0.48\ \text{rad}.$$

If the parameters of unbalance don't change during the 3 - 4 turns round, the axis of rotation is used for the parametric statement of problem. It permits to shorten the time of calculation of initial data about 10 times.

Yuri Menshikov Efficiency of suggested algorithm was also shown on other tests [1,3,12,15].

REFERENCES

1. W. Lees and M. I. Friswell, "The Evaluation of Rotor Imbalance in Flexibly Mounted Machines," Journal of Sound and Vibration, Vol. 208, No. 5, 1997, pp. 671-683.
2. M. G. Smart, M. I. Friswell and A. W. Lees, "Estimating Turbogenerator Foundation Parameters: Model Selection and Regularization," Proceedings of the Royal Society of London A, Vol. 456, 2000, pp. 1583-1607.

3. M. S. Darlow, "Balancing of High-Speed Machinery: Theory, Methods and Experimental Results," Mechanical Systems and Signal Processing, Vol. 1, No. 1, 1982, pp. 105-134.
4. Yu. L. Menshikov and N. V. Polyakov, "Operative Evaluation of Unbalance Characteristics of a Deformable Rotor," Proceedings of 8th International Symposium on Technical Diagnostics (IMEKO), Dresden, 23-25 September 1992, pp. 399-408.
5. Yu. L. Menshikov and N. V. Polyakov, "The New Statement of Problem of Unbalance Identification," Proceedings of ICTAM 2004, Warsaw, 15-21 August 2004.
A. N. Tikhonov and V. Y. Arsenin, "The Methods of Solution of the Incorrectly Formulated Problems," Science, Moscow, 1979. [Citation Time(s):4]
6. V. Goncharskij, A. C. Leonov and A. G. Yagola, "About One Regularized Algorithm for Ill-Posed Problems with Approximate Given Operator," Journal of Computational Mathematics and Mathematical Physics, Vol. 12, No. 6, 1972, pp. 1592-1594.
7. Yu. L. Menshikov, "Useful Hypothesis in Inverse Problems of Interpretation," Proceedings of International Conference Inverse Problems and Applications (IPA2013), Linkoping University, Linköping, 2-6 April 2013.
8. Yu. L. Menshikov, "Solution Estimations of Measurement's Inverse Problem," Proceedings of 4th Inverse Problems, Design and Optimization Symposium (IPDO-2013), Book of Abstracts, Albi, 26-28 June 2013.
9. Yu. L. Menshikov, "Synthesis of Adequate Mathematical Description as Solution of Special Inverse Problems," European Journal of Mathematical Sciences, Vol. 2, No 3, 2013, pp. 256-271.
10. Yu. L. Menshikov, "The Reduction of Initial Date Inaccuracy in Ill-Posed Problems," Proceedings of 15th IMACS World Congress on Scientific Computation, Modelling and Applied Mathematics, Vol. VI, Berlin, 24-29 August 1997, pp. 577-582.
11. Yu. L. Menshikov, "Identification of Optimal Mathematical Model of External Impacts," Proceedings of MATHMOD 03, Vienna, 5-7 February 2003.
12. Yu. L. Menshikov, "Some New Statements of Inverse Problems of Recognition," 10th International Conference on Mathematical Methods in Electromagnetic Theory, Kharkov, 14-17 September 2004, pp. 464-466.
13. Yu. L. Menshikov, "Recognition of Rotor Machines Unbalance," Differential Equations and Their Appendices in Physics. The Collection of the Proceedings, Dnepropetrovsk, 1988, pp. 44-56.
14. Yu. L. Menshikov, "Identification of Rotor Unbalance in Minimax Statement," 13th International Congress on Sound and Vibration, Vienna, 2-6 July 2006, pp. 345-352.

CHAPTER 8

Mean Square Solutions of Second-Order Random Differential Equations by Using the Differential Transformation Method

Ayad R. Khudair, S. A. M. Haddad, Sanaa L. Khalaf

Department of Mathematics, Faculty of Science, University of Basrah, Basrah, Iraq

ABSTRACT

The differential transformation method (DTM) is applied to solve the second-order random differential equations. Several examples are represented to demonstrate the effectiveness of the proposed method. The results show that DTM is an efficient and accurate technique for finding exact and approximate solutions.

Keywords: Random Differential Equations, Stochastic Differential Equation, Differential Transformation Method

1. INTRODUCTION

The ordinary differential equations which contain random constant or random variables are well known topics which are called the random ordinary differential equations. The subject of second-order random differential equations is one of much current interests due to the great importance of many applications in engineering, biology and physical phenomena (see, e.g. Chil'es and Delfiner [1] , Cort'es et al. [2] , Soong [3] and references therein). Recently, several first-order random differential models have been solved by using the Mean Square Calculus [2] [4] - [11] . Variety scientific problems have been modeled by using the nonlinear second-or- der random differential equations. However, most of these equations cannot be solved analytically. Thus, accurate and efficient numerical techniques are needed. There are several semi-numerical techniques which have been considered to obtain exact and approximate solutions of linear and nonlinear differential equations, such as adomian decomposition method (ADM) [12] , variational iteration method (VIM) [13] and homotopy perturbation method (HPM) [14] . We observe that semi-numerical methods are very prevalent in the current literature, cf. [12] - [14] .

The object of this work is to describe how to implement the differential transformation method (DTM) for finding exact and approximate solutions of the second-order random differential equations. To this end, the second-order random differential equations and the concept of the differential transformation method are presented in Section 2. In Section 3, we consider the statistical functions of the mean square solution of the second- order random differential equation. Section 4 is devoted to numerical examples.

2. DIFFERENTIAL TRANSFORM METHOD

The differential transform method (DTM) has been used by Zhou [15] . This method is effective to obtain exact and approximate solutions of linear and nonlinear differential equations. To describe the basic ideas of DTM, we consider the second order random differential equation,

$$L[x(t)]+N[x(t),A]=g(t), \tag{1}$$

$$x(0)=y_0, \quad \left.\frac{dx(t)}{dt}\right|_{t=0}=y_1 \tag{2}$$

where $L[x(t)] = \frac{d^2x(t)}{dt^2}$, $x(t)$ is an unknown function, $N[x(t), A]$ is a nonlinear operator, $g(t)$ is the source in homogeneos term, and A, y_0 and y_1 are random variables.

We now write the differential transform of function as

$$X(k) = \frac{1}{k!} \frac{d^k x(t)}{dt^k}\bigg|_{t=0} \tag{3}$$

In fact, $x(t)$ is a differential inverse transform of the form

$$x(t) = \sum_{k=0}^{\infty} X(k) t^k \tag{4}$$

It is clear from (3) and (4) that the concept of differential transform is derived from Taylor series expansion. That is

$$x(t) = \sum_{k=0}^{\infty} \frac{1}{k!} \frac{d^k x(t)}{dt^k}\bigg|_{t=0} t^k \tag{5}$$

Original function	Transformed function
$x(t) \pm y(t)$	$X(k) \pm Y(k)$
$cx(t)$	$cX(k)$
$y(t) = \frac{dx(t)}{dt}$	$Y(k) = (k+1)X(k+1)$
$y(t) = \frac{d^2x(t)}{dt^2}$	$Y(k) = (k+1)(k+2)X(k+2)$
$u(t) = x(t)y(t)$	$U(k) = \sum_{s=0}^{k} X(s)Y(k-s)$
$x(t) = t^s$	$X(k) = \begin{cases} 1 & k = s \\ 0 & k \neq s \end{cases}$
$y(t) = t\frac{dx(t)}{dt}$	$Y(k) = kX(k)$
$y(t) = t\frac{d^2x(t)}{dt^2}$	$Y(k) = k(k+1)X(k+1)$
$y(t) = t^2\frac{d^2x(t)}{dt^2}$	$Y(k) = k(k-1)X(k)$

Differential transform for some functions.

Notes that, the derivatives in differential transform method does not evaluate symbolically.

In keeping with Equations (3) and (4), let $X(k)$, $Y(k)$ and $U(k)$, respectively, are the transformed functions of $x(t)$, $y(t)$ and $u(t)$. The fundamental mathematical operations of differential transformation are listed in the following table.

3. STATISTICAL FUNCTIONS OF THE MEAN SQUARE SOLUTION

Before proceeding to find the computation of the main statistical functions of the mean square solution of Equations (1) and (2) we briefly clarify some concept, notation, and results belonging to the so-called L_P calculus. The reader is referred to the books by Soong [3] , Loeve [16] , and Wong and Hajek [17] . Throughout the paper, we deal with the triplet Probabilistic space (Ω, F, P). Thus, suppose $L_2 = L_2(\Omega, F, P)$ is the set of second order random variables. Then the random variable $X:\Omega \rightarrow R \in L_2$, if $E\left[X^2\right] < \infty$, where $E[\square]$ is an expectation operator. The norm on is denoted by $\|\square\|_2$. For example, for the random variable X we define $\|X\|_2 = \left(E\left[X^2\right]\right)^{\frac{1}{2}}$, in such way that $\left(L_2, \|X\|_2\right)$ is a Banach space. In addition, let T (real interval) represent the space of times, we say that $\{X(t), t \in T\}$ is a second order stochastic process, if the random variable $X(t) \in L_2$ for each $t \in T$.

A sequence of second order random variables $\{X_n, n \geq 0\}$ converges to $X(t) \in L_2$, if

$$\lim_{n \to \infty} \|X_n - X\| = \lim_{n \to \infty} \left(E\left[X_n - X^2\right]^{\frac{1}{2}} \right) = 0.$$

To proceed from (4), we truncate the expansion of at the term as follows

$$x_N(t) = \sum_{k=0}^{N} X(k) t^k \tag{6}$$

By using the independence between A, y_0 and y_1 we have

$$E\left[x_N(t)\right] = \sum_{k=0}^{N} E\left[X(k)\right] t^k \tag{7}$$

$$V\left[x_N(t)\right] = \sum_{j=0}^{N}\sum_{i=0}^{N} \operatorname{cov}\left(X(i), X(j)\right) t^{i+j} \tag{8}$$

where

$$\operatorname{cov}\left(X(i), X(j)\right) = E\left(X(i)X(j)\right) - E\left[X(i)\right]E\left[X(j)\right],\ \forall i, j = 0, 1, \cdots, N.$$

The following Lemma guarantee the convergent of the sequence $E\left[X_N(t)\right]$ to $E\left[X(t)\right]$ and the sequence $V\left[X_N(t)\right]$ to $V\left[X(t)\right]$ if the sequence the $X_N(t)$ converges to $X(t)$.

Lemma [5] : Let $\{X_N\}$ and $\{Y_N\}$ be two sequences of 2-r.vs X and Y, respectively, i.e., $\lim_{N\to\infty} X_N = X$ and $\lim_{N\to\infty} Y_N = Y$ then

$$\lim_{N\to\infty} E\left[X_N Y_N\right] = E\left[XY\right].$$

If $X_N = Y_N$, then $\lim_{N\to\infty} E\left[X_N^2\right] = E\left[X^2\right]$, $\lim_{N\to\infty} E\left[X_N\right] = E\left[X\right]$ and $\lim_{N\to\infty} V\left[X_N\right] = V\left[X\right]$.

4. NUMERICAL EXAMPLES

In this section, we adopt several examples to illustrate the using of differential transform method for approximating the mean and the variance.

Example 1: Consider random initial value problem $\frac{d^2x(t)}{dt^2} + A^2x(t) = 0$, $x(0) = Y_0$ and $\left.\frac{dx(t)}{dt}\right|_{t=0} = Y_1$

where $A^2 \sim Be(\alpha = 2, \beta = 1)$ and independently of the initial conditions Y_0 and Y_1 which satisfy $E[Y_o] = 1$, $E\left[Y_o^2\right] = 2$, $E\left[Y_1\right] = 1$, $E\left[Y_1^2\right] = 3$ and $E[Y_0Y_1] = 0$.

The approximate mean and variance are

$$E[x(t)] = 1 + t - \frac{1}{3}t^2 - \frac{1}{9}t^3 + \frac{1}{48}t^4 + \frac{1}{240}t^5 - \frac{1}{1800}t^6 - \frac{1}{12600}t^7 + \frac{1}{120960}t^8 + \cdots$$

$$V[x(t)] = 1 - 2t + \frac{4}{3}t^2 + \frac{8}{9}t^3 - \frac{19}{72}t^4 - \frac{67}{540}t^5 + \frac{103}{4050}t^6 + \frac{41}{4725}t^7 - \frac{247}{172800}t^8 - \cdots$$

Figure 1 explain the Graph of the expectation approximation solution by using DTM with n = 18, while Figure 2 explain the Graph of variance approximation solution by using DTM with n = 18.

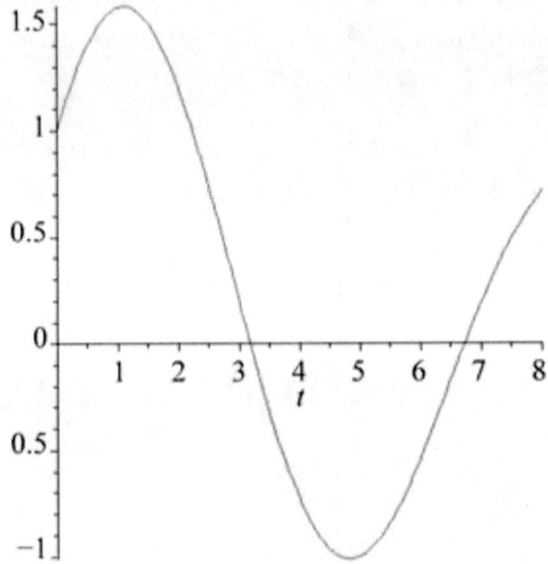

Figure 1. Graphs of the expectation approximation solution of the DTM with n = 18.

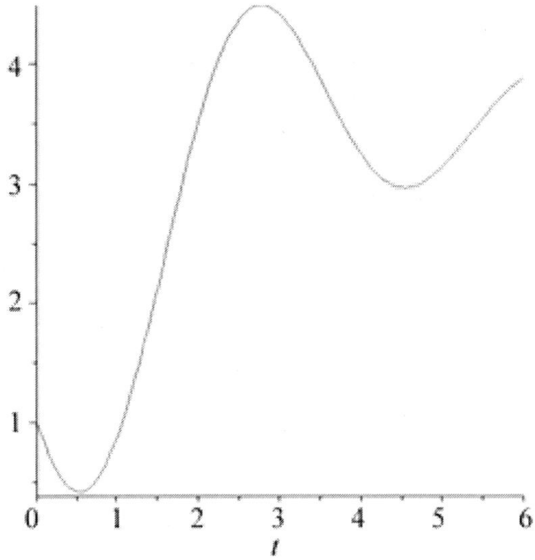

Figure 2. Graphs of variance approximation solution of the DTM with n = 18.

Example 2: Consider random initial value problem $\frac{d^2x(t)}{dt^2} + Atx(t) = 0$, $X(0) = Y_0$ and $\left.\frac{dx(t)}{dt}\right|_{t=0} = Y_1$ where A is a Beta r.v. with parameters $\alpha = 2$ and $\beta = 3$, i.e. $A \square Be(\alpha = 2, \beta = 3)$ and the initial conditions Y_o and Y_1 are independent r.v.'s such as $E[Y_o] = 1$, $E[Y_o^2] = 2$, $E[Y_1] = 2$, $E[Y_1^2] = 5$.

The approximate mean and variance are

$$E[x(t)] = 1 + 2t - \frac{1}{15}t^3 - \frac{1}{15}t^4 + \frac{1}{900}t^6 + \frac{1}{1260}t^7 - \frac{1}{113400}t^9 - \frac{1}{198450}t^{10} + \cdots$$

$$V[x(t)] = 1 + t^2 - \frac{2}{15}t^3 - \frac{1}{15}t^5 + \frac{2}{225}t^6 + \frac{1}{450}t^7 + \frac{83}{25200}t^8 - \frac{83}{283500}t^9 - \frac{2}{18375}t^{10} - \cdots$$

Figure 3 explain the Graph of the expectation approximation solution by using DTM with n = 18, while Figure 4 explain the Graph of variance approximation solution by using DTM with n = 18.

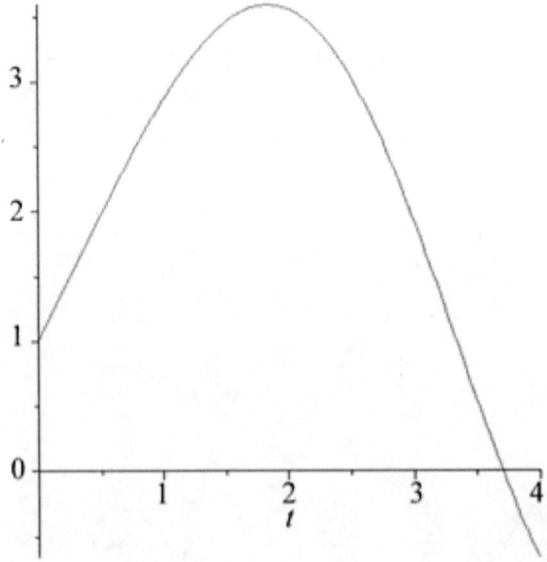

Figure 3. Graphs of the expectation approximation solution of the DTM with n = 18.

Example 3: Consider the problem $\frac{d^2x(t)}{dt^2}+2A\frac{dx(t)}{dt}+A^2x(t)=0$, and $\left.\frac{dx(t)}{dt}\right|_{t=0}=Y_1$ whereA is a Beta r.v. with parameters $\alpha=2$ and $\beta=1$, i.e. $A \sim Be(\alpha=2,\beta=1)$ and independently of the initial conditions Y_o and Y_1 which are independent r.v.' satisfy $E[Y_o]=1$, $E[Y_o^2]=2$, $E[Y_1]=1$, $E[Y_1^2]=1$.

The approximate mean and variance are

$$E[x(t)]=1+t-\frac{11}{12}t^2+\frac{23}{60}t^3-\frac{13}{120}t^4+\frac{59}{2520}t^5-\frac{83}{20160}t^6+\frac{37}{60480}t^7-\frac{143}{1814400}t^8+\cdots$$

$$V[x(t)]=1-\frac{1}{2}t^2+\frac{4}{15}t^3+\frac{103}{720}t^4-\frac{649}{2520}t^5+\frac{8959}{50400}t^6-\frac{3133}{37800}t^7+\frac{517}{17280}t^8-\cdots$$

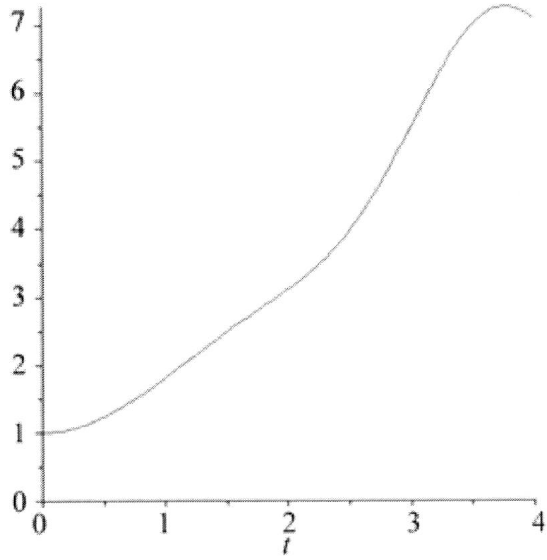

Figure 4. Graphs of variance approximation solution of the DTM with n = 18.

Figure 5 explain the Graph of the expectation approximation solution by using DTM with n = 18, while Figure 6 explain the Graph of variance approximation solution by using DTM with n = 18.

Example 4: Consider the problem $\frac{d^2x(t)}{dt^2}+Atx(t)=0$, $X(0)=Y_0$ and $\left.\frac{dx(t)}{dt}\right|_{t=0}=Y_1$ where A is a uniform r.v. with parameters $\alpha=0$ and $\beta=1$, i.e. $A \sim U(\alpha=0,\beta=1)$ and independently of the initial conditions Y_o and Y_1 which are independent r.v.'s satisfy $E[Y_o]=1$, $E[Y_o^2]=2$, $E[Y_1]=1$, $E[Y_1^2]=1$.

The approximate mean and variance are

$$E[x(t)]=1+t-\frac{11}{12}t^2+\frac{23}{60}t^3-\frac{13}{120}t^4+\frac{59}{2520}t^5-\frac{83}{20160}t^6+\frac{37}{60480}t^7-\frac{143}{1814400}t^8+\cdots$$

$$V[x(t)]=1+\frac{1}{432}t^6-\frac{1}{1440}t^8+\frac{131}{1008000}t^{10}-\frac{41}{2177280}t^{12}+\frac{71471}{31294771200}t^{14}\cdots$$

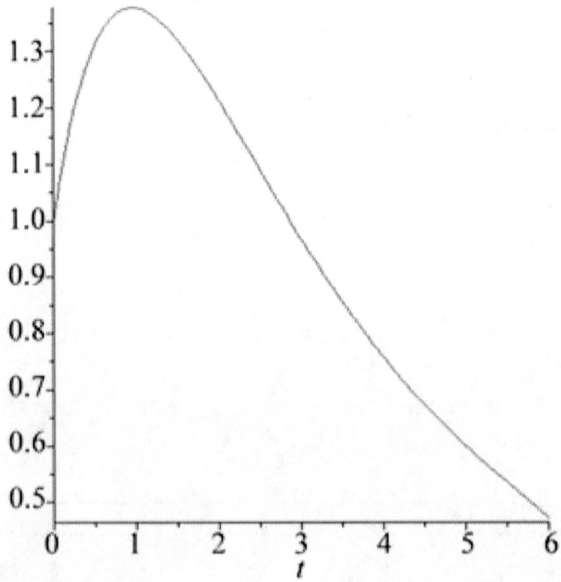

Figure 5. Graphs of the expectation approximation solution of the HAM with n = 18.

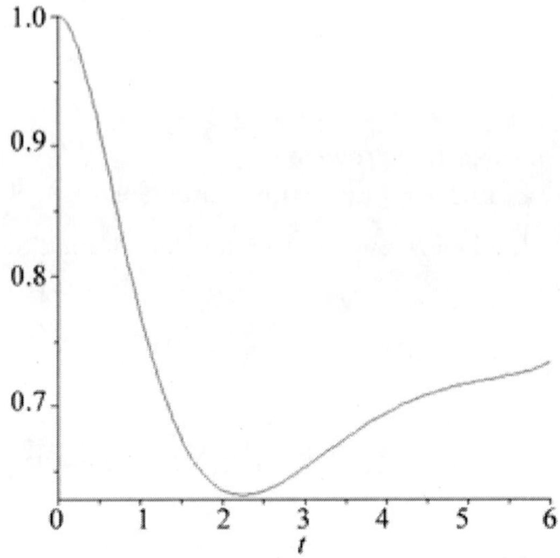

Figure 6. Graphs of variance approximation solution of the HAM with n = 18.

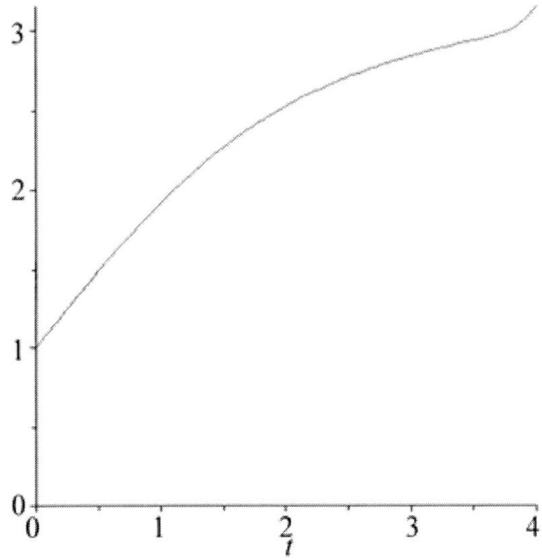

Figure 7. Graphs of the expectation approximation solution of the DTM with n = 18.

Figure 7 explain the Graph of the expectation approximation solution by using DTM with n = 18, while Figure 8 explain the Graph of variance approximation solution by using DTM with n = 18.

Example 5: Consider the problem $\frac{d^2x(t)}{dt^2}+Ax(t)=0$, $X(0)=Y_0$ and $\left.\frac{dx(t)}{dt}\right|_{t=0}=Y_1$ where A is auniform r.v. with parameters $\alpha=0$ and $\beta=2$, i.e. $A \sim U(\alpha=0,\beta=2)$ and independently of the initial conditions and Y_1 which are independent r.v.'s satisfy $E[Y_o]=1$, $E[Y_o^2]=4$, $E[Y_1]=1$, $E[Y_1^2]=2$.

The approximate mean and variance are

$$E[x(t)]=1+t-\frac{1}{2}t^2-\frac{1}{6}t^3+\frac{1}{18}t^4+\frac{1}{90}t^5-\frac{1}{360}t^6-\frac{1}{2520}t^7+\frac{1}{12600}t^8+\frac{1}{113400}t^9-\cdots$$

$$V[x(t)]=3-2t^2+\frac{13}{12}t^4+\frac{1}{18}t^5-\frac{61}{270}t^6-\frac{2}{135}t^7+\frac{599}{22680}t^8+\frac{101}{56700}t^9-\cdots$$

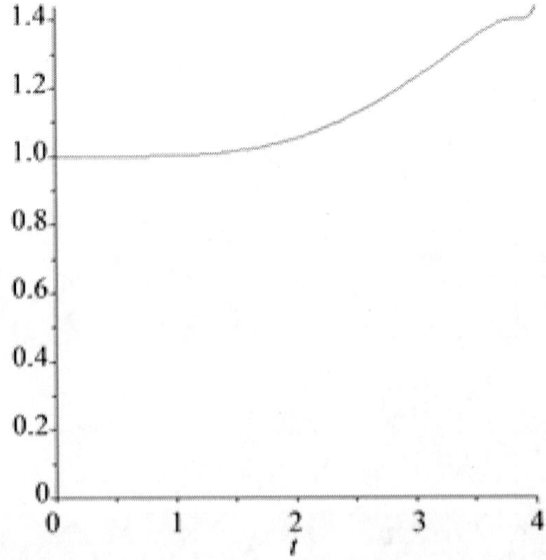

Figure 8. Graphs of variance approximation solution of the DTM with n = 18.

Figure 9 explain the Graph of the expectation approximation solution by using DTM with n = 18, while Figure 10 explain the Graph of variance approximation solution by using DTM with n = 18.

Example 6: Consider the problem $\frac{d^2x(t)}{dt^2} + Ax(t) = -x(t) + \sin(t)$, $x(0) = Y_0$ and $\left.\frac{dX(t)}{dt}\right|_{t=0} = Y_1$ where

A is a uniform r.v. with parameters $\alpha = 1$ and $\beta = 2$, i.e. $A \square U(\alpha = 1, \beta = 2)$ and independently of the

initial conditions Y_o and Y_1 which satisfy $E[Y_o] = 1$, $E[Y_o^2] = 2$, $E[Y_1] = 1$, $E[Y_1^2] = 6$ and $E[Y_0 Y_1] = 0$.

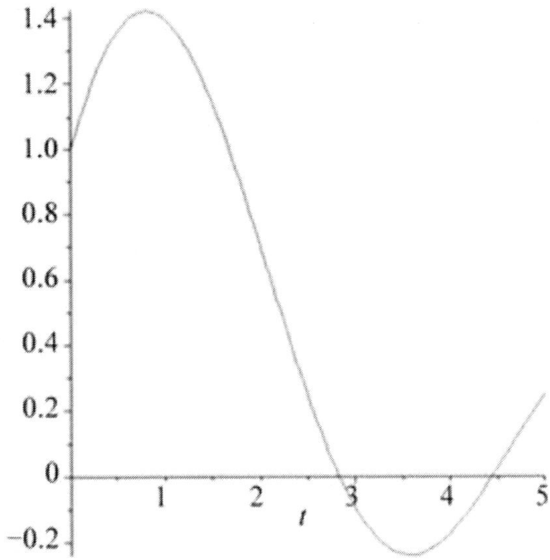

Figure 9. Graphs of the expectation approximation solution of the DTM with n = 18.

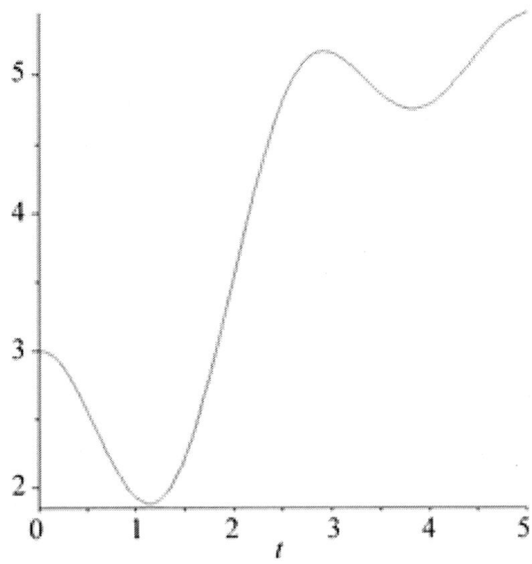

Figure 10. Graphs of variance approximation solution of the DTM with n = 18.

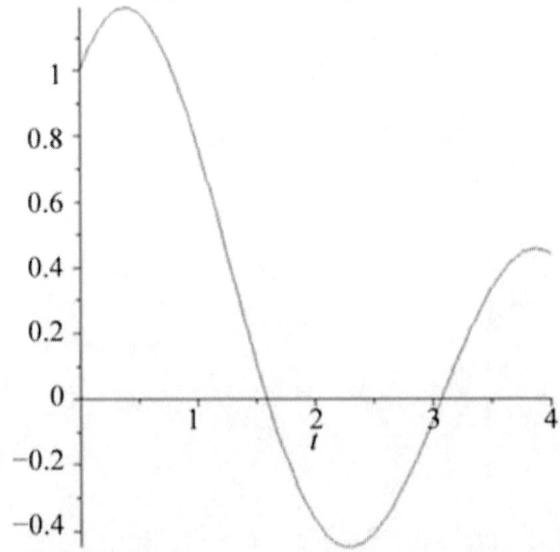

Figure 11. Graphs of the expectation approximation solution of the DTM with n = 18.

The approximate mean and variance are

$$E\left[x(t)\right]=1+t-\frac{5}{4}t^2-\frac{1}{4}t^3+\frac{19}{72}t^4+\frac{17}{720}t^5-\frac{13}{576}t^6-\frac{11}{8640}t^7+\frac{211}{201600}t^8+\cdots$$

$$V\left[x(t)\right]=1-2t+\frac{5}{2}t^2+\frac{10}{3}t^3-\frac{293}{144}t^4-\frac{67}{40}t^5+\frac{185}{288}t^6+\frac{12221}{30240}t^7-\frac{198419}{1814400}t^8-\cdots$$

Figure 11 explain the Graph of the expectation approximation solution by using DTM with n = 18, while Figure 12 explain the Graph of variance approximation solution by using DTM with n = 18.5.

ConclusionIn this paper, we successfully applied the differential transform method to solve the second-order random differential Equations (1)-(2) with coefficients which depend on a random variable A which has been assumed to be independent of the random initial conditions y_0 and y_1. This includes the computation of approximations of the mean and variance functions to the random solution. These approximations not only agree but also improve those provided by the Adomian Decomposition Method [12] , Variational Iteration Method [13] and Homotopy Perturbation Method [14] as we have illustrated through different examples. Otherwise, the differential transform method is very effective and powerful tools for the second-order random differential equation because it is a direct way without using linearization, perturbation or restrictive assumptions.

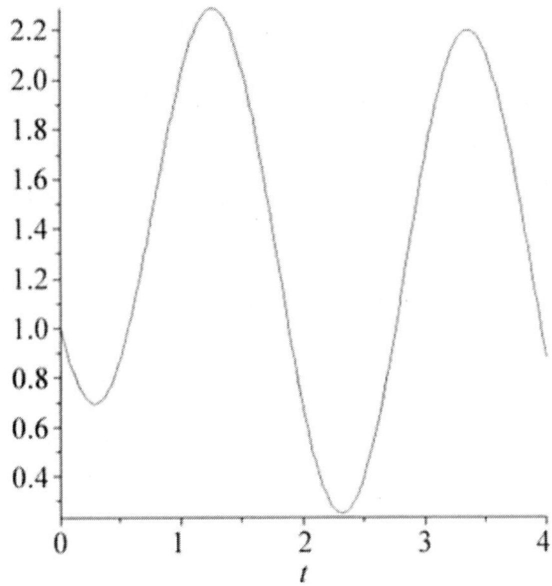

Figure 12. Graphs of variance approximation solution of the DTM with n = 18.

REFERENCES

1. Chiles, J. and Delfiner, P. (1999) Geostatistics: Modelling Spatial Uncertainty. John Wiley, New York.
2. Cortes, J.C., Jodar, L. and Villafuerte, L. (2009) Random Linear-Quadratic Mathematical Models: Computing Explicit Solutions and Applications. Mathematics and Computers in Simulation, 79, 2076-2090.
3. Soong, T.T. (1973) Random Differential Equations in Science and Engineering. Academic Press, New York.
4. Cortés, J.C., Jódar, L. and Villafuerte, L. (2007) Mean Square Numerical Solution of Random Differential Equations: Facts and Possibilities. Computers and Mathematics with Applications, 53, 1098-1106.
5. Cortés, J.C., Jódar, L. and Villafuerte, L. (2007) Numerical Solution of Random Differential Equations: A Mean Square Approach. Mathematical and Computer Modelling, 45, 757-765.
6. Cortés, J.C., Jódar, L. and Villafuerte, L. (2007) Mean Square Numerical Solution of Random Differential Equations: Facts and Possibilities. Computers and Mathematics with Applications, 53, 1098-1106.

7. Cortés, J.C., Jódar, L. and Villafuerte, L. (2009) Random Linear Quadratic Mathematical Models: Computing Explicit Solutions and Applications. Mathematics and Computers in Simulation, 79, 2076-2090.
8. Villafuerte, L., Braumann, C.A., Cortés, J.C. and Jódar, L. (2010) Random Differential Operational Calculus: Theory and Applications. Computers & Mathematics with Applications, 59, 115-125.
9. Calbo, G., Cortés, J.C. and Jódar, L. (2010) Mean Square Power Series Solution of Random Linear Differential Equations. Computers and Mathematics with Applications, 59, 559-572.
10. Cortés, J.C., Jódar, L., Villanueva, R.-J. and Villafuerte, L. (2010) Mean Square Convergent Numerical Methods for Nonlinear Random Differential Equations. Lecture Notes in Computer Science, 5890, 1-21.
11. Cortés, J.C., Jódar, L., Villafuerte, L. and Company, R. (2011) Numerical Solution of Random Differential Models. Mathematical and Computer Modelling, 54, 1846-1851.
12. Khudair, A.R., Ameen, A.A. and Khalaf, S.L. (2011) Mean Square Solutions of Second-Order Random Differential Equations by Using Adomian Decomposition Method. Applied Mathematical Sciences, 5, 2521-2535.
13. Khudair, A.R., Ameen, A.A. and Khalaf, S.L. (2011) Mean Square Solutions of Second-Order Random Differential Equations by Using Variational Iteration Method. Applied Mathematical Sciences, 5, 2505-2519.
14. Khalaf, S.L. (2011) Mean Square Solutions of Second-Order Random Differential Equations by Using Homotopy Perturbation Method. International Mathematical Forum, 6, 2361-2370.
15. Zhou, J.K. (1986) Differential Transformation and Its Applications for Electrical Circuits. Huazhong University Press, Wuhan.
16. Lo'eve, M. (1963) Probability Theory. Van Nostrand, Princeton, New Jersey.
17. Wong, E. and Hajek, B. (1985) Stochastic Processes in Engineering System. Springer Verlag, New York.

CHAPTER 9

Oscillation Properties of Third Order Neutral Delay Differential Equations*

Elmetwally M. Elabbasy[1], Osama Moaaz[1], Ebtesam Sh. Almehabresh[2]

[1]*Department of Mathematics, Faculty of Science, Mansoura University, Mansoura, Egypt*

[2]*Department of Mathematics, Faculty of Education, AL Asmarya Islamic University, Zliten, Libya*

ABSTRACT

Oscillation criteria are established for third-order neutral delay differential equations with deviating arguments. These criteria extend and generalize those results in the literature. Moreover, some illustrating examples are also provided to show the importance of our results.

Keywords: Oscillation, Third Order, Neutral Delay, Differential Equations

1. INTRODUCTION

This article is concerned with the oscillation and the asymptotic behavior of solutions of the third-order neutral delay differential equations with deviating argument of the form

$$\left(r(t)\left[z''(t)\right]^{\alpha}\right)' + \int_c^d q(t,\xi)x^{\alpha}\left(g(t,\xi)\right)\mathrm{d}\xi = 0, t \ge t_0, \tag{E}$$

where $z(t) = x(t) + \int_a^b p(t,\eta)x(\tau(t,\eta))\mathrm{d}\eta.$ We assume that:

(H) $p,\tau \in C([t_0,\infty)\times[a,b],R)$ $r \in C([t_0,\infty),(0,\infty))$; $q,g \in C([t_0,\infty)\times[c,d],R)$,

α is a quotient of odd positive integers, $\tau(t,\eta) \le t$, $0 \le \int_a^b p(t,\eta)\mathrm{d}\eta \le p < 1$, $g(t,\xi) \le t$, $\lim_{t\to\infty}\tau(t,\eta) = \lim_{t\to\infty} g(t,\xi) = \infty$ and $q(t,\xi) > 0$.

A function $x(t) \in C([t_x,\infty)), t_x \ge t_0$ is called a solution of (E), if it has the properties $z(t) \in C^1([t_x,\infty))$, $z'(t) \in C^1([t_x,\infty))$, $r(t)\left[z''(t)\right]^{\alpha} \in C^1([t_x,\infty))$ and satisfies (E) on $[t_x,\infty)$. We consider only those solutions $x(t)$ of (E) which satisfy for all $T \ge t_x$. We assume that (E) possesses such solution. A solution of (E) is called oscillatory if it has arbitrarily large zeros on $[t_x,\infty)$; otherwise, it is called nonoscillatory.

In the recent years, great attention in the oscillation theory has been devoted to the oscillatory and asymptotic properties of the third-order differential equations (see [1] - [14]). Baculikova et al. [2] [3] , Dzurina et al. [4] and Mihalikova et al. [11] studied the oscillation of the third-order nonlinear differential equation

$$\left(r(t)\left[\left(x(t) + p(t)x(\tau(t))\right)''\right]^{\alpha}\right)' + q(t)f\left(x(g(t))\right) = 0, t \ge t_0,$$

under the condition

$$\int_{t_0}^{\infty} r^{-1/\alpha}(s) = \infty.$$

Li et al. [10] considered the oscillation of

$$\left(r_2(t)\left[r_1(t)\left(x(t) + p(t)x(\tau(t))\right)'\right]'\right)' + q(t)x(g(t)) = 0, t \ge t_0,$$

under the assumption

$$\int_{t_0}^{\infty} r_1^{-1}(s) = \infty \quad \text{and} \quad \int_{t_0}^{\infty} r_2^{-1}(s) < \infty.$$

The aim of this paper is to discuss asymptotic behavior of solutions of class of third order neutral delay differential Equation (E) under the condition

$$\int_{t_0}^{\infty} r^{-1/\alpha}(s) < \infty. \tag{1}$$

By using Riccati transformation technique, we established sufficient conditions which insure that solution of class of third order neutral delay differential equation is oscillatory or tends to zero. The results of this study extend and generalize the previous results.

2. MAIN RESULTS

In this section, we will establish some new oscillation criteria for solutions of (E).

Theorem 2.1. Assume that conditions (1) and (H) are satisfied. If for some function $\rho \in C^1([t_0,\infty),(0,\infty))$, for all sufficiently large $t_1 \geq t_0$ and for $t_3 \geq t_2 \geq t_1$, one has

$$\limsup_{t\to\infty} \int_{t_3}^{t} \left(\rho(s) q^*(s)(1-p)^\alpha G(s) - \frac{1}{(\alpha+1)^{\alpha+1}} \frac{r(s)(\rho'(s))^{\alpha+1}}{\rho^\alpha(s)} \right) ds = \infty, \tag{2}$$

where

$$G(s) = \left(\frac{\int_{t_2}^{g(s,c)} \left(\int_{t_1}^{v} r^{-\frac{1}{\alpha}}(u) du \right) dv}{\int_{t_1}^{s} r^{-\frac{1}{\alpha}}(u) du} \right)^\alpha, q^*(s) = \int_c^d q(s,\xi) d\xi, \tag{3}$$

and

$$\int_{t_0}^{\infty}\int_{v}^{\infty}\left[\frac{1}{r(u)}\int_{u}^{\infty}\int_{c}^{d}q(s,\xi)\mathrm{d}\xi\mathrm{d}s\right]^{\frac{1}{\alpha}}\mathrm{d}u\mathrm{d}v=\infty. \tag{4}$$

If

$$\limsup_{t\to\infty}\int_{t_2}^{t}\left(\delta(s)q^{*}(s)(1-p)^{\alpha}\left(g(s,c)-t_1\right)^{\alpha}-\frac{1}{(\alpha+1)^{\alpha+1}}\frac{1}{\left(\delta(s)r(s)\right)^{\alpha}}\right)\mathrm{d}s=\infty \tag{5}$$

where

$$\delta(t)=\int_{t}^{\infty}\frac{1}{r^{\frac{1}{\alpha}}(s)}\mathrm{d}s, \tag{6}$$

then every solution $x(t)$ of (E) is either oscillatory or converges to zero as $t\to\infty$.

Proof. Assume that $x(t)$ is a positive solution of (E). Based on the condition (1), there exist three possible cases

(1) $z(t)>0, z'(t)>0, z''(t)>0, \left(r(t)\left[z''(t)\right]^{\alpha}\right)'\le 0;$

(2) $z(t)>0, z'(t)<0, z''(t)>0, \left(r(t)\left[z''(t)\right]^{\alpha}\right)'\le 0;$

(3) $z(t)>0, z'(t)>0, z''(t)<0, \left(r(t)\left[z''(t)\right]^{\alpha}\right)'\le 0,$

for $t\ge t_1$, t_1 is large enough. We consider each of three cases separately. Suppose first that $z(t)$ has the property (1). We define the function $\omega(t)$ by

$$\omega(t)=\rho(t)\frac{r(t)\left(z''(t)\right)^{\alpha}}{\left(z'(t)\right)^{\alpha}}. \tag{7}$$

Then, $\omega(t)>0$ for $t\geq t_1$. Using $z'(t)>0$, we have

$$\begin{aligned}x(t)&=z(t)-\int_a^b p(t,\eta)x(\tau(t,\eta))\mathrm{d}\eta\\&\geq z(t)-\int_a^b p(t,\eta)z(\tau(t,\eta))\mathrm{d}\eta\\&\geq z(t)-z(\tau(t,b))\int_a^b p(t,\eta)\mathrm{d}\eta\\&\geq(1-p)z(t).\end{aligned} \tag{8}$$

Since

$$z'(t)\geq\int_{t_1}^{t}\frac{\left(r(s)\left[z''(s)\right]^{\alpha}\right)^{\frac{1}{\alpha}}}{r^{\frac{1}{\alpha}}(s)}\mathrm{d}s\geq\left(r(t)\left[z''(t)\right]^{\alpha}\right)^{\frac{1}{\alpha}}\int_{t_1}^{t}\frac{1}{r^{\frac{1}{\alpha}}(s)}\mathrm{d}s,$$

we have that

$$\left(\frac{z'(t)}{\int_{t_1}^{t}r^{-\frac{1}{\alpha}}(s)\mathrm{d}s}\right)'\leq 0. \tag{9}$$

Thus, we get

$$\begin{aligned}z(t)&=z(t_2)+\int_{t_2}^{t}\left(\frac{z'(t)}{\int_{t_1}^{s}r^{-\frac{1}{\alpha}}(u)\mathrm{d}u}\int_{t_1}^{s}r^{-\frac{1}{\alpha}}(u)\mathrm{d}u\right)\mathrm{d}s\\&\geq\frac{z'(t)}{\int_{t_1}^{t}r^{-\frac{1}{\alpha}}(u)\mathrm{d}u}\int_{t_2}^{t}\left(\int_{t_1}^{s}r^{-\frac{1}{\alpha}}(u)\mathrm{d}u\right)\mathrm{d}s,\end{aligned} \tag{10}$$

for $t\geq t_2>t_1$. Differentiating (7), we obtain

$$\omega'(t)=\rho'(t)\frac{r(t)(z''(t))^{\alpha}}{(z'(t))^{\alpha}}+\rho(t)\frac{\left(r(t)[z''(t)]^{\alpha}\right)'}{(z'(t))^{\alpha}}-\alpha\rho(t)\frac{r(t)(z''(t))^{\alpha+1}}{(z'(t))^{\alpha+1}}.$$

It follows from (E), (7) and (8) that

$$\omega'(t)\le-\rho(t)\int_c^d q(t,\xi)\mathrm{d}\xi(1-p)^{\alpha}\frac{z^{\alpha}(g(t,c))}{(z'(t))^{\alpha}}+\frac{\rho'(t)}{\rho(t)}\omega(t)-\alpha\frac{\omega^{\frac{\alpha+1}{\alpha}}(t)}{(\rho(t)r(t))^{\frac{1}{\alpha}}},$$

that is

$$\omega'(t)\le-\rho(t)\int_c^d q(t,\xi)\mathrm{d}\xi(1-p)^{\alpha}\frac{z^{\alpha}(g(t,c))}{(z'(g(t,c)))^{\alpha}}\frac{(z'(g(t,c)))^{\alpha}}{(z'(t))^{\alpha}}$$
$$+\frac{\rho'(t)}{\rho(t)}\omega(t)-\frac{\alpha}{(\rho(t)r(t))^{\frac{1}{\alpha}}}\omega^{\frac{\alpha+1}{\alpha}},$$

which follows from (9) and (10) that

$$\omega'(t)\le-\rho(t)\int_c^d q(t,\xi)\mathrm{d}\xi(1-p)^{\alpha}\left(\frac{\int_{t_2}^{g(t,c)}\left(\int_{t_1}^{s}r^{-\frac{1}{\alpha}}(u)\mathrm{d}u\right)\mathrm{d}s}{\int_{t_1}^{t}r^{-\frac{1}{\alpha}}(u)\mathrm{d}u}\right)^{\alpha}$$
$$+\frac{\rho'(t)}{\rho(t)}\omega(t)-\frac{\alpha}{(\rho(t)r(t))^{\frac{1}{\alpha}}}\omega^{\frac{\alpha+1}{\alpha}}(t).$$

Hence, we have

$$\omega'(t)\le-\rho(t)\int_c^d q(t,\xi)\mathrm{d}\xi(1-p)^{\alpha}\left(\frac{\int_{t_2}^{g(t,c)}\left(\int_{t_1}^{s}r^{-\frac{1}{\alpha}}(u)\mathrm{d}u\right)\mathrm{d}s}{\int_{t_1}^{t}r^{-\frac{1}{\alpha}}(u)\mathrm{d}u}\right)^{\alpha}$$
$$+\frac{1}{(\alpha+1)^{\alpha+1}}\frac{r(t)(\rho'(t))^{\alpha+1}}{\rho^{\alpha}(t)}.$$

Integrating the last inequality from $t_3(>t_2)$ to t, we get

$$\omega(t_3)\geq\int_{t_3}^{t}\left(\rho(s)q^*(s)(1-p)^{\alpha}G(s)-\frac{1}{(\alpha+1)^{\alpha+1}}\frac{r(s)(\rho'(s))^{\alpha+1}}{\rho^{\alpha}(s)}\right)ds, \tag{11}$$

which contradicts (2). Assume now that $z(t)$ has the property (2). Using the similar proof ([1] , Lemma 2), we can get $\lim_{t\to\infty}x(t)=0$ due to condition (4). Thirdly, assume that $z(t)$ has the property (3). From $\left(r(t)[z''(t)]^{\alpha}\right)'<0$, $r(t)[z''(t)]^{\alpha}$ is decreasing. Thus we get

$$r(s)[z''(s)]^{\alpha}\leq r(t)[z''(t)]^{\alpha}, s\geq t\geq t_1.$$

Dividing the above inequality by $r(s)$ and integrating it from t to l, we obtain

$$z'(l)\leq z'(t)+r^{\frac{1}{\alpha}}(t)z''(t)\int_t^l r^{-\frac{1}{\alpha}}(s)ds.$$

Letting $l\to\infty$, we get

$$0\leq z'(t)+r^{\frac{1}{\alpha}}(t)z''(t)\int_t^{\infty} r^{-\frac{1}{\alpha}}(s)ds,$$

that is

$$-\frac{r^{\frac{1}{\alpha}}(t)z''(t)}{z'(t)}\int_t^{\infty} r^{-\frac{1}{\alpha}}(s)ds\leq 1. \tag{12}$$

Define function ψ by

$$\psi(t)=\frac{r(t)\left(z''(t)\right)^{\alpha}}{\left(z'(t)\right)^{\alpha}},t\ge t_1. \tag{13}$$

Then $\psi(t)<0$ for $t\ge t_1$. Hence, by (12) and (13), we obtain

$$-\delta(t)\psi^{\frac{1}{\alpha}}(t)\le 1. \tag{14}$$

Differentiating (13), we get

$$\psi'(t)=\frac{\left(r(t)\left[z''(t)\right]^{\alpha}\right)'}{\left(z'(t)\right)^{\alpha}}-\alpha r(t)\frac{\left(z''(t)\right)^{\alpha+1}}{\left(z'(t)\right)^{\alpha+1}}.$$

Using $z'(t)>0,$ we have (8). From (E) and (8), we have

$$\psi'(t)\le-\int_c^d q(t,\xi)\mathrm{d}\xi(1-p)^{\alpha}\frac{z^{\alpha}(g(t,c))}{\left(z'(t)\right)^{\alpha}}-\alpha r(t)\left(\frac{z''(t)}{z'(t)}\right)^{\alpha+1}. \tag{15}$$

In view of (3), we see that

$$z(t)\ge z'(t)(t-t_1). \tag{16}$$

Hence,

$$\left(\frac{z(t)}{(t-t_1)}\right)'\le 0,$$

which implies that

$$\frac{z(g(t,c))}{z(t)}\ge\frac{(g(t,c)-t_1)}{(t-t_1)}. \tag{17}$$

By (13) and (15)-(17), we get

$$\psi'(t) \le -\int_c^d q(t,\xi)\mathrm{d}\xi(1-p)^\alpha \left(g(t,c)-t_1\right)^\alpha - \alpha r^{-\frac{1}{\alpha}}(t)\psi^{\frac{\alpha+1}{\alpha}}(t).$$

Multiplying the last inequality by $\delta(t)$ and integrating from $t_2(>t_1)$ to t, we obtain

$$0 \ge \psi(t)\delta(t) - \psi(t_2)\delta(t_2) + \int_{t_2}^t \delta(s)q^*(s)(1-p)^\alpha \left(g(s,c)-t_1\right)^\alpha \mathrm{d}s$$
$$+\int_{t_2}^t \frac{\alpha\psi^{\frac{\alpha+1}{\alpha}}(s)\delta(s)}{r^{\frac{1}{\alpha}}(s)}\mathrm{d}s - \int_{t_2}^t \frac{\psi(s)}{r(s)}\mathrm{d}s,$$

which follows that

$$1+\psi(t_2)\delta(t_2) \ge \int_{t_2}^t \left(\delta(s)q^*(s)(1-p)^\alpha\left(g(s,c)-t_1\right)^\alpha - \frac{1}{(\alpha+1)^{\alpha+1}}\frac{1}{\left(\delta(s)r(s)\right)^\alpha}\right)\mathrm{d}s,$$

which contradicts (5). This completes the proof. W

3. EXAMPLES

The following examples illustrate applications of our result in this paper.

Example 3.1. For $t \ge 1$ and $\lambda > 0$, consider the third-order differential equation

$$\left(t^{\frac{4}{3}}\left(x(t)+\int_0^1 p_1 x\left(\frac{t}{2}\right)\mathrm{d}\eta\right)''\right)' + \int_0^1 \frac{2\lambda}{t^{\frac{5}{3}}}\xi x(t-\xi)\mathrm{d}\xi = 0. \tag{18}$$

Let $\rho(t)=1,\ \alpha=1,\ a=0,\ b=1,\ c=0,\ d=1,\ r(t)=t^{\frac{4}{3}},\ p(t,n)=p_1$ such that $0 \le \int_0^1 p_1 \mathrm{d}\eta \le p < 1,\ \tau(t,\eta)=\frac{t}{2},\qquad g(t,\xi)=t-\xi.$ Note that,

$$\int_{t_0}^{\infty} r^{-\frac{1}{\alpha}}(s)\mathrm{d}s = \int_1^{\infty} s^{-\frac{4}{3}}\mathrm{d}s = 3 < \infty, \delta(t) = 3t^{-\frac{1}{3}},$$

and

$$\int_1^{\infty}\int_v^{\infty} u^{-\frac{4}{3}} \int_u^{\infty}\int_0^1 \frac{2\lambda\xi}{s^{\frac{5}{3}}}\mathrm{d}\xi \mathrm{d}s\mathrm{d}u\mathrm{d}v = \infty.$$

Furthermore

$$\limsup_{t\to\infty}\int_{t_3}^{t}\left(\rho(s)q^{*}(s)(1-p)^{\alpha}G(s) - \frac{1}{(\alpha+1)^{\alpha+1}}\frac{r(s)(\rho'(s))^{\alpha+1}}{\rho^{\alpha}(s)}\right)\mathrm{d}s$$

$$= \limsup_{t\to\infty}\int_{t_3}^{t}\left(\frac{\lambda}{6}(1-p)s^{-\frac{5}{3}}\left(\frac{9s^{\frac{2}{3}} - 6t_1^{-\frac{1}{3}}s + \beta}{s^{-\frac{1}{3}} - t_1^{-\frac{1}{3}}}\right)\right)\mathrm{d}s = \infty,$$

such that $q^{*}(s)$, $G(s)$ are defined as in (3) and $\beta = 6t_1^{-\frac{1}{3}}t_2 - 9t_2^{\frac{2}{3}}$,

$$\limsup_{t\to\infty}\int_{t_2}^{t}\left(\delta(s)q^{*}(s)(1-p)^{\alpha}(g(t,c)-t_1)^{\alpha} - \frac{1}{(\alpha+1)^{\alpha+1}}\frac{1}{(\delta(s)r(s))^{\alpha}}\right)\mathrm{d}s$$

$$= \limsup_{t\to\infty}\int_{t_2}^{t}\left(3\lambda(1-p)s^{-2}(s-t_1) - \frac{1}{12}s^{-1}\right)\mathrm{d}s = \infty.$$

Using our result, every solution of (18) is either oscillatory or converges to zero as $t\to\infty$ if $\lambda > 1/36(1-p)$.

Example 3.2. For $t \geq 1$ and $\mu > 0$, consider the third-order differential equation

$$\left(t^{5}\left[\left(x(t) + \int_1^2 \frac{\eta}{t+1}x\left(\frac{t+\eta}{3}\right)\mathrm{d}\eta\right)''\right]^{3}\right)' + \int_0^1 \frac{2\mu}{3}t^{-2}\xi x^{3}\left(t - \frac{\xi}{2}\right)\mathrm{d}\xi = 0. \tag{19}$$

Let $\rho(t)=1,\ \alpha=3,\ a=1,\ b=2,\ c=0,\ d=1,\ r(t)=t^5,\ p(t,n)=\frac{\eta}{t+1}$ such that $0\le\int_1^2(\eta/(t+1))\mathrm{d}\eta\le\frac{3}{4}<1,\ \tau(t,\eta)=(t+\eta)/2,\ q(t,\xi)=\frac{2\mu}{3}t^{-2}\xi,\ g(t,\xi)=\left(t-\frac{\xi}{2}\right).$ Note that,

$$\int_{t_0}^{\infty}r^{-\frac{1}{\alpha}}(s)\mathrm{d}s=\int_1^{\infty}s^{-\frac{5}{3}}\mathrm{d}s=\frac{3}{2}<\infty,\delta(t)=\frac{3}{2}t^{-\frac{2}{3}},$$

and

$$\int_1^{\infty}\int_v^{\infty}\left[u^{-5}\int_u^{\infty}\int_0^1\frac{2\mu\xi}{3s^2}\mathrm{d}\xi\mathrm{d}s\right]^{\frac{1}{3}}\mathrm{d}u\mathrm{d}v=\infty.$$

Furthermore

$$\limsup_{t\to\infty}\int_{t_3}^{t}\left(\rho(s)q^*(s)(1-p)^{\alpha}G(s)-\frac{1}{(\alpha+1)^{\alpha+1}}\frac{r(s)(\rho'(s))^{\alpha+1}}{\rho^{\alpha}(s)}\right)\mathrm{d}s$$

$$=\lim_{t\to\infty}\text{supp}\int_{t_3}^{t}\left(\frac{\mu}{(3)^4(4)^3}s^{-2}\left(\frac{9s^{\frac{1}{3}}-3t_1^{\frac{2}{3}}s+\beta}{s^{\frac{2}{3}}-t_1^{\frac{2}{3}}}\right)^3\right)\mathrm{d}s=\infty,$$

such that $q^*(s)$, $G(s)$ are defined as in (3) and $\beta=3t_1^{\frac{2}{3}}t_2-9t_2^{\frac{1}{3}}$,

$$\limsup_{t\to\infty}\int_{t_2}^{t}\left(\delta(s)q^*(s)(1-p)^{\alpha}(g(t,c)-t_1)^{\alpha}-\frac{1}{(\alpha+1)^{\alpha+1}}\frac{1}{(\delta(s)r(s))^{\alpha}}\right)\mathrm{d}s$$

$$=\limsup_{t\to\infty}\int_{t_2}^{t}\left(\frac{\mu}{128}s^{-\frac{8}{3}}(s-t_1)^3-\frac{1}{864}s^{-13}\right)\mathrm{d}s=\infty.$$

Using our result, every solution of (19) is either oscillatory or converges to zero as $t\to\infty$ if $\lambda>0$ for some $t_1\in\left(0,\frac{1}{12}\right).$

Example 3.3. For $t\ge1$ and $\gamma>0,$ consider the third-order differential equation

$$\left(t^{\frac{5}{4}}\left(x(t)+\int_{1}^{2}\frac{1}{3}x(t-\eta)\mathrm{d}\eta\right)''\right)'+\int_{0}^{1}\frac{3\gamma}{t^{\frac{7}{4}}}\xi x(t)\mathrm{d}\xi=0. \tag{20}$$

Let $\rho(t)=1,\quad \alpha=1,\quad b=2,\quad c=0,\quad d=1,\quad r(t)=t^{\frac{5}{4}},$ such that $0\le\int_{0}^{1}\frac{1}{3}\mathrm{d}\eta\le\frac{1}{3}<1$, $\tau(t,\eta)=t-\eta$, $q(t,\xi)=3\gamma\xi/t^{\frac{7}{4}}$, $g(t,\xi)=t$. Note that,

$$\int_{t_0}^{\infty}r^{-\frac{1}{\alpha}}(s)\mathrm{d}s=\int_{1}^{\infty}s^{-\frac{5}{4}}\mathrm{d}s=4<\infty,\delta(t)=4t^{-\frac{1}{4}},$$

and

$$\int_{1}^{\infty}\int_{v}^{\infty}u^{-\frac{5}{4}}\int_{u}^{\infty}\int_{0}^{1}\frac{3\gamma\xi}{2s^{\frac{7}{4}}}\mathrm{d}\xi\mathrm{d}s\mathrm{d}u\mathrm{d}v=\infty.$$

Furthermore

$$\limsup_{t\to\infty}\int_{t_3}^{t}\left(\rho(s)q^{*}(s)(1-p)^{\alpha}G(s)-\frac{1}{(\alpha+1)^{\alpha+1}}\frac{r(s)(\rho'(s))^{\alpha+1}}{\rho^{\alpha}(s)}\right)\mathrm{d}s$$

$$=\limsup_{t\to\infty}\int_{t_3}^{t}\left(\frac{\gamma}{3}s^{-\frac{7}{4}}\left(\frac{4s^{\frac{3}{4}}-3t_1^{-\frac{1}{4}}s+\beta}{s^{-\frac{1}{4}}-t_1^{-\frac{1}{4}}}\right)\right)\mathrm{d}s=\infty,$$

such that $q^{*}(s)$, $G(s)$ are defined as in (3) and $\beta=3t_1^{-\frac{1}{4}}t_2-4t_2^{\frac{3}{4}}$,

$$\limsup_{t\to\infty}\int_{t_2}^{t}\left(\delta(s)q^{*}(s)(1-p)^{\alpha}(g(t,c)-t_1)^{\alpha}-\frac{1}{(\alpha+1)^{\alpha+1}}\frac{1}{(\delta(s)r(s))^{\alpha}}\right)\mathrm{d}s$$

$$=\limsup_{t\to\infty}\int_{t_2}^{t}\left(4\gamma s^{-2}(s-t_1)-\frac{1}{16}s^{-1}\right)\mathrm{d}s=\infty.$$

Using our result, every solution of (20) is either oscillatory or converges to zero as $t \to \infty$ if $\lambda > \frac{1}{64}$.

REFERENCES

1. Baculikova, B. and Dzurina, J. (2010) Oscillation of Third-Order Functional Differential Equations. EJQTDE, 43, 1-10.
2. Baculikova, B. and Dzurina, J. (2010) Oscillation of Third-Order Neutral Differential Equations. Mathematical and Computer Modelling, 52, 215-226.
3. Baculikova, B. and Dzurina, J. (2010) On the Asymptotic Behavior of a Class of Third Order Nonlinear Neutral Differential Equations. Central European Journal of Mathematics, 8, 1091-1103.
4. Dzurina, J., Thandapani, E. and Tamilvanan, S. (2010) Oscillation of Solutions to Third-Order Half-Linear Neutral Differential Equations. EJDEs, 2012, 1-9.
5. Elabbasy, E.M., Barsoum, M.Y. and Moaaz, O. (2015) Boundedness and Oscillation of Third Order Neutral Differential Equations with Deviating Arguments. Journal of Applied Mathematics and Physics, 3, 1367-1375.
6. Elabbasy, E.M., Hassan, T.S. and Moaaz, O. (2012) Oscillation Behavior of Second-Order Nonlinear Neutral Differential Equations with Deviating Arguments. Opuscula Mathematica, 32, No. 4.
7. Elabbasy, E.M. and Moaaz, O. (2012) On the Asymptotic Behavior of Third-Order Nonlinear Functional Differential Equations. Serdica Mathematical Journal, 42, 157-174.
8. Elabbasy, E.M. and Moaaz, O. (2016) On the Oscillation of Third Order Neutral Differential Equations. Asian Journal of Mathematics and Applications, 2016, Article ID: 0274.
9. Elabbasy, E.M. and Moaaz, O. (2016) Oscillation Criteria for Third Order Nonlinear Neutral Differential Equations with Deviating Arguments. International Journal of Science and Research, 5, No. 1.
10. Li, T., Thandapani, E. and Graef, J.R. (2012) Oscillation of Third-Order Neutral Retarded Differential Equations. International Journal of Pure and Applied Mathematics, 75, 511-520.
11. Mihalikova, B. and Kostikova, E. (2009) Boundedness and Oscillation of Third Order Neutral Differential Equations. Tatra Mountains Mathematical Publications, 43, 137-144.
12. Saker, S.H. and Dzurina, J. (2010) On the Oscillation of Certain Class of Third-Order Nonlinear Delay Differential Equations. Mathematica Bohemica, 135, 225-237.
13. Saker, S.H. (2006) Oscillation Criteria of Third-Order Nonlinear Delay Differential Equation. Mathematica Slovaca, 56, 433-450.
14. Moaaz, O. (2014) Oscillation Theorems for Cartain Second-Order Differential Equations. Lambert Academic Publishing, Germany.

CHAPTER 10

Review Study of Detection of Diabetes Models through Delay Differential Equations

Dimplekumar Chalishajar, David H. Geary, Geoffrey Cox

Department of Applied Mathematics, Virginia Military Institute, Lexington, VA, USA

ABSTRACT

Mathematical models based on advanced differential equations are utilized to analyze the glucose-insulin regulatory system, and how it affects the detection of Type I and Type II diabetes. In this paper, we have incorporated several models of prominent mathematicians in this field of work. Three of these models are single time delays, where either there is a time delay of how long it takes insulin produced by the pancreas to take effect, or a delay in the glucose production after the insulin has taken effect on the body. Three other models are two-time delay models, and based on the specific models, the time delay takes place in some sort of insulin production delay or glucose production delay. The intent of this paper is to use these multiple delays to analyze the glucose-insulin regulatory system, and how if it is not properly working at any point, the high risk of diabetes becomes a reality.

Keywords: Delay Differential Equations, MATLAB, Diabetes, Insulin-Glucose Model

1. INTRODUCTION

The human body needs to maintain a level of glucose concentration that is not too high in order for it to function properly. In order for glucose to be produced and utilized, insulin production and utilization must occur. If insulin is not produced, or not enough of it is produced, then proper glucose levels cannot be maintained, which leads to the disease of diabetes. Diabetes mellitus is a condition in which there is too much glucose in our blood. The pancreatic endocrine hormones insulin and glucagon are responsible for regulating and maintaining your body's glucose concentration level. Insulin is produced in the pancreas, and is released from β cells. These β cells stimulate the cells to absorb enough glucose from the blood for the fuel or energy that they need. Insulin also stimulates the liver to absorb and store any left-over glucose. Glucagon is released when the glucose concentration level is too low. This stimulates the liver cells to release the left-over glucose into the blood, so cells will have enough energy to carry out their functions. If a person's blood glucose concentration level is constantly above the normal range for humans, they suffer from the chronic condition, known as diabetes.

There are two type of diabetes. Type I diabetes, or insulin-dependent diabetes, is an autoimmune disease that occurs when the insulin-producing β cells in the pancreas are attacked and destroyed by other cells in the body, causing the pancreas to produce little or no insulin at all. The cure for Type I diabetes is to have insulin replacement treatment for the rest of their life. This type of diabetes is known as juvenile diabetes, because it normally develops before the age of 40, often times before then. It is the least common of the two (10%).

Type II, or non-insulin dependent diabetes, is a metabolic disorder which develops when the body can still produce insulin, but not enough, or when the insulin that is produced does not work properly, or better known as insulin resistance. Type II diabetes, is more likely to occur if there is a history of it in your family, and is often associated with obesity. Symptoms for Type II diabetes develop slowly over time, if at all, and occur mostly in people over the age of 40. Life expectancy for the Type I diabetes is reduced by about 15 years, and 10 years for Type II diabetes. If diabetes is left untreated or improperly managed, it can lead to heart disease, blindness, kidney disease, and non-traumatic limb amputations.

Due to these factors and harmful effects of diabetes, many researchers have been motivated to study the glucose-insulin endocrine metabolic regulatory system with hopes to better understand the mechanistic functions and causes of the dysfunction of the metabolic system. Many types of models have been made to better understand the role of the insulin-glucose regulatory system, including ordinary differential equations, and delay differential equation, both of which are included in this paper.

In the glucose-insulin endocrine metabolic regulatory system, the two pancreatic endocrine hormones, insulin and glucagon, are the primary regulatory factors. Numerous in-vivo and in-vitro experiments indicate that insulin secretion consists of two oscillations occurring with different time scales. One is rapid oscillations (5 - 15 min) and the second is ultradian oscillations (50 - 150 min). The mechanisms underlying both types of oscillations are still not fully understood. The rapid oscillations may arise from an intra-pancreatic pacemaker mechanism resulting in periodic secretary bursting of insulin from pancreatic β cells. This rapid oscillation is usually super imposed on the slow ultradian oscillation. The ultradian oscillations of insulin secretion are assumed to result from instability in the glucose-insulin endocrine hormones of the metabolic regulatory system.

In this paper, we propose a number of delay differential equation (DDE) models by introducing two and three explicit time delays in order to model the glucose-insulin levels in the metabolic regulatory system. The two and three time delay models not only confirm many experimental observations, but also demonstrate robustness, and produce simulation profiles in better agreement with physiological data. As a result, we suspect that one of the possible causes of ultradian insulin secretion oscillations is the time delay of the insulin secretion simulated by the elevated glucose concentration. Our analysis shows that three time delay model is more accurate than two time delay.

In this paper we use several proposed Delay Differential Equation models (DDE) to show the delays of the metabolic system. We use single delay models in addition of multiple delay models to form a new proposed model combining the theories of the previous models used.

2. DELAY DIFFERENTIAL EQUATIONS

A delay differential equation (DDE) is an equation where the evolution of the system at a certain time, solution t, depends on the state of the system at an earlier time; say$^{(t-\tau)}$. This is distinct from the ordinary differential equation (ODE) where the derivatives depend only on the current value of the independent variable. The solution of the ODEs therefore requires knowledge of not only the current state, but also of the state of a certain time previously.

2.1. Differential Delay Equation Methods

Many approaches of DDE solutions start from the problem in the formulations

$$y'(t)=f\left(t,\,y(t),y(t-\tau)\right);\ t_o\le t\le t_f$$
$$y(t)=\phi(t);\ t\le t_o \tag{2.1}$$

The function $y(t)$ represents the same physical quantity that evolves overtime. The derivative $y'(t)$ depends on past values of $y(t)$, $\phi(t)$ represents the initial function, and τ is the delay term.

The theory of DDEs has become an important area of investigation stimulated by their numerous applications to problem in mechanics, electric, and electronic engineering, medicines, biology, ecology, etc. Since many systems arising from realistic models heavily depend on history (which is characterized by the effect of finite (or infinite) delay on all equations), so there is a need to study partial functional delay differential system with the delay argument; and many evolution process characterized the fact that at certain moments of time they experience a change of state abruptly, this process is subject to short time perturbation whose duration is negligible in comparison with the duration of the process.

There are several approaches to finding numerical solutions to (2.1). These include the direct application of linear multiple step method and Runge-Kutta (4th order mainly) method for ODEs. If the delay is non-constant, the methods are combined with interpolation.

The first approaches to the numerical solution of DDEs of the above form were characterized by the direct application of formulas for ODEs, called linear multi-step methods. For example, Euler's forward method, which is

$$y_{(n+1)}=y_n+h_{(n+0)}f\left(t_n,y_n,y_q\right), \tag{2.2}$$

for the same integer $q<n$.

Beuman's method ([1] , Beuman et al., 2003) avoids the need for interpolation in solving the numerical solution of (2.1). To simplify first we consider the case of the constant delay τ. Then the discontinuity points are $\varepsilon_k=t_o+k_r$. In the first interval $\left[t_o,t_o+\tau\right]$, (2.1) has the form

$$y'(t) = f(t-\tau, y_1(t), \phi(t-\tau)),$$
$$y(t_o) = \phi(t_o)$$

In the second interval $[t_o+\tau, t_o+2\tau], y_1(t) = y(t-\tau)$ and $y_2(t) = y(t)$ can be defined, so the DDE (2.1) can be written as the two dimensional system of ODEs.

$$y_1'(t) = f(t-\tau,\ y_1(t), \phi(t-\tau)),$$
$$y_2'(t) = f(t_1, y_2(t), y(t))$$
$$y_1(t_o+\tau) = \phi(t_o)$$
$$y_2(t_o+\tau) = y(t_o+\tau).$$

Hence in general, over the interval $[t_o+(k+1)\tau, t_o+k\tau]$ the DDE (2.1) can be written as the k-dimensional system of ODEs

$$y_i'(t) = f(t-(k-i)\tau, y_p(t), y_{i-1}(t));\ i = 1,2,\cdots,k$$
$$y(t_o+k_1\tau) = y(t_o+(i-1)\tau);\ i = 1,2,\cdots,k \tag{2.3}$$

where,

$$y_o(t) = \phi(t-k\tau);\ y_i(t) = y(t-(k-i)\tau);\ i = 1,2,\cdots,k$$

passing from k to k + 1 means shifting the interval of integration from $[t_o+(k+1)\tau, t_o+k\tau]$ to $[t_o+k\tau, t_o+(k+1)\tau]$, and extending the solution from $[t_o, t_o+k\tau]$ to $[t_o+k\tau, t_o+(k+1)\tau]$ by adding the component $y_k(t) = y(t)$ in the current interval. Therefore, a standard numerical method for solving ODE can be used to solve (2.2), for increasing k, larger system, for each step k, the numerical solution of the k-dimensional system (2.2) is intended to provide an approximate value of $y_k(t_o+k\tau) = y(t_o+k\tau)$ to be used as the initial value of the new component $y_{k+1}(t) = y(t)$ in the next step. The process ends when $t_o < k\tau < t_f$, for some k. So it is necessary to solve a system of ever-increasing dimension and to calculate repeatedly the same pieces of the solution related to the previous intervals. However, due to the presence of the delay argument that is storing and interpolating the computed solution throughout the interval $[t_o, t_f - \tau]$, reducing to a system of ODEs avoids standard complications ([1] , Beuman et al., 2003).

2.2. An Example of a Delay Differential Equation

$$x'(t) = 5x(t-3)$$
$$x(t) = 2; \quad -3 \le t \le 0$$

Since $0 \le t \le 3$ implies $-3 \le t-3 \le 0$, we have $x(t-3)=2$ and the differential equation becomes

$$x'(t) = 5x(t-3)$$
$$x'(t) = 5 \times 2 = 10$$

Therefore,

$$x(t) = \int x'(t)\,dt$$
$$x(t) = \int 10\,dt + C_1 = 10t + C_1$$

Using $x(0)=2$,

$$x(0) = 10(0) + C_1 = 2$$
$$C_1 = 2$$

Therefore,

$$x(t) = 10t + 2; \quad 0 \le t \le 3$$

In the next subinterval, $3 \le t \le 6$ or $0 \le t-3 \le 3$, we have

$$x(t-3) = 10(t-3) + 2$$

The differential equation becomes

$$x'(t) = 5x(t-3)$$
$$x'(t) = 5[10(t-3)+2] = 50t - 150 + 10$$
$$x'(t) = 50t - 140$$
$$x(t) = \int x'(t)\,\mathrm{d}t$$
$$x(t) = \int (50t - 140)\,\mathrm{d}t$$
$$x(t) = (50t^2/2) - 140t + C_2 = 25t^2 - 140t + C_2$$

Since $x(3) = 32$,

$$x(3) = 32 = 25(3)^2 - 140(3) + C_2$$

$$32 = 225 - 420 + C_2$$

$$32 = -195 + C_2$$

$$C_2 = 227$$

Therefore,

$$x(t) = 25t^2 - 140t + 22$$

2.3. Solving Differential Delay Equations in MATLAB

Using the routine dde23 ([2] , Thompson et al., 2001) in MATLAB, it is easier to solve a large class of delay differential equations (DDE). The problem is restricted to solving systems of equations of the form

$$y'(x) = f(t, y(x), y(x-\tau_1), y(x-\tau_2), \cdots, y(x-\tau_k))$$

for constant delays τ_j such that $\tau = \min(\tau_1, \tau_2, \cdots, \tau_k)$. The equations hold on $a < x < b$, which requires the history $y(x) = S(x)$ to be given for $x \leq a$. DDE 23 uses explicit Runge-Kutta triple methods in solving Ordinary

Differential Equations (ODE), in particular the Runge-Kutta triple BS (2,3), and extends these materials to solve DDEs. A triple of s stages involves three formulas. If you have an approximation y_n to $y(x)$ at x_n and wish to compute an approximation at $x_{x+1} = x_n + c_i h_n$. For $i = 1, 2, \cdots, s$, the stages $f_{ni} = f(x_{ni}, y_{ni})$ are defined in terms of $x_{ni} = x_n + c_i h_n$ and

$$y_{ni} = y_n + h_n \sum_{j=1}^{i-1} a_{ij} f(n_j)$$

The first formula used in the triple is an approximation used to advance the integration, written in terms of the increment function

$$\varphi(x_n, y_n) = \sum_{i=1}^{s} b_i f(n_i) \tag{2.4}$$

The second formula is used only for selection the step size

$$y_{n+1} = y_n + h_n \sum_{i=1}^{s} b_i^4 f(n_i) = y_n + h_n{}_\Phi(x_n, y_n) \tag{2.5}$$

The third formula is often described as a continuous extension of the first formula, and has the form

$$y_{n+\sigma} = y_n + h_n \sum_{i=1}^{s} b_i(\sigma) f(n_i) \tag{2.6}$$

The coefficients, $b_i(\sigma)$, are polynomials in σ, so this represents a polynomial approximation to $y(x_n + \sigma h_n)$ for $0 \le \sigma \le 1$. It is assumed that this formula yields the value y_n when $\sigma = 0$ and $y_{n+1}s$ when $\sigma = 1$.

When the delay term is greater than the step size and we have an available approximation $S(x)$ to $y(x)$ for all $x \le x_n$. All the $x_{ni} - \tau_j \le x_n$ and $f_{ni} = f(x_{ni}, y_{ni}, S(x_{ni} - \tau_1), \cdots, (x_{ni} - \tau_k))$ is an explicit recipe for the stage and the formulas are explicit, the function $S(x)$ is the initial history for $x \le a$.

After taking the step to x_{n+1} then use (2.6) to define $S(x)$ on $[x_n, x_{n-1}]$ as $S(x_n + \sigma h_n) = y_{n+\sigma}$.

When the step size is greater than the smallest delay, the "history" term $S(n)$ is evaluated in the span of the current step and the formulas are defined implicitly. In this situation the formulas are evaluated with simple iteration. On reading x_n, $S(x)$ is defined for $x \le x_n$. Its definition is extended to $[x_n, x_n + h_n]$ and the resulting function is called $S^{(o)}(x)$. A typical stage of simple iteration begins with the approximate solution $S^m(x)$. The next iteration is computed with the explicit formula

$$S^{(m+1)}(x_n + \sigma h_n) = y_n + h_n \varphi\left(x_n, y_n, \sigma; S^{(m)}(x)\right)$$

DDE 23 predicts $S^{(o)}(x)$ to be the constant y_o for the first step ([3] , Thompson et al., 2001).

A typical introduction of DDE 23 has the form

sol = DDE23(ddefile, lags, history, tspan);

where ddefile is the name of the function for evaluation the DDE, lags is the constant delays, supplied as an array. history can be specified as the name of a function that evaluates the solution and returns it as a column vector, and tspan is the interval of integration. See ([4] , Wang et al., 2007) chapter Numerical Results for some examples of using DDE 23 to solve some simple delay differential equation models.

3. MATHEMATICAL MODELS FOR DIABETES

3.1. Modeling through Ordinary Differential Equations

The insulin-glucose model that is used in this paper was originally developed by Sturis. The purpose of the model was to show the slow oscillations that take place inside of the pancreas and liver, and could give some insight on how they can lead to diabetes if not functioning properly ([5] , Tolic et al., 2000).

The secretion of insulin in the glucose-insulin endocrine metabolic system occurs in an oscillatory manner over a range of 50 - 150 minutes.

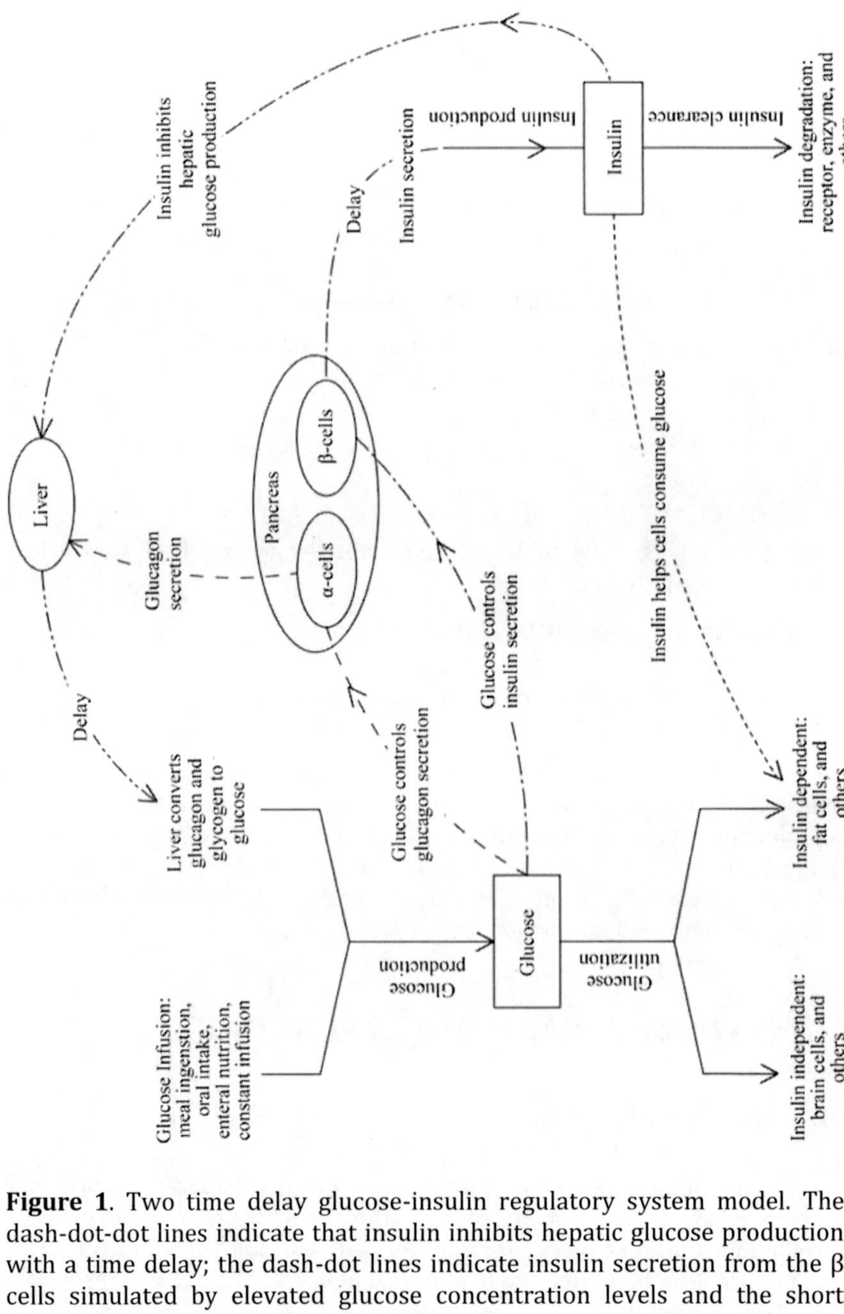

Figure 1. Two time delay glucose-insulin regulatory system model. The dash-dot-dot lines indicate that insulin inhibits hepatic glucose production with a time delay; the dash-dot lines indicate insulin secretion from the β cells simulated by elevated glucose concentration levels and the short dashed line indicates the insulin induced glucose utilization in cells with time delay; the dashed lines indicate low glucose concentration levels triggering α cells in the pancreas to release glucagon.

Two time delays exist in the time. One is due to the electric action inside of the β cells upon glucose stimulation to release insulin. The second time delay represents the delayed effect of insulin on hepatic glucose ([6] , Li et al., 2006).

The two time delay model of the glucose-insulin regulatory system can be seen in Figure 1.

There are three feedback loops in this model: glucose stimulates pancreatic insulin secretion, insulin stimulates glucose uptake and inhibits hepatic glucose production, and glucose enhances its own uptake. The system also includes two significant time delays: the first is the fact that the physiological action of insulin after the utilization of glucose is correlated with the concentration of insulin. The second is associated with the time lag between the appearance of insulin in the plasma and its inhibitory effect on the hepatic glucose production ([5] , Tolic et al., 2000).

The model has three main variables: the amount of glucose in the plasma and intercellular space, G, the amount of insulin in the plasma, I_p, and the amount of insulin in the intercellular space, I_i. In addition to the three main variables, there are three others, x_1, x_2, x_3 that represent the above-mentioned delay between insulin in plasma and its effect on the hepatic glucose production ([5] , Tolic et al., 2000).

The equations describing the dynamics of the model are

$$\frac{dG}{dt} = G_{in} - f_2(G) - f_3(G) f_4(I) + f_5(I) \tag{3.1}$$

$$\frac{dI_p}{dt} = f_1(G) - E\left(\frac{I_p}{V_p} - \frac{I_i}{V_i}\right) - \frac{I_p}{t_p} \tag{3.2}$$

$$\frac{dI_i}{dt} = E\left(\frac{I_p}{V_p} - \frac{I_i}{V_i}\right) - \frac{I_p}{t_p} \tag{3.3}$$

$$\frac{dx_1}{dt} = \frac{3}{t_d}\left(I_p - x_1\right) \tag{3.4}$$

$$\frac{dx_2}{dt} = \frac{3}{t_d}\left(x_1 - x_2\right) \tag{3.5}$$

$$\frac{dx_3}{dt} = \frac{3}{t_d}(x_2 - x_3) \tag{3.6}$$

Insulin production: Insulin can only be produced by β cells in the pancreas. The β cell secretion is in in response to the elevated glucose concentration. There are other functions of the β cell, however glucose this the most common and important provocation for insulin release. The formula to show the pancreatic insulin production of β cells is given as

$$f_1(G) = R_m \Big/ \left(1 + \exp\left(\left(C_1 - G/V_g\right)/a_1\right)\right) \tag{3.7}$$

$$f_1'(G) = \left[\left(R_m/V_g a_1\right)\left(\exp\left(\left(C - G/V_g\right)/a_1\right)\right)\right] \Big/ \left[\left(1 + \exp\left(\left(C_1 - G/V_g\right)/a_1\right)\right)\right]^2$$

Insulin degradation and clearance: The liver and the kidney are the primary sites of insulin degradation and clearance. Insulin that is not cleared by the liver or the kidney is eventually removed from other tissues (i.e. muscles). Insulin degradation is process of the kidney and liver that controls insulin action by inactivating and removing the excess of the hormone. Insulin degradation is directly proportional to the insulin concentration. Insulin clearance is the process of regulating the cellular response to the hormone by decreasing insulin availability.

Glucose production: The formation of glucose in the body is caused by two sources. The first is the consumption of carbohydrates, or compounds of carbon, oxygen and hydrogen, that form sugars, starches and fibers. These carbohydrates are absorbed into the bloodstream through hydrolysis. The consumption of carbohydrates takes place through meal ingestion, oral intake (coffee with sugar), etc. The second source of glucose production is the liver. When level of glucose concentration in the bloodstream drops, the β cells stop releasing insulin. However, α cells, also released by the pancreas begin to release another hormone, glucagon. When glucagon is released it takes control of metabolic pathways (or a sequence of chemical reactions undergone by a compound or class of compounds in a living organism—definition off internet) in the liver, and forces the liver to release glucose.

Glucose Utilization: Glucose utilization, like glucose production consists of two parts, insulin-independent utilization, and insulin-dependent utilization. Insulin-independent glucose consumers are the brain and nerve cells. This glucose intake of the brain and nerve cells is given as

$$f_2(G) = U_b / \left(1 - \exp\left(-G / \left(C_2 - V_g\right)\right)\right) \tag{3.8}$$

$$f_2'(G) = \left(U_b / C_2 V_g\right)\left[\exp\left(-G / C_2 V_g\right)\right]$$

This function represents the dependency on the glucose concentration alone ([6] , Li et al., 2006). Glucose utilization for muscles and fat tissues depend not only on glucose concentrations levels, but also insulin concentration levels. The $f_3(G)$ shows the glucose dependent term describing glucose utilization

$$f_3(G) = G / \left(C_3 V_g\right) \tag{3.9}$$

$$f_3'(G) = 1 / \left(C_3 V_g\right)$$

The insulin dependent term is given as

$$f_4(I) = U_0 + \left(U_m - U_0\right) / \left(1 + \exp\left(-\beta \ln\left(I / C_4\left(1/V_i + 1/Et_i\right)\right)\right)\right) \tag{3.10}$$

$$f_4'(I) = \frac{\left(\beta\left(U_m - U_0\right)/I\right)\left[\exp\left(-\beta \ln\left(I / C_4\left(1/V_i + 1/Et_i\right)\right)\right)\right]}{\left[1 + \exp\left(-\beta \ln\left(I / C_4\left(1/V_i + 1/Et_i\right)\right)\right)\right]^2}$$

Production of glucose controlled by insulin concentration (I) is stated as hepatic glucose production. In healthy people the pancreas has the ability to constantly measure levels of glucose in the bloodstream. The pancreas then responds to the glucose levels and releases a certain amount of insulin to counteract the glucose intake. The formula to show the above reaction is

$$f_5(G) = R_g / \left(1 + \exp\left(\alpha\left(I/V_p - C_5\right)\right)\right) \tag{3.11}$$

$$f_5'(I) = -R_g\left(\alpha/V_p\right)\left(\exp\left(\alpha\left(I/V_p - C_5\right)\right)\right) / \left(1 + \exp\left(\alpha\left(I/V_p - C_5\right)\right)\right)$$

3.2. Analysis of Insulin Glucose Feedback Model

In order to fully understand certain feature of the model, the insulin-glucose feedback model has been simplified. The simplified model has two important properties: it is analytically tractable, and it shows the same main characteristics as the original model (i.e. self-sustained oscillations and values of the state variables in the same range as the original model) ([5] , Tolic et al., 2000).

The function $f_2(G)$ is approximated with a constant and the function $f_1(G)$ by a first-order polynomial. The function $f_4(I)$ is also well approximated by a first-order polynomial in the range of obtained values of intercellular insulin. The function $f_5(I)$ shows significantly greater variation in its second derivative than all the other functions, in the range of glucose and insulin concentration. For values of I in the range 65 - 115 mU in $f_5(I)$ changes from concave to convex, and is well approximated by a third-order polynomial.

The function $f_1(G)$ is replaced with the first-order Taylor expansion around the mean value of G, $f_2(G)$ with a constant, $f_4(I)$ with the first-order expansion around the mean value of I, and $f_5(I)$ with the third- order expansion around the mean value of I. Note that the mean values were obtained by a simulation of glucose infusion at the rate of 216 mg·min^{-1} using the original model ([5] , Tolic et al., 2000) and Table 1 and Table 2. The standard model to for the detection of diabetes without delay is studied and analyzed. It is well accepted, reasonable and reliable model, refer [7]

The simplified model is

$$dI_p/dt = aI_p + bI_i + cG + d \tag{3.12}$$

$$dI_i/dt = eI_p + fI_i \tag{3.13}$$

$$dG/dt = gI_iG + hG + kI + lI^2 + mI^3 + p \tag{3.14}$$

$$dx_1/dt = rI_p - rx_1 \tag{3.15}$$

Table 1. Parameters of the functions in Equations (7)-(11). Values (data) are obtained experimentally in laboratory.

Parameter	Value	Parameter	Value
V_g (l)	10	U_0 (mg/min)	40
R_m (mU/min)	210	U_m (mg/min)	940
a_1 (mg/l)	300	β	1.77
C_1 (mg/l)	2000	C_4 (mU/l)	80
U_b (mg/min)	72	R_g (mU/l)	180
C_2 (mg/l)	144	α (l/mU)	0.29
C_3 (mg/l)	1000	C_5 (mU/l)	26
V_p (l)	3	t_p (min)	6
V_i (l)	11	t_i (min)	100
E (l/min)	0.2	t_d (min)	36

Table 2. Commonly accepted experimental values used in these models.

Parameter	Value	Parameter	Value
a (min^{-1})	−0.233	h (min^{-1})	2.64×10^{-3}
b (min^{-1})	0.0182	k (mg·mU^{-1}·min^{-1})	17.5
c (mg·mU^{-1}·min^{-1})	4.79×10^{-3}	l (mg·mU^{-2}·min^{-1})	−0.315
d (mU·min^{-1})	−43.9	n (mg·mU^{-3}·min^{-1})	1.48×10^{-3}
e (min^{-1})	0.0667	p (mg·min^{-1})	80.5
f (min^{-1})	−0.0282	r (min^{-1})	0.0833
g (mU^{-1}·min^{-1})	-9.44×10^{-5}		

$$dx_2/dt = rx_1 - rx_2 \tag{3.16}$$

$$dx_3/dt = rx_2 - rx_3 \tag{3.17}$$

The time evolution of plasma glucose and insulin concentration that result from the simplified model does not differ significantly from the results of the original model.

In the simulation of an exogenous insulin infusion, equation, dI_p/dt, was replaced by

$$dI_p/dt = m\left(1 + A\sin\left(2\pi t/T\right)\right) + aI_p + bI_i \tag{3.18}$$

where m = 21 mU/min. A was set to zero for the constant infusion and to 0.3 for the oscillatory infusion. The period T = 120 min. Since the glucose infusion rate in the experiments equals 420 mg/min, using Table 1 and Table 2 (experimental data). The parameter p in Equation (3.14) was set to 285 mg/min to compensate for the difference between the rate of the glucose infusion of 420 and 216 mg/min ([5] , Tolic et al., 2000).

From an exact solution of the two first order differential equations (3.13) and (3.18), the numerical solutions to (3.15)-(3.17) were computed. For large t, the solutions are periodic functions with period T and may be expressed in the form

$$y_j = Y_j + A_j \sin\left(\left(2\pi t/T\right) + \varphi_j\right),\ j = 1,\cdots,5 \tag{3.19}$$

The only equation in the simplified model with non-linear terms is Equation (3.14). It can be expressed in the form

$$dG(t)/dt + P(t)G(t) = Q(t) \tag{3.20}$$

where

$$P(t) = -gI_i(t) - h \tag{3.21}$$

and

$$Q(t) = kI(t) + l(I)^2 + n(I)^3 + p \tag{3.22}$$

Equation (20) has the general solution

$$G(t) = [1/\mu(t)]\left[\mu(t_o)G_o + \int_{t0}^{t}\mu(\zeta)Q(\zeta)\mathrm{d}(\zeta)\right] \tag{3.23}$$

where

$$\mu(t) = \exp\left(H\left(\cos\left(2\pi t/T + \varphi_i\right) - \cos\left(\varphi_i\right)\right)\right)\exp(-Jt) \tag{3.24}$$

$$J = g\langle I_i \rangle + h \tag{3.25}$$

and

$$H = gA_i\left(T/2\pi\right) \tag{3.26}$$

3.3. Modeling through Delay Differential Equations

In section 3.1 we have discussed the dynamics of the insulin-glucose feedback system through Equations (1)-(6). The hepatic glucose production time delay is stimulated by introducing three auxiliary variables, x_1, x_2, x_3, which is caused the third order delay. We demonstrate how the auxiliary variables stimulate the time delay as follows: For the sake of convenience we would like to assume the first order delay (i.e. x_1).

$$x_1'(t) = (3/t_d)\left(I_p(t) - x_1(t)\right);\ t_d > 0$$

Then

$$x_1'(t - t_d)t_d = I_p(t - t_d) - x_1(t - t_d)$$

$$I_p(t-t_d)=x_1(t-t_d)+x_1'(t-t_d)t_d$$

When $I_p(t-t_d)\approx x_1(t)$

$$x_1(t)=x_1(t-t_d)+x_1'(t-t_d)t_d+Ot_d^2$$

Using Taylor's Series Expansion for the function of one variable, we get

$$I_p(t-t_d)=I_p(t_d)+(t-t_d)I_p'(t-t_d)t_d+(O)(t_d)^2$$

I: Explicit Single Time Delay Differential Equation Model

In [8] , Engleborghs et al. (2001) replaced the auxiliary variables $x_i, i=1,2,3$, and introduced a single time delay in the Negative Loop Model and proposed the flowing DDE Model

$$\mathrm{d}G(t)/\mathrm{d}t=G_{in}-f_2(G(t))-f_3(G(t))f_4(I_i(t))+f_5(I(t-\tau))$$

$$\mathrm{d}I_p(t)/\mathrm{d}t=f_1(G(t))-(I_i/t_i)\mathrm{d}I_i(t)/\mathrm{d}t$$

$$\mathrm{d}x_1/\mathrm{d}t=(3/t_d)(I_p-x_1)$$

$$\mathrm{d}x_2/\mathrm{d}t=(3/t_d)(x_1-x_2)$$

$$\mathrm{d}x_3/\mathrm{d}t=(3/t_d)(x_2-x_3)$$

where the functions f_i, and their parameters are assumed to be the same as those in the model (Section 3.1.1). G(t) is the amount of glucose, $I_p(t)$ and $I_i(t)$ are the amounts of insulin in the plasma and the intercellular space, respectively. V_p is the plasma insulin distribution volume, V_i is the effective volume of the intercellular space, E is the diffusion transfer rate, t_p and t_i are the insulin degradation time constants in the plasma and intercellular space, respectively. G_{in} indicates the (exogenous)glucose supply rate to the plasma, and $x_1(t), x_2(t),\ x_3(t)$ are three auxiliary variables associated with certain delays of the insulin on the hepatic glucose production in time t_d. The positive constant delay τ mimics the hepatic glucose production delay.

This model consists of two negative feedback loops, describing the effects of insulin on glucose utilization and glucose production, respectively. Both loops include the stimulatory effect of glucose on insulin secretion. The model mimics the delayed insulin-dependent glucose uptake by splitting the insulin into two separate compartments, plasma and interstitial space. The hepatic glucose production time delay is simulated by introducing three variables x_1, x_2, x_3 (third order delay).

Unfortunately, this model overlooked the time delay from glucose stimulated insulin release to the delayed insulin-dependent glucose uptake. Due to the complex chemical reactions of the β cells, it takes a few minutes (5 - 15 mins) for the insulin being ready to help cells utilizing glucose after the plasma glucose concentration level is elevate. As confirmed by, Sturis et al. (1995) [9] , Tolic (2000) [5] , the effects of time delay are significant (Li et al., 2006) [6] .

II. Single Time Delay Proposed by Drozdov and Khanina (1995) [10]

The rate of insulin secretion f_1 is assumed to depend only on the glucose concentration c_z. $f_1(c_z)$ is as follows

$$f_1(c_z) = 210/(1 + \exp(a + bc_z)).$$

The numerator in the right-hand side of the numerator in $f_1(c_z)$ characterizes the rate of insulin secretion for very large concentrations of glucose. Parameters a and b are determined using experimental data (let a = 5.21 and b = −0.03).

The function f_2 is suggested as

$$f_2 = (9/(1 + \exp(7.76 - 1.772\ln(c_x)))) + 0.4$$

The rate of glucose production f_3 is assumed to depend only on the insulin concentration c_x, but at the moment $t - \tau$, where τ is a delay in glucose production. The parameter τ lies in the interval from a few minutes to an hour. The function f_3 is proposed as

$$f_3 = 160/\left(1+\exp\left(0.29c_x - 7.5\right)\right)$$

The delay in this proposed model is a delay in the $\mathrm{d}z(t)/\mathrm{d}t$ term ($\mathrm{d}x_3/\mathrm{d}t$ term in other models). The rate of glucose production in the $\mathrm{d}z(t)/\mathrm{d}t$ term is assumed to only depend on the insulin concentration, at the moment $t-\tau$, where τ is a delay in glucose production, which affects the amount of glucose produced by the liver. ([10] , Drozdov et al., 1995).

The model is as follows

$$\mathrm{d}x(t)/\mathrm{d}t = f_1\left(0.1z(t)/V_3\right) - \left(\left(E/V_1\right) + \left(1/T_1\right)x(t) + \left(E/V_2\right)y(t)\right)$$

$$\mathrm{d}y(t)/\mathrm{d}t = \left(E/V_1\right)x(t) - \left(\left(E/V_2\right) + \left(1/T_2\right)\right)y(t)$$

$$\mathrm{d}z(t)/\mathrm{d}t = f_3\left(x(t-\tau)/V_1\right) - \left(0.1z(t)/V_3\right)f_2\left(y(t)/V_2\right) + \left(L - p_0\right)$$

where V_1 (Volume of plasma) = 3, V_2(Volume of interstitial liquid) = 11, V_3 (Volume of the glucose compartment) = 10

III: Explicit Two Time Delay Differential Equation Model

This is a single delay DDE model used for insulin therapies for both Type I and Type II diabetes mellitus and insulin degregation rate assumes Michaelis-Menten kinetics. The model equations proposed by Wang, Li, and Kuang (2009) [11] , is as follows

$$\mathrm{d}G(t)/\mathrm{d}t = G_{in} - f_2\left(G(t)\right) - f_3\left(G(t)\right)f_4\left(I_i(t)\right) + f_5\left(I(t-\tau)\right)$$

$$\mathrm{d}I_p(t)/\mathrm{d}t = bf_1\left(G\left(t-\tau_1\right)\right) - \left(\mathrm{d}/\mathrm{d}t\right)I_i(t)$$

$$\mathrm{d}x_1/\mathrm{d}t = \left(3/t_d\right)\left(I_p - x_1\right)$$

$$\mathrm{d}x_2/\mathrm{d}t = \left(3/t_d\right)\left(x_1 - x_2\right)$$

$$\mathrm{d}x_3/\mathrm{d}t = \left(3/t_d\right)\left(x_2 - x_3\right)$$

The major analytical results of this model were to obtain a sufficient condition for global asymptotical stability induced by a Liapunov function with the hepatic glucose production time delay satisfying the cases when $t = 0$ and $t > 0$. This result has the following consequences: 1) If the hepatic glucose production time delay, τ, and the insulin transfer time between the plasma and interstitial compartments, t_i and t_p, are sufficiently small, then the solution converges to a steady state. 2) If the hepatic glucose production value, R_g, is very small, the oscillation will not be sustained.

For Type I diabetes, b = 0, (means no insulin is secreted from the pancreas). For Type IIdiabetes, $0 < t \leq 1$. In this approach, we keep one explicit time delay τ_1 as that in the two time delay model $G'(t)$ and $I'(t)$, but mimic the hepatic glucose production time delay by variable chain as in the model in Section 3.1.1. ([11] , Wang et al., 2009).

IV: Alternative Explicit Two Time Delay Differential Equation Model (Insulin Therapies Related Model)

The idea of insulin therapy is to mimic the normal reaction of the beta cells in the pancreas when they are stimulated by an increase in glucose concentration of the blood, for example after meal ingestion. Numerous experiences have demonstrated that insulin is released from beta cells in two oscillatory models: pulsatile oscillations and ultradin oscillations. Insulin therapies must mimic the insulin secretion at these two time scales and are normally introduced based on clinical experiences, although mathematical models have been proposed for some specific situations ([4] , Wang et al., 2007).

In the paper by Wang, the following generic model is put forward to simulate the pancreatic insulin secretion with insulin infusion after blood glucose concentration increases for Type I diabetes. We model the effect of time delay t_2 in glucose utilization by $f_3(G(t))$ and $f_4(I(t-t_0))$. This results in the following alternative model with two explicit delays.

$$dG(t)/dt = G_{in} - f_2(G(t)) - f_3(G(t))f_4(I(t-\tau_2)) + f_5(I(t-\tau_0))$$

$$dI_p(t)/dt = f_1(G(t)) - d(I_i(t))/dt$$

$$dx_1/dt = (3/t_d)(I_p - x_1)$$

$$dx_2/dt = (3/t_d)(x_1 - x_2)$$

$$dx_3/dt = (3/t_d)(x_2 - x_3)$$

The notations in models (III) and (IV) are the same as the models discussed previously. Data are used from Table 1 and Table 2 to set up the MATLAB codes.

```
function res = CGC_2TD_Eqns_IV(t,SOL,DSOL)
global Vp Vi tpti td E Gin di Vg

I  = SOL(1); % Insulin variable from SOL vector
G  = SOL(2); % Glucose variable from SOL vector
T2 = DSOL(1,1); %tau2
T0 = DSOL(1,2); %tau0

dIdt = f1(G)-di*I;
dGdt = Gin-f2(G)-f3(G)*f4(T2)+f5(T0);

res = [dIdt; dGdt]; % return solution vector
end
```

V: Alternative Explicit Two Time Delay Differential Equation Model (Two-Time Delay Model proposed by Li, Kuang, Mason (2006)) [6] .

Li, Kuang, and Mason consider two time delays: τ_1 and τ_2. The first, τ_1, is to denote the total time delay form the time that the glucose concentration level is elevated to the moment that the insulin has been transported to the interstitial space and becomes "remote insulin". Therefore insulin secretion can be approximated by $f_1(G(t-\tau_1))$ with time delay $\tau_1 < 0$.

The second time delay, τ_0, has to do with the delay of effect of hepatic glucose production measured from the time that the insulin has become "remote insulin" to the moment that a significant change of hepatic glucose production occurs. τ_0 in the model represents glucose utilization delay. The model is as follows ([6] , Li et al., 2006)

$$dG(t)/dt = G_{in} - f_2(G(t)) - f_3(G(t))f_4(I(t)) + f_5(I(t-\tau_0))$$

$$dI_p(t)/dt = f_1G(t-\tau_1) - dI_i(t)/dt$$

$$dx_1/dt = (3/t_d)(I_p - x_1)$$

$$dx_2/dt = (3/t_d)(x_1 - x_2)$$

$$dx_3/dt = (3/t_d)(x_2 - x_3)$$

This model suggests that the time delay from insulin secretion stimulated by glucose to the insulin becoming "remote insulin" is not negligible; especially the delay of insulin secretion triggered by elevated glucose. Therefore, we suspect that one of the possible causes of ultradian insulin secretion oscillations is the time delay of the insulin secretion stimulated by the elevated glucose concentration.

In this two time delay model, τ_1 denotes the total delay from the time that the glucose concentration level is elevated to the moment that the insulin has been transferred to the interstitial space, thus becoming 'remote insulin'. τ_0 has to do with the delay of hepatic glucose production measured from the time that insulin has become 'remote insulin' to the moment that a significant change of hepatic glucose production occurs. G_i is the glucose infusion rate; $d_i > 0$ is the insulin clearance rate. Functions $f_1 - f_5$ are defined previously.

Based on our numerical simulations and analysis, we suspect that time total time delay, τ_1, is critically responsible for the oscillation. The total time delay is measured from the moment that the glucose concentration level starts to increase to the moment that the insulin has been transported to the interstitial space.

```
function res = CGC_2TD_Eqns_V(t,SOL,DSOL)
global Vp Vi tpti td E Gin di Vg

I  = SOL(1); % Insulin variable from SOL vector
G  = SOL(2); % Glucose variable from SOL vector
T1 = DSOL(2,1); %tau1
T0 = DSOL(1,2); %tau0

dIdt = f1(T1)-di*I;
dGdt = Gin-f2(G)-f3(G)*f4(I)+f5(T0);

res = [dIdt; dGdt]; % return solution vector
end
```

VI: Another Alternative Two Time Delay Differential Equation Model, (Palumbo, Panunzi, DeGaetano [2004]) [12]

To our knowledge, there are no mathematical models considering β cell activity in the form of DDEs. Our aim is to describe the activity of β cell through the flowing two time delay DDE model.

$$\mathrm{d}G(t)/\mathrm{d}t = G_{in} - f_2(G(t)) - f_3(G(t))f_4(I(t-\tau_g)) + f_5(I(t))$$

$$\mathrm{d}I_p(t)/\mathrm{d}t = bf_1(G(t-\tau_i)) - \mathrm{d}/\mathrm{d}t(I_i(t))$$

$$\mathrm{d}x_1/\mathrm{d}t = (3/t_d)(I_p - x_1)$$

$$\mathrm{d}x_2/\mathrm{d}t = (3/t_d)(x_1 - x_2)$$

$$\mathrm{d}x_3/\mathrm{d}t = (3/t_d)(x_2 - x_3)$$

where τ_g is the apparent delay with which the pancreas varies secondary insulin release in response to varying plasma glucose concentration. τ_i is the delay with which insulin acts in stimulating glucose uptake by peripheral tissues ([12] , Palumbo et al., 2004). This model is very similar to the one given in (V).

There are two possibilities of delay of insulin release by the pancreas τ_g in response to variation of glucose concentration in plasma τ_i: (1) $\tau_g = \tau_i = \pm\infty$, (2) $\tau_g \neq \tau_i$.

This model describes the dynamics of the glucose-insulin feedback system which involves an equation for the number of β cells. Additionally, it provides a description of the mechanisms that link glucose metabolism to changes in the electrical activity of β cell that induces insulin secretion. Codes for the numerical simulations of this model are as follows:

```
function res = CGC_2TD_Eqns_VI(t,SOL,DSOL)
global Vp Vi tpti td E Gin di Vg

I  = SOL(1); % Insulin variable from SOL vector
G  = SOL(2); % Glucose variable from SOL vector
T1 = DSOL(2,1); %tau1
T2 = DSOL(1,2); %tau2

dIdt  = f1(T1)-di*I;
dGdt   = Gin-f2(G)-f3(G)*f4(T2)+f5(I);

res = [dIdt; dGdt]; % return solution vector
end
```

VII: Multiple Time Delay Model

The final model, which has not been considered, is a multiple delay differential equation, made from combinations of the previous four models given. The model is as follows:

$$dG(t)/dt = G_{in} - f_2(G(t)) - f_3(G(t))f_4(I(t-\tau_2)) + f_5(I(t-\tau_0))$$

$$dI_p(t)/dt = f_1(G(t-\tau_1)) - d/dt(I_i(t))$$

$$dx(t)/dt = f_1(0.1z(t)/V_3) - ((E/V_1) + (1/T_1)x(t) + (E/V_2)y(t))$$

$$dy(t)/dt = (E/V_1)x(t) - ((E/V_2) + (1/T_2))y(t)$$

$$dz(t)/dt = f_3(x(t-\tau)/V_1) - (0.1z(t)/V_3)f_2(y(t)/V_2) + (L - p_0)$$

This model is made up from all of the previous sections, taking common delays from each and making one model out of the previous five, and adding one additional delay in dx_3/dt. The two time delays in $dG(t)/dt$ come from the model proposed by Wang, Li, and Kuang in 2007 (Section IV) [4]. Similarly in 2006, Li, Kuang and Mason [6], made an earlier model (Section V) that proposed the time delay in $dG(t)/dt$ in f_5 and the time delay in $dI_p(t)/dt$. We connected the two proposed models to form the basis for our final model. The time delays in $dG(t)/dt$ and $dI_p(t)/dt$ are also related form the model proposed by Palumbo, Panunzi, and DeGaetano in 2004 (Section VI) [12]. The last three equations $(dx(t)/dt, dy(t)/dt, dz(t)/dt)$ with the time delay in the $dz(t)/dt$ equation, were taken from the model proposed by Drozdov and Khanina ([10] , Drozdov et al., 1995).

This function represents all the delays we have presented so far in this paper. The time delays in the $dG(t)/dt$ function τ_0 and τ_2. In this equation the τ_0 is seen in the $f_5(I)$, or the glucose production controlled by insulin concentration, and represents the hepatic glucose production delay. The τ_2 seen in the $f_4(I)$, or the insulin- dependent glucose uptake (dependent upon glucose because multiplied by $f_3(G)$), is the time delay for the insulin-dependent glucose utilization by the cells ([4] , Wang et al., 2007).

The delay in the $dI_p(t)/dt$ term is represented by τ_1, which is the total time delay form the time that the glucose concentration level is elevated to the moment that the insulin has been transported to the interstitial space and becomes "remote insulin" as stated in Section V. This delay can be added to our model, because the time delay $t-\tau_0$ is shared between the model in Section IV and Section V. The last delay in our system is the delay in the $dz(t)/dt$ term, given by Drozdov and Khanina ([10] , Drozdov et al., 1995). This delay represents the delay in glucose production, which affects the amount of glucose produced by the liver.

```
function res = CGC_3TD_Eqns(t,SOL,DSOL)
global Vp Vi tpti td E Gin di Vg
I  = SOL(1); % Insulin variable from SOL vector
G  = SOL(2); % Glucose variable from SOL vector
T1 = DSOL(2,1); %tau1
T2 = DSOL(1,2); %tau2
T0 = DSOL(1,3); %tau0
dIdt = f1(T1)-di*I;
dGdt  = Gin-f2(G)-f3(G)*f4(T2)+f5(T0);
res = [dIdt; dGdt]; % return solution vector
```

Remark: The parameter values assumed in Figures 1-3 are experimental values. Such values may have significant effects on the results studied in this paper.

Summary: In this paper, we studied delay differential equation (DDE) models by introducing two and three explicit time delays and model the glucose-insulin endocrine metabolic regulatory system. The three-time delay model not only confirms many exciting experimental observations, but also demonstrates robustness, and produces simulation profiles in better agreement with observed data. As a result, we suspect that one of the possible causes of ultradian insulin secretion oscillations is the time delay of the insulin secretion simulated by the elevated glucose concentration.

Like the two-time delay models, three-time delay models can also be generalized to get more accuracy in the detection of diabetes.

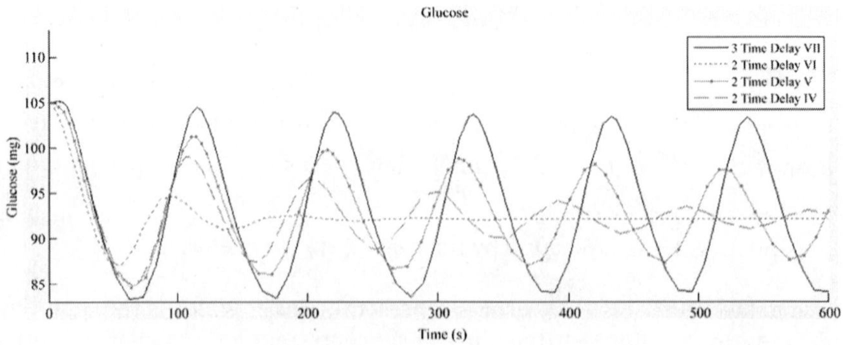

Figure 2. Glucose profiles [models (IV)-(VII)]. $\tau_1 = 4, \tau_2 = 7, \tau_0 = 12, G_{in} = 1.08, d_i = 0.06$.

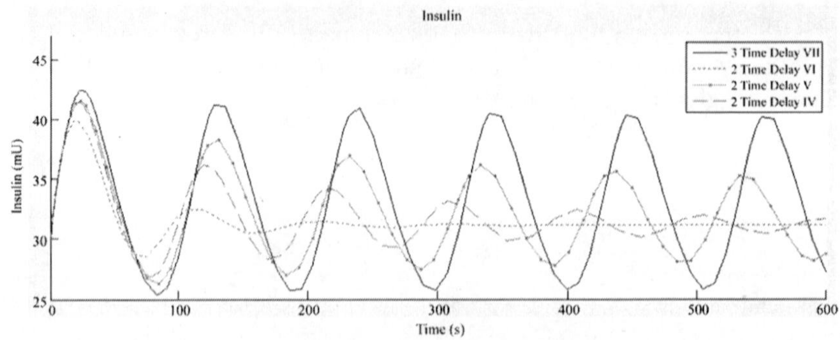

Figure 3. Insulin profiles [models (IV)-(VII)]. $\tau_1 = 4, \tau_2 = 7, \tau_0 = 12, G_{in} = 1.08, d_i = 0.06$.

ACKNOWLEDGEMENTS

We are thankful to Dr. Roger Thelwell (James Madison University, VA) for working with us throughout the process to set up the MATLAB codes.

REFERENCES

1. Beuman, A. and Zennaro, M. (2003) Numerical Methods for Delay Differential Equations. Oxford University Press, NY, USA.
2. Thompson, S. and Shampine, L.F. Solving Delay Differential Equations with dde23. http://www.radford.edu/thompson/webddes/tutorial
3. Thompson, S. and Shampine, L.F. (2001) Solving Delay Differential Equations in MATLAB. Applied Numerical Mathematics, 37, 441-458.
4. Wang, H.Y., Li, J.X. and Kuang, Y. (2007) Mathematical Modeling & Qualitative Analysis of Insulin Therapies. Mathematical Biosciences, 210, 17-33.
5. Tolic, I.M., Mosekilde, E. and Sturis, J. (2000) Modeling the Insulin-Glucose Feeback System: The Significance of Pulsatile Insulin Secretion. Journal of Theoretical Biology, 207, 361-375.
6. Li, J.X., Kuang, Y. and Mason, C.C. (2006) Modeling the Glucose-insulin Regulatory System and Ultradin Insulin Secretory Oscillations with Two Explicit Time Delays. Journal of Theoretical Biology, 242, 722-735.
7. Chalishajar, D. and Stanford, A. (2014) Mathematical Analysis of Insulin-Glucose Feedback System of Diabetes. International Journal of Engineering and Applied Sciences, 5, 36-58.
8. Englrborgh, K., Lemaire, V., Belair, J. and Roose, D. (2001) Numerical Bifurcation Analysis of Delay Differential Equations Arising from Physiological Modeling. Journal of Mathematical Biology, 42, 361-385.
9. Sturis, J., Scheen, A.J., Leproult, R., Polonsky, J.S. and Van Cauter, E. (1995) 24-Hours Glucose Profilesduring Continuous or Oscillatory Insulin Infusion. Journal of Clinical Investigation, 95, 1464-1471.
10. Drozdov, A. and Khanina, H. (1995) A Model for Ultradin Oscillations of Insulin and Glucose. Mathematical and Computer Modelling, 22, 23-38.
11. Wang, H.Y., Li, J.X. and Kuang, Y. (2009) Enhanced Modeling of the Glucose-Insulin System and Its Applications in Insulin Therapies. Journal of Biological Dynamics, 3, 22-38.
12. Palumbo, P., Panunzi, S. and De Gaetano, A. (2004) Qualitative Behavior of a Family of Delay-Differential Models of the Glucose-Insulin System. Istituto Di Analisi Dei Sistemi Ed Informatica, 620, 3-28.

CHAPTER 11

On Applications of Generalized Functions in the Discontinuous Beam Bending Differential Equations

Dimplekumar Chalishajar, Austin States, Brad Lipscomb

Department of Applied Mathematics, Virginia Military Institute (VMI), Lexington, USA

ABSTRACT

This paper discusses the mathematical modeling for the mechanics of solid using the distribution theory of Schwartz to the beam bending differential Equations. This problem is solved by the use of generalized functions, among which is the well known Dirac delta function. The governing differential Equation is Euler-Bernoulli beams with jump discontinuities on displacements and rotations. Also, the governing differential Equations of a Timoshenko beam with jump discontinuities in slope, deflection, flexural stiffness, and shear stiffness are obtained in the space of generalized functions. The operator of one of the governing differential Equations changes so that for both Equations the Dirac Delta function and its first distributional derivative appear in the new force terms as we present the same in a Euler-Bernoulli beam. Examples are provided to illustrate the abstract theory. This research is useful to Mechanical Engineering, Ocean Engineering, Civil Engineering, and Aerospace Engineering.

Keywords: Mechanics of Solids, Discontinuities in a Beam Bending Differential Equations, Generalized Functions, Jump Discontinuities

1. INTRODUCTION

This article introduces the method for computing lateral deflections of plane beams undergoing symmetric bending. Reviewers should be acquainted with: integration of ordinary differential Equations, and statics of plane beams under symmetric bending.

Our primary objective is to apply the discontinuous beam bending differential Equation to different application obviously representing beam bending. One of the most common types of structural components is a beam, recommended more in Civil and Mechanical Engineering. A beam resembles as a bar-like structural that is used to support transverse loading and carry it to the supports. Beams resist against transverse loads through a bending action, which creates compressive longitudinal stresses on one side of a beam and tensile stress on the other side. With these two combinations between compressive longitudinal stress and tensile stress, an internal bending moment starts to occur. In the case of a discontinuous load, we begin applying beam-bending differential Equation for each part of the beam.

All models use some sort of approximation to the underlying physics because beams are three-dimensional bodies.

Transverse loading being resisted on a preferred longitudinal plane is known as a

plane beam. Because the classical beam theory is the simplest and most associated

model for plane beams, it presents assumptions such as:

1) Planar symmetry: The longitudinal axis appears to be straight with a cross section of the beam being longitudinal plane of symmetry. Each sections that lie on the plane, both resultant of the transverse loads. The resultant of the transverse loads acting on each section lies on the plane.
2) Cross sectional variation: The cross section remains constant or varies.
3) Normality: The plane sections are originally normal to the longitudinal axis of the beam remain plane and normal to the reformed longitudinal axis upon bending.

4) Strain energy: Transverse shear and axial forces are ignored, while only internal strain energy from another object accounts for the bending moment deformations.

5) Linearization: The infinitesimal deformation of the beam is brought into the mix due to the consideration of transverse deflections, rotations and deformations.

6) Material model: The heterogeneous beams are fabricated with several elastic and isotropic materials, such as reinforced concrete.

Transverse shear and axial force are ignored, while only internal strain energy from another object accounts for the bending moment deformations. The assumption of infinitesimal deformation is brought into the mix due to the consideration of transverse deflections, rotations and deformations. The assumption is made that heterogeneous beams are fabricated with several elastic and isotropic materials.

Now we will begin our discussion on classical beam theory, also known as The Euler- Bernoulli Beam Theory.

Beam coordinates system:

The coordinate system throughout the beam undergoes a transverse loading at a point on the top surface will shorten. As for the other, it will elongate. This causes a neutral surface between the top and the bottom.

Beam motion:

We associate beam motion as the loading on a x, y plane beam is structured in to two dimensional displacement field $[u(x,y)]$ and $[v(x,y)]$ u and v are respected as the axial and transverse displacement components with respect to a beam point.

Beam loading:

$P(x)$ is denoted as the transverse force per unit length occurs on the plane beam in a positive y direction. We can determine the strong and the weak loading points based on what beam is being used. For instance, support on a simply supported beam is found on the end points that prohibit transverse displacements. In contrast, one side of a cantilever beam does not have an end support, resulting with one being clamped on and the other being free. Airplane wings, diving boards, and stabilizers are prime examples of cantilever beams.

2. SINGULAR LOADING CONDITIONS

This section will explain the equivalent distributed force for a family of singular loading condition by using Schwartz's distribution theory.

Definition 1. Let $q(x)$ be a distributed force. The nth order moment of $q(x)$ that x_0 is given by

$$M^n(x_0) = \int_{-\infty}^{+\infty} (x - x_0)^n q(x) \mathrm{d}x \tag{1}$$

Definition 2. Let $q(x)$ be a distributed force in the small open segment (interval). $(x_0 - \epsilon, x_0 + \epsilon)$ Also consider

$$M^n(\epsilon) = \begin{cases} \int_{-\infty}^{\infty} (x - x_0)^n q_\epsilon(x) \mathrm{d}x \neq 0; m = n \\ \int_{-\infty}^{\infty} (x - x_0)^m q_\epsilon(x) \mathrm{d}x = 0; m \neq n \end{cases} \tag{2}$$

Then

$$M_0^n = \lim_{\epsilon \to 0} \int_{x_0 - \epsilon}^{x_0 + \epsilon} (x - x_0)^n q_\epsilon(x) \mathrm{d}x \tag{3}$$

In 1959, Timoshenko and Woinowsky-Krieger [1] studied the concentrated double moment of the beam, which is the limiting situation of two opposite movements acting on two different separated points. They proved the result in a deflection with a discontinuous slope at the point of the concentrated double movement.

In this article, we want to study the equivalent distributed force in the loading function of a point moment of order n by using the distribution theory, refer [2] . We will show that the loading function for this loading condition is expressed by

$$q(x) = \frac{M^2}{2} \delta^{(2)}(x - x_0) \tag{4}$$

where M^2 is the value of double movement and $\delta^{(2)}$ is the second distributional derivative of δ .

Theorem 1. The equivalent distributed force of a unit moment of order n applied at $x = x_0$ is

$$q_n(x) = \frac{\left(-1^n\right)}{n!}\delta^{(n)}(x - x_0) \tag{5}$$

where $\delta^{(n)}$ is the nth distributional derivative of the Dirac Delta function.

Corollary 1. The equivalent distributed force for an upward concentrated force of magnitude p is

$$q_p(x) = p\delta(x - x_0) \tag{6}$$

This was obtained by Timoshenko (1976) [3] and Shames (1989) [4] , where the limiting case of a load distributed over a very short portion of a beam. The shearing forces of an Euler-Bernoulli beam can be applied to act as another proof of this representation by using the discontinuity of a concentrated force, appears in Section 3.4.

Corollary 2. The equivalent distributed force of a clockwise concentrated moment of magnitude M is

$$q_M(x) = M\delta^{(1)}(x - x_0) \tag{7}$$

This result was founded by Shames in (1989) [4] , in which this loading is considered to be the limiting case of two concentrated forces M/ϵ, ϵ apart, when ϵ goes to zero. The bending moment of an Bernoulli beam can be applied to act as another proof of this representation by using the discontinuity a concentrated moment introduces. It appears in Section 3.3.

Corollary 3. The equivalent distributed force of a concentrated double moment is given by Equation (4). As Timoshenko and Woinowsky-Krieger (1959) [1] mention, this loading results in a deflection with a discontinuous slope at the point of double moment. We see later that, in an Euler-Bernoulli beam with a jump discontinuity in slope, this forcing function appears.

3. A MATHEMATICAL EXPLANATION FOR CORNER CONDITION IN CLASSICAL PLATE THEORY AND EQUALLY DISTRIBUTED FORCE FOR DISTRIBUTED MOMENTS

In this section we obtain the equivalent distributed force of a distributed moment. We then give a mathematical explanation for corner condition in classical plate theory.

It can be shown that the force function of a distributed moment, $m(x)$, can be expressed in terms of m and $\delta^{(1)}$

$$q(x) = \left(m * \delta^{(1)}\right)(x) \tag{8}$$

But distribution theory shows that for any function f

$$\left(f * \delta^{(n)}\right)(x) = f^{(n)}(x) \tag{9}$$

Hence,

$$q(x) = m^{(1)}(x) \tag{10}$$

Accordingly, for a distributed moment the first distributional derivative of $m(x)$, is the forcing function. Imagine a beam that has a length of L under a distributed moment $m_0(x)$, (see Figure 1(a)). The moment can be written as

$$m(x) = \begin{cases} m_0(x); & 0 < x < L \\ 0; & L \leq x \text{ or } x \leq 0 \end{cases} \tag{11}$$

Hence,

$$m(x) = m_0(x)\left[H(x) - H(x-L)\right] \tag{12}$$

where H is Heaviside's function. When substituting Equation (12) into Equation (10), the output is

$$\begin{aligned} q(x) = m^{(1)}(x) &= m_0^{(1)}(x)\left[H(x) - H(x-L)\right] + m_0(x)\left[\delta(x) - \delta(x-L)\right] \\ &= m'_x\left[H(x) - H(x-L)\right] + m_0(0)\delta(x) - m_0(L)\delta(x-L) \end{aligned} \tag{13}$$

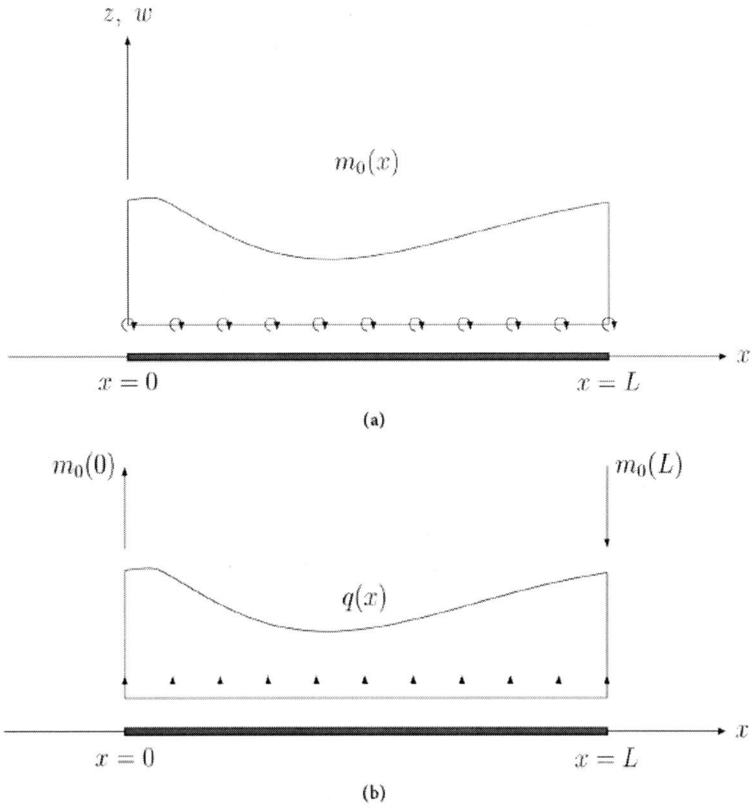

Figure 1. (a) A beam under a distributed moment. (b) The equivalent force system.

The distributed moment is equivalent to the distributed force $m_0'(x)$ on both end points of the beam given by $(0, L)$. Also distributed moment is equivalent to two concentrated forces $m_0(0)$ and $-m_0(L)$ at $x = 0$ and $x = L$, respectively, as shown in Figure 1(b). Similarly, if $m(x)$ is a partially distributed moment in (x_1, x_2) is equivalent to a distributed force in this interval (x_1, x_2) and two concentrated forces at $x = 0 = x_1$ and $x = L = x_2$.

$$Q_x' = \left(\frac{\partial M_{xy}}{\partial y}\right)_{x=a} \tag{14}$$

Corner condition: Timoshenko and Woinowky-Krieger (1959) [1] were unable to explain the corner condition mathematically, although they mentioned it physically as below:

- Corner conditions definition—physically: The polygonal loaded plates will usually produce concentrated reaction at corner points with the distributed reaction along the edges.
- Corner conditions definition—mathematically: In classical plate theory Q_x' Equation (14) consists of a system of distributed forces and two concentrated forces as the corner points.

Note: Corner conditions phenomenon does not appear in sheer deformation theo- ries.

4. REPRESENTING POINT LOADS AND MOMENTS THROUGH JUMP DISCONTINUITIES, DEFLECTION, AND FLEXURAL STIFFNESS USING THE EULER-BERNOULLI BEAM THEORY

The classical method of solving the differential Equation of Euler-Bernoulli beam with jump discontinuity in slope, deflection and flexural stiffness, is to solve the problem on both sides of the discontinuities and then apply boundary and continuity conditions. Here we will solve a problem of differential Equation in the space of generalized functions we solve the problem as a single beam using generalized functions therefore we will consider only one point of jump discontinuity and then generalize this idea with n singular points of an Euler-Bernoulli beam.

Euler-Bernoulli beam theory provides the following displacement field assumptions:

$$u_1(x,y,z) = -z\frac{\mathrm{d}w(x)}{\mathrm{d}x}$$

$$u_2(x,y,z) = 0$$

$$u_3(x,y,z) = w(x) \tag{15}$$

where u_1, u_2, u_3 are displacement components along the x, y, and z axes respectively. The beam lies along the x-axis and the loads are applied vertically along the z-axis. We can use Equation (15) and the foundation of virtual work, the governing equilibrium Equation can be declared as:

$$q(x) = \frac{\mathrm{d}^2}{\mathrm{d}x^2}\left(EI\frac{\mathrm{d}^2 w}{\mathrm{d}x^2}\right)$$

$$\frac{q(x)}{EI} = \frac{\mathrm{d}^2}{\mathrm{d}x^2}\left(\frac{\mathrm{d}^2 w}{\mathrm{d}x^2}\right) = \frac{\mathrm{d}^4 w}{\mathrm{d}x^4}$$

$$\frac{q(x)}{EI} = \frac{\mathrm{d}^4 w}{\mathrm{d}x^4} \tag{16}$$

where EI is the flexural stiffness and q is a distributed force and is called the loading function.

The first three derivatives of w are continuous and the fourth derivative is piecewise continuous, only when q is a piecewise continuous function. On the other hand, there are some equivalent distributed conditions for which the loading function cannot be declared as a classical function. A general case of these conditions were studied in Section 2. However, displacement of the beam or its derivatives can sometimes have discontinuities that are separate from the loading condition. That introduces the focus of this section.

The beam shown in Figure 2 is of length L and the boundary conditions are arbitrary at $x=0,L$. The flexural stiffness of the beam is changed periodically at $x=x_0$. This point will also house discontinuities in slope and deflection. The most general case uses a combination of an internal hinge with a rotational spring along with a shear-free connection with a translational spring. The constants of the rotational and translation springs are K_r and K_t, respectively. Now we take

$$w\left(x_0^+\right)-w\left(x_0^-\right)=\Delta \tag{17}$$

$$\frac{\mathrm{d}}{\mathrm{d}x}\left[w\left(x_0^+\right)-w\left(x_0^-\right)\right]=\frac{\mathrm{d}}{\mathrm{d}x}(\Delta)=\theta \tag{18}$$

The beam has two components to it, segments AB and BC. Hence, why the Heaviside's function is applied.

$$w(x)=w_1(x)+\left[w_2(x)-w_1(x)\right]H(x-x_0) \tag{19}$$

where w is the deflection of the beam, w_1 is the deflection of the segment AB, and w_2 is the deflection of the segment BC. After the calculation given in the appendix, the governing differential Equation of the beam is

$$\frac{\overline{\mathrm{d}w}}{\mathrm{d}x}=\frac{\mathrm{d}w_1}{\mathrm{d}x}+\left(\frac{\mathrm{d}w_2}{\mathrm{d}x}-\frac{\mathrm{d}w_2}{\mathrm{d}x}\right)H(x-x_0)+\Delta\delta(x-x_0)$$

$$\frac{\overline{\mathrm{d}^2w}}{\mathrm{d}x^2}=\frac{\mathrm{d}^2w_1}{\mathrm{d}x^2}+\left(\frac{\mathrm{d}^2w_2}{\mathrm{d}x^2}-\frac{\mathrm{d}^2w_2}{\mathrm{d}x^2}\right)H(x-x_0)+\theta\delta(x-x_0)+\Delta\delta^{(1)}(x-x_0)$$

$$\frac{\overline{\mathrm{d}^3w}}{\mathrm{d}x^3}=\frac{\mathrm{d}^3w_1}{\mathrm{d}x^3}+\left(\frac{\mathrm{d}^3w_2}{\mathrm{d}x^3}-\frac{\mathrm{d}^3w_2}{\mathrm{d}x^3}\right)H(x-x_0)+\left(\frac{\mathrm{d}^2w_2}{\mathrm{d}x^2}-\frac{\mathrm{d}^2w_2}{\mathrm{d}x^2}\right)\delta(x-x_0)$$
$$+\theta\delta^{(1)}(x-x_0)+\Delta\delta^{(2)}(x-x_0)$$

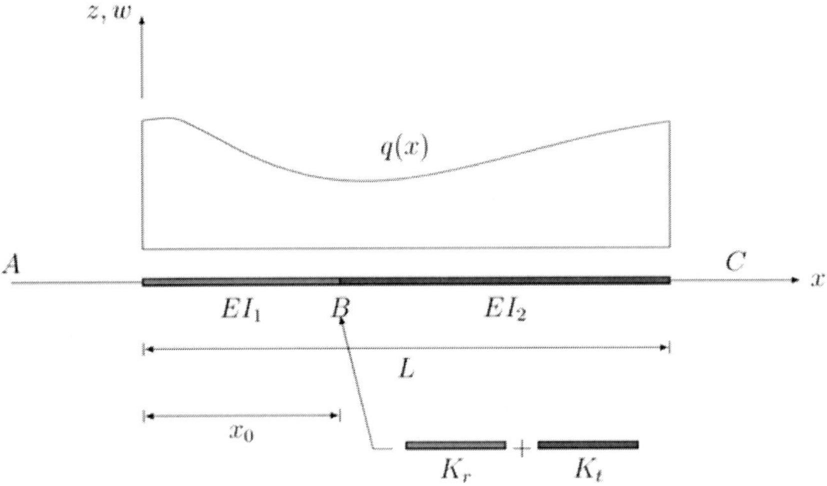

Figure 2. A beam with a jump discontinuity in slope, deflection, and flexural stiffness with arbitrary boundary conditions under a distributed force.

$$\frac{\bar{d}^4 w}{dx^4} = \frac{q(x)}{EI} + \frac{q(x)}{EI}\left(\frac{1}{\alpha}-1\right)H(x-x_0) + \frac{K_t\Delta}{EI}\left(\frac{1}{\alpha}-1\right)\delta(x-x_0)$$
$$+\frac{K_r\theta}{EI}\left(\frac{1}{\alpha}-1\right)\delta^{(1)}(x-x_0) + \theta\delta^{(2)}(x-x_0) + \Delta\delta^{(3)}(x-x_0) \tag{20}$$

where the distributional differentiation is denoted by the bar. As can be seen having jump discontinuities in slope and deflection is equivalent to having double and triple point moments $M^2 = 2\theta$, $M^3 = 6\Delta$ at the point of jump discontinuities. Also, continuity conditions can be written as

$$EI\frac{d^2 w(x_0^-)}{dx^2} = K_r\theta, \quad EI\frac{d^3 w(x_0^-)}{dx^3} = K_t\Delta \tag{21}$$

when you apply the four boundary conditions at $x = 0, L$ as well as the continuity conditions from Equation (21), you are able to gather the deflection w. As can be seen only the force term changes; the form of the operator of the differential Equation is the same as that of Equation (16).

4.1. Solution Procedure

Kanwal [5] proposed a method, that we are about to use, in order to solve a differential Equation in the space of a generalized function. The general solution is

$$w(x) = w_h(x) + w_p(x) \tag{22}$$

w_h and w_p are solutions to the following differential Equations

$$\frac{\mathrm{d}^4 w_h}{\mathrm{d}x^4} = \frac{q(x)}{EI} \tag{23}$$

$$\frac{\bar{\mathrm{d}}^4 w_h}{\mathrm{d}x^4} = \frac{q(x)}{EI}\left(\frac{1}{\alpha}-1\right)H(x-x_0) + \frac{K_t\Delta}{EI}\left(\frac{1}{\alpha}-1\right)\delta(x-x_0) + \frac{K_r\theta}{EI}\left(\frac{1}{\alpha}\right)\delta^{(1)}(x-x_0) + \theta\delta^{(2)}(x-x_0) + \Delta\delta^{(3)}(x-x_0) \tag{24}$$

when finding w_p, we assume that

$$w_p(x) = W(x)H(x-x_0) \tag{25}$$

Hence,

$$\frac{\bar{\mathrm{d}}^4 w_p}{\mathrm{d}x^4} = \frac{\mathrm{d}^4 W(x)}{\mathrm{d}x^4}H(x-x_0) + \frac{\mathrm{d}^3 W(x_0)}{\mathrm{d}x^3}H(x-x_0) + \frac{\mathrm{d}^2 W(x_0)}{\mathrm{d}x^2}\delta^{(1)}(x-x_0) + \frac{\mathrm{d}W(x_0)}{\mathrm{d}x}\delta^{(2)}(x-x_0) + W(x_0)\delta^{(3)}(x-x_0) \tag{26}$$

Equating the coefficient of the generalized functions in Equation (24) and Equation (26), we are able to obtain

$$\frac{\mathrm{d}^4 W(x)}{\mathrm{d}x^4} = \frac{q(x)}{EI}\left(\frac{1}{\alpha}-1\right) \tag{27}$$

$$\frac{d^3W(x_0)}{dx^3}=\frac{K_t\Delta}{EI}\left(\frac{1}{\alpha}-1\right),\quad \frac{d^2W(x_0)}{dx^2}=\frac{K_r\Delta}{EI}\left(\frac{1}{\alpha}-1\right) \tag{28}$$

$$\frac{dW(x_0)}{dx}=\theta,\quad W(x_0)=\Delta \tag{29}$$

After solving Equation (27) and applying the initial conditions (29), we are able to obtain

$$W=W(x,\Delta,\theta) \tag{30}$$

Solving Equation (23) for w_h, we have four integration constants. Applying the four boundary conditions at $x=0,L$ for $w=w_h+w_p$ and the continuity conditions (21), we obtain the beam deflection. Obviously, this is not an efficient method and has no superiority over the classical method. A more efficient method is proposed here for calculating the beam deflection.

4.2. Auxiliary Beam Method

Suppose w represents the deflection of an Euler-Bernoulli beam with jump discontinuities in slope, deflection, and flexural stiffness at the point $x=x_0$. The deflection is defined as:

$$\bar{w}=w(x)-\Delta H(x-x_0)-\theta(x-x_0)H(x-x_0)-\frac{K_r\theta}{2EI}\left(\frac{1}{\alpha}-1\right)(x-x_0)^2H(x-x_0)$$
$$-\frac{K_t\Delta}{6EI}\left(\frac{1}{\alpha}\right)(x-x_0)^3H(x-x_0) \tag{31}$$

$w(x)$ is the classical function. Using Equation (31) in Equation (20) allows us to obtain

$$\frac{d^4\bar{w}(x)}{dx^4}=\frac{q(x)}{EI}+\frac{q(x)}{EI}\left(\frac{1}{\alpha}\right)H(x-x_0) \tag{32}$$

We also have, from Equation (31)

$$\overline{w}(0) = w(0),$$

$$\overline{w}(L) = w(L) - \Delta - \theta(L - x_0) - \frac{K_r\theta}{2EI}\left(\frac{1}{\alpha} - 1\right)(L - x_0)^2 - \frac{K_t\Delta}{2EI}\left(\frac{1}{\alpha} - 1\right)(L - x_0)^3 \tag{33}$$

$$\frac{d\overline{w}(0)}{dx} = \frac{dw(0)}{dx},$$

$$\frac{d\overline{w}(L)}{dx} = \frac{dw(L)}{dx} - \theta - \frac{K_r\theta}{EI}\left(\frac{1}{\alpha} - 1\right)(L - x_0) - \frac{K_t\Delta}{EI}\left(\frac{1}{\alpha} - 1\right)(L - x_0)^2 \tag{34}$$

$$\frac{d\overline{w}^2(0)}{dx^2} = \frac{d^2w(0)}{dx^2},$$

$$\frac{d\overline{w}^2(L)}{dx^2} = \frac{d^2w(L)}{dx^2} - \frac{K_r\theta}{EI}\left(\frac{1}{\alpha} - 1\right) - \frac{K_t\Delta}{EI}\left(\frac{1}{\alpha} - 1\right)(L - x_0) \tag{35}$$

The continuity conditions for the auxiliary beam are:

$$\frac{d^2\overline{w}(x_0^-)}{dx^2} = \frac{d^2w(x_0^-)}{dx^2} = \frac{K_r\theta}{EI}, \quad \frac{d^3\overline{w}(x_0^-)}{dx^3} = \frac{d^3w(x_0^-)}{dx^3} = \frac{K_t\Delta}{EI} \tag{36}$$

Therefore, instead of solving two differential Equations for the two beam segments and applying eight boundary and continuity conditions, only one differential Equation with six boundary and continuity Equations is solved. To clarify the method, three examples are solved in the next section.

Three examples are presented and solved in order to show the efficiency of The Euler-Bernoulli Beam Theory with jump discontinuities.

Example 1. This is an example of a internal hinged beam under a uniform distributed force. The beam shown in Figure 3 is clamped at $x = 0$ and simply supported at $x = L$.

The flexural stiffness is constant.

$$q(x) = -q_0, \quad \alpha = 1, \quad K_r = K_t = 0, \quad \Delta = 0 \tag{37}$$

From Equation (32), the G.D.E of the auxiliary beam is

$$\frac{d^4\overline{w}}{dx^4} = -\frac{q_0}{EI} \tag{38}$$

$$\frac{d^3\overline{w}}{dx^3} = -\frac{q_0}{EI}x + a_3 \tag{39}$$

$$\frac{d^2\overline{w}}{dx^2} = -\frac{q_0}{EI}\frac{x^2}{2} + a_3 x + a_2 \tag{40}$$

$$\frac{d\overline{w}}{dx} = -\frac{q_0}{EI}\frac{x^3}{6} + a_3\frac{x}{2} + a_2 x + a_1 \tag{41}$$

Lastly,

$$\overline{w} = -\frac{q_0}{24EI}x^4 + a_3 x^3 + a_2 x^2 + a_1 x + a_0 \tag{42}$$

It is known that for this beam,

$$w(0) = \frac{dw(0)}{dx} = w(L) = \frac{d^2 w(L)}{dx^2} = 0 \tag{43}$$

Using Equation (33) through Equation (35) we are able to find,

$$\overline{w}(0) = \frac{dw(0)}{dx} = 0, \quad \overline{w}(L) = (\lambda - 1)\theta L, \quad \frac{d^2 w(L)}{dx^2} = 0, \quad \frac{d^2 w(\lambda L)}{dx^2} = 0 \tag{44}$$

Using Equation (27) and Equation (29) we are able to obtain a_0, a_1, a_2 and a_3

$$a_0 = a_1 = 0, \quad a_2 = \frac{-q_0}{16EI} L^2 + \frac{3}{2} \frac{(\lambda - 1)\theta}{L}, \quad a_3 = \frac{q_0}{8EI} L\left(1 - \frac{1}{6}\right) - \frac{(\lambda - 1)\theta}{2L^2} \tag{45}$$

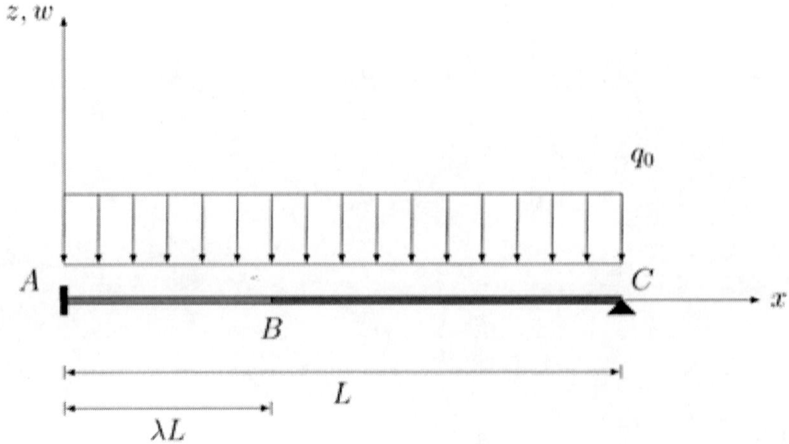

Figure 3. A clamped, simply supported beama with an internal hinge under a uniform distri- buted force.

Now, using Equation (42) and Equation (44) we obtain

$$\bar{w}(x) = -\frac{q_0}{24EI}\left[x^4 - 2(\lambda + 1) L x^3 + 6\lambda L^2\right] \tag{46}$$

We also find that

$$\theta = \frac{(4\lambda - 1) q_0 L^3}{24(1 - \lambda) EI} \tag{47}$$

Therefore, from Equation (31) we can derive

$$w(x) = -\frac{q_0}{24EI}\left[x^4 - 2(\lambda + 1) L x^3 + 6\lambda L^2 x^2 - \frac{(4\lambda - 1)}{(1 - \lambda)} L^3 (x - \lambda L) H(x - \lambda L)\right] \tag{48}$$

Example 2.

This example uses a beam that has a jump discontinuity at $x = \lambda L$, shown in Figure 4.

For this beam we know,

$$q(x) = -\frac{q_0}{L}x, \quad \alpha = 1, \quad K_r = K_t = 0, \quad \theta = 0 \tag{49}$$

Using Equation (49) we can find the G.D.E of the auxiliary beam

$$\frac{d^4\overline{w}}{dx^4} = -\frac{q_0}{EIL}x \tag{50}$$

$$\frac{d^3\overline{w}}{dx^3} = -\frac{q_0}{2EIL}x^2 + a_3 \tag{51}$$

$$\frac{d^2\overline{w}}{dx^2} = -\frac{q_0}{6EIL}x^3 + a_3x + a_2 \tag{52}$$

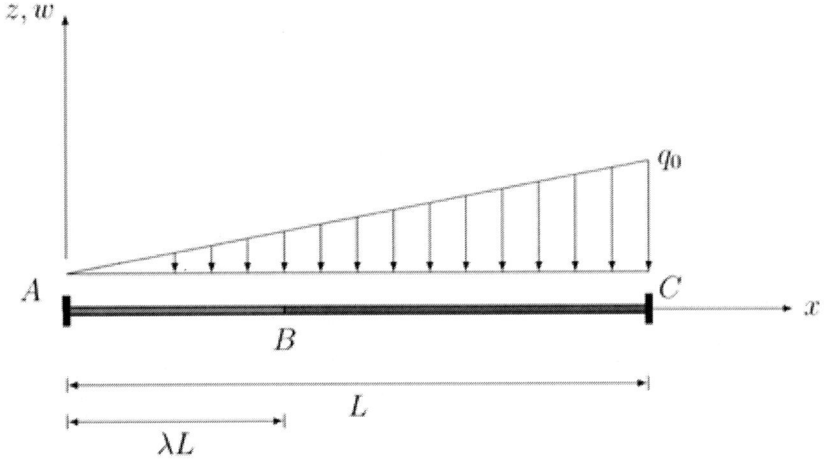

Figure 4. A double clamped beam with an internal shear-free connection under a linearly vary- ing distributed force.

$$\frac{\mathrm{d}\overline{w}}{\mathrm{d}x}=-\frac{q_0}{24EIL}x^4+a_3\frac{x^2}{2}+a_2x+a_1 \tag{53}$$

Thus,

$$\overline{w}(x)=-\frac{q_0}{120EIL}x^5+a_3x^3+a_2x^2+a_1x+a_0 \tag{54}$$

We know the following is true for this beam,

$$w(0)=\frac{\mathrm{d}w(0)}{\mathrm{d}x}=w(L)=\frac{\mathrm{d}w(L)}{\mathrm{d}x}=0 \tag{55}$$

$$\frac{\mathrm{d}^2w(x_0^-)}{\mathrm{d}x^2}=\frac{\mathrm{d}^3w(x_0^-)}{\mathrm{d}x^3} \tag{56}$$

Now, using Equation (33) through Equation (35) we have

$$\overline{w}(0)=\frac{\mathrm{d}\overline{w}(0)}{\mathrm{d}x}=0,\quad \overline{w}(L)=w(L)-\Delta=-\Delta,\quad \frac{\mathrm{d}\overline{w}(L)}{\mathrm{d}x}=\frac{\mathrm{d}w(L)}{\mathrm{d}x}=0 \tag{57}$$

$$\frac{\mathrm{d}^3\overline{w}(xL)}{\mathrm{d}x^3}=0 \tag{58}$$

Now using Equation (54), Equation (57) and Equation (58), we obtain

$$a_0=\frac{q_0}{24EIL}x^2\left[\frac{-14}{5}x^3+4x^2-8\right]$$

$$a_2=\frac{-q_0}{3EIL}x^2$$

$$a_3=\frac{q_0}{2EIL}x^2$$

In a similar way as Example 1, we can find $\bar{w}(x)$ to be

$$\bar{w}(x) = -\frac{q_0 x^2}{24EIL}\left[2x^3 - 20\lambda^2 L^2 x - 5L^3\left(1-6\lambda^2\right)\right] \tag{59}$$

$$\Delta = \frac{q_0 L^4\left(10\lambda^2 - 3\right)}{24EI} \tag{60}$$

Now, from Equation (31)

$$w(x) = -\frac{q_0 x^2}{24EIL}\left[2x^3 - 20\lambda^2 L^2 x - 5L^3\left(1-6\lambda^2\right)\right] + \frac{q(0)L^4\left(10\lambda^2 - 3\right)}{24EI} H(x-\lambda L) \tag{61}$$

Example 3. A Simply supported beam under a uniform distributed force with jump discontinuity in flexural stiffness at $x = \lambda L$ is shown in Figure 5.

For this beam

$$\Delta = \theta = 0, \quad \alpha = 2, \quad q(x) - q_0 \tag{62}$$

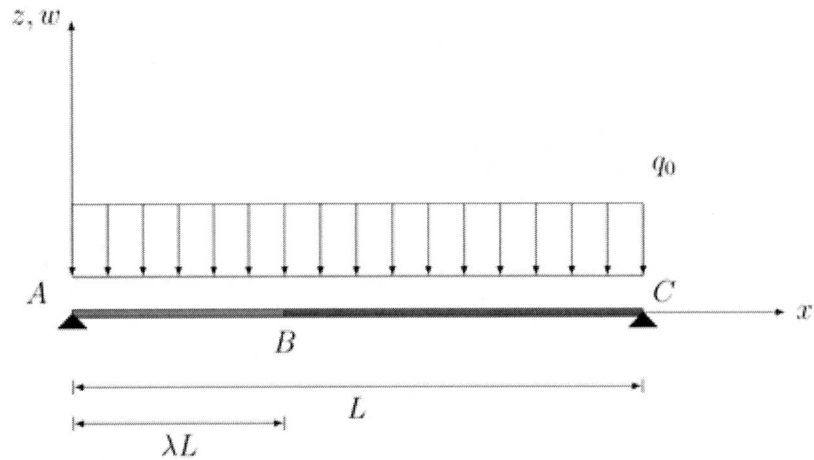

Figure 5. A simply supported beam with a jump discontinuity in flexual stiffness under a uniform distributed force.

and the deflection of the beam is defined as:

$$\bar{w}(x)=w(x)-\frac{M_0}{2EI}\left(\frac{1}{\alpha}-1\right)(x-\lambda L)^2 H(x-\lambda L) -\frac{V_0}{6EI}\left(\frac{1}{\alpha}-1\right)(x-\lambda L)^3 H(x-\lambda L) \tag{63}$$

The governing differential Equation of the beam is

$$\frac{\mathrm{d}^4\bar{w}}{\mathrm{d}x^4}=\frac{-q_0}{EI}+\frac{q_0}{2EI}H(x-\lambda L) \tag{64}$$

The boundary and continuity conditions are

$$\bar{w}(0)=0,\quad \bar{w}(L)=\frac{M_0}{4EI}(l-\lambda)^2 L^2 \tag{65}$$

$$\frac{\mathrm{d}^2\bar{w}(0)}{\mathrm{d}x^2}=0,\quad \frac{\mathrm{d}^2\bar{w}(L)}{\mathrm{d}x^2}=\frac{M_0}{2EI}+\frac{V_0 L}{2EI}(1-\lambda) \tag{66}$$

$$\frac{\mathrm{d}^2\bar{w}(\lambda L)}{\mathrm{d}x^2}=\frac{M_0}{EI},\quad \frac{\mathrm{d}^3\bar{w}(\lambda L)}{\mathrm{d}x^3}=\frac{V_0}{EI} \tag{67}$$

When you combine the boundary and continuity Equations, you will get

$$\bar{w}(0)\frac{\mathrm{d}^2 w(0)}{\mathrm{d}x^2}=0 \tag{68}$$

$$-\bar{w}(L)+\frac{(1-\lambda)^2}{4}\frac{\mathrm{d}^2\bar{w}(\lambda L)}{\mathrm{d}x^3}=0 \tag{69}$$

$$-\frac{\mathrm{d}^2\bar{w}(L)}{\mathrm{d}x^2}+\frac{\mathrm{d}^2\bar{w}(\lambda L)}{2\mathrm{d}x^2}+\frac{L(1-\lambda)}{2}\frac{\mathrm{d}^3\bar{w}(\lambda L)}{\mathrm{d}x^3}=0 \tag{70}$$

From Equation (54) we are able to obtain,

$$\bar{w}(x) = -\frac{q_0 x^4}{24EI} + \frac{q_0 (x-\lambda L)^4}{48EI} H(x-\lambda L) + a_0 + a_1 x + a_2 x^2 + a_3 x^3 \tag{71}$$

When applying the boundary and continuity conditions to Equation (61), we are able to obtain

$$\bar{w}(x) = -\frac{q_0}{48EI}\left[2x^4 - 4Lx^3 + L^3\left[(1-\lambda)^4 - 6\lambda(1-\lambda)^3 + 2\right]x - (x-\lambda L)^4 H(x-\lambda L)\right] \tag{72}$$

From Equation (57) we are able to obtain,

$$M_0 = \frac{\lambda(1-\lambda)}{2} q_0 L^2, \quad V_0 = \left(\frac{1}{2}\right) q_0 L \tag{73}$$

Therefore, from Equation (64) we obtain

$$w(x) = -\frac{q_0}{48EI} 2x^4 - 4Lx^3 + L^3\left[(1-\lambda)^4 - 6\lambda(1-\lambda)^3 + 2\right]x + \left[-(x-\lambda L)^4 + 6L^2\lambda(1-\lambda)(x-\lambda L) + 4L\left(\frac{1}{2}-\lambda\right)(x-\lambda L)^3\right] H(x-\lambda L) \tag{74}$$

Assuming $\lambda = 1$ gives us the deflection of a simply-supported beam with a constant flexural stiffness EI under a uniform distributed force

$$w(x) = -\frac{q_0}{24EI} x^4 + \frac{q_0 L}{12EI} x^3 + \frac{q_0 L^3}{24EI} x \tag{75}$$

For n point loads (moment or force) one has to use $(n+1)$ differential Equation and has to apply $(n+1)$ boundary and continuity conditions. By using Macauly's bracket we have only one expression for the bend moment and loading function using the singularity function matter. In this case only one differential Equation with four boundary conditions are required to be solved.

In the case of n jump discontinuities, if one uses the auxiliary beam methods, shown in this article, instead of solving $(n+1)$ differential Equations and applying $4(n+1)$ boundary and continuity conditions need to be solved. In almost all practical problems, we do not have all three kinds of discontinuities at the same point (ie if a beam has n internal hinges, then the number of continuity Equations is reduced to n).

In Section 3 for finding the governing differential Equation of an Euler-Bernoulli beam with jump discontinuities, the beam was partitioned to continuous beam segments. The next section will use the same ideas in order to find the equivalent distributed forces for point forces and point moments.

4.3. Equivalent Force Function for Concentrated Force and Moment: A Nonclasical Approach

Here we will use that a concentrated force and a concerntrated moment represnt jump discontinities into shearing orce (the third derivative of the beam deflection) and the bending moment (the second derivative of the beam deflection) respectively, of an Euler-Bernoullie beam. As mentioned earlier, in Section 2, the classical proof of Equation (6) and Equation (7) is based on considering the singular loading condition as a distributed force over a very short length of the beam. The concentrated force and the concentrated moment introduced jump discontinuities into the sheering force, third derivative of the Equation, and the bending moment, second derivative of the Equation, of an Euler-Bernoulli beam. In this section we will study a non classical approach for a concentrated force of magnitude P, Equation (6) and a concentrated moment of M Equation (7). Figure 6 shows a beam with a concentrated force P_0 and a concentrated moment M_0 applied at $x = x_0$. The beam AC may be assumed to be composed of two beam segments, AB and BC. The deflections of the two beam segments AB and BC are denoted by w_1 and w_2 respectively. There is no loading for $0 < x < x_0$ and $x_0 < x < L$; hence, we have

$$\frac{d^4 w_1}{dx^4} = 0; \quad x \in (0, x_0) \tag{76}$$

$$\frac{d^4 w_2}{dx^4} = 0; \quad x \in (x_0, L) \tag{77}$$

w is the deflection of the beam and can be written as

$$w(x) = w_1(x) + [w_2(x) - w_1(x)]H(x - x_0) \tag{78}$$

We know that the magnitudes of w_1 and w_2 are equal at $x = x_0$, the same is true for their first derivatives. Therefore,

$$\frac{d^2w(x)}{dx^2} = \frac{d^2w_1(x)}{dx^2} + \left[\frac{d^2w_2(x)}{dx^2} - \frac{d^2w_1(x)}{dx^2}\right]H(x - x_0) \tag{79}$$

From the figure above, we can write

$$V_2(x_0^+) - V_1(x_0^-) = P_0$$

$$M_2(x_0^+) - M_1(x_0^-) = M_0 \tag{80}$$

Then,

$$\frac{d^3w_2(x_0^+)}{dx^3} - \frac{d^3w_1(x_0^-)}{dx^3} = \frac{P_0}{EI}$$

$$\frac{d^2w_2(x_0^+)}{dx^2} - \frac{d^2w_1(x_0^-)}{dx^2} = \frac{M_0}{EI} \tag{81}$$

If you differentiate both sides from Equation (82) with respect to x, you get

$$\frac{d^3w(x)}{dx^3} = \frac{d^3w_1(x)}{dx^3} + \left[\frac{d^3w_2(x)}{dx^3} - \frac{d^3w_1(x)}{dx^3}\right]H(x - x_0) + \frac{M_0}{EI}\delta(x - x_0) \tag{82}$$

also

$$\frac{\bar{\mathrm{d}}^4 w(x)}{\mathrm{d}x^4} = \frac{\mathrm{d}^4 w_1(x)}{\mathrm{d}x^4} + \left[\frac{\mathrm{d}^4 w_2(x)}{\mathrm{d}x^4} - \frac{\mathrm{d}^4 w_1(x)}{\mathrm{d}x^4}\right] H(x - x_0)$$
$$+ \frac{M_0}{EI}\delta(x - x_0) + \frac{P_0}{EI}\delta(x - x_0) \tag{83}$$

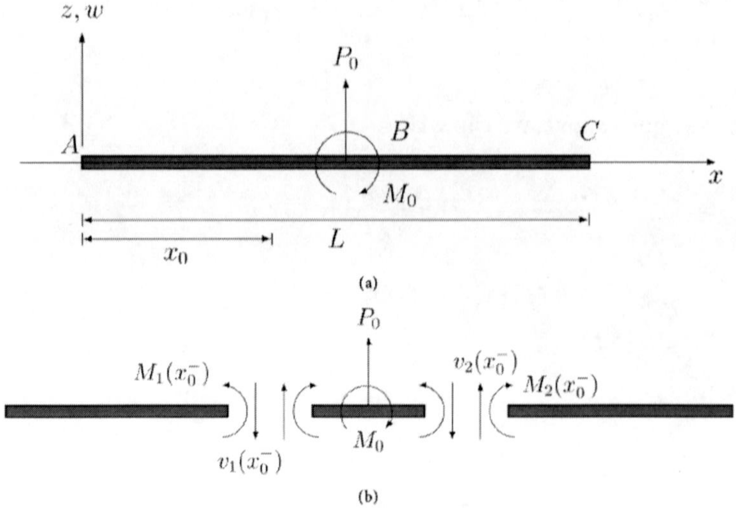

Figure 6. (a) A beam under a concentrated force and a concentrated moment. (b) Moment and shear discontinuity at the point of the action of concentrated loads.

Now, from Equation (76), Equation (77) and Equation (86) we obtain

$$\frac{\bar{\mathrm{d}}^4 w(x)}{\mathrm{d}x^4} = \frac{P_0\delta(x - x_0) + M_0\delta'(x - x_0)}{EI} \tag{84}$$

Therefore, the equivalent force function is

$$q(x) = q_P(x) + q_M(x) = P_0\delta(x - x_0) + M_0\delta(x - x_0) \tag{85}$$

5. TIMOSHENKO BEAM WITH JUMP DISCONTINUITIES

Much like the work introduced on Euler-Bernoulli beams; we can also acquire jump discontinuities in slope, deflection, flexural stiffness, and shear stiffness to a system of differential Equations of a Timoshenko beam [6] . Differentiation between Euler- Bernoulli beams and Timoshenko beams are the distant shear deformation in the Timoshenko beams. Shear deformation is where a force is being applied on one part and another force is being applied on another part, but in the opposite direction. Along the x, y, and z-axes, are displacement components u_1, u_2, and u_3. The displacement field is provided by the Equations

$$u_1(x,y,z) = z\phi(x)$$

$$u_1(x,y,z) = 0$$

$$u_1(x,y,z) = w^T(x) \tag{86}$$

u_1, u_2, and u_3 are displacement components in the (x,y,x) plane. Also, ϕ is the rotation of the Timoshenko beam about the y-axis and the superscript T shows the deflection of the Timoshenko beam. The Governing system of differential Equations can be written as,

$$\frac{d}{dx}\left(EI\frac{d\phi}{dx}\right) - GA'\left(\phi + \frac{dw}{dx}\right) = 0$$

$$\frac{d}{dx}\left[GA'\left(\phi + \frac{dw}{dx}\right)\right] + q(x) = 0 \tag{87}$$

the shear modulus, G, is a distinguished ratio of the shear stress over the shear strain. The shear stress is the force applied to a certain amount of area it is applied to. The shear strain on the other hand, is the rate of change in displacement of the strain combine with $A' = K_s A$,

$$\Omega = \frac{GA'}{EI} \tag{88}$$

the ratio of shear and flexural stiffness has now been defined.

We can substitute Equation (88) into Equation (87) yields,

$$\Omega\frac{dw^T}{dx}-\frac{d^2\phi}{dx^2}+\Omega\phi=0$$

$$\frac{d\phi}{dx}+\frac{d^2w^T}{dx^2}+\frac{q(x)}{GA'}=0 \tag{89}$$

Next, ponder a Timoshenko beam has jump discontinuities in the slope, deflection, shear stiffness, and flexural stiffness at $x=x_0$ on its length L. We refer to the heavy- side function as used for the Euler-Bernoulli beam to initiate the two Equations of deflection and rotation of the Timoshenko beam, we can write

$$w^T(x)=w_1^T(x)+\left[w_2^T(x)-w_1^T(x)\right]H(x-x_0)$$

$$\phi(x)=\phi_1(x)+\left[\phi_2(x)-\phi_1(x)\right]H(x-x_0) \tag{90}$$

Because deflection and rotation both have jump discontinuities at $x=x_0$, we write

$$w_2^T(x_0)-w_1^T(x_0)=\Delta^T,\quad \phi_2(x_0)-\phi_1(x_0)=\theta^T \tag{91}$$

It is known that

$$M_1(x_0)=EI_1\frac{d\phi_1(x_0)}{dx},\quad M_2(x_0)=EI_2\frac{d\phi_2(x_0)}{dx} \tag{92}$$

and

$$V_1(x_0)=GA'\left[\phi_1(x_0)+\frac{dw_1(x_0)}{dx}\right],\quad V_2(x_0)=GA'\left[\phi_2(x_0)+\frac{dw_2(x_0)}{dx}\right] \tag{93}$$

Also, for an infinitesimal element including the discontinuity point at equilibrium implies

$$M_1(x_0)=M_2(x_2)=K_r\theta^T$$

$$V_1(x_0)=V_2(x_2)=K_t\Delta^T \tag{94}$$

As we can see from Equation (94), $K_t$and K_r are the stiffness of the translational and rotational springs at $x=x_0$. So after reviewing and comparing Equation (92), Equation (93) and Equation (94),

$$\frac{d\phi_2(x_0)}{dx}-\frac{d\phi_1(x_0)}{dx}=\frac{K_r\theta^T}{EI}\left(\frac{1}{\alpha}-1\right)$$

$$\frac{dw_2(x_0)}{dx}-\frac{dw_1(x_0)}{dx}=\frac{K_t\Delta^T}{EI}\left(\frac{1}{\beta}-1\right)-\theta^T \tag{95}$$

where

$$EI_1=EI,\quad EI_2=\alpha EI$$

$$GA_1'=GA,\quad GA_2'=\beta GA' \tag{96}$$

Differentiating Equation (90), we obtain

$$\frac{\bar{d}w^T}{dx}=\frac{dw_1^T}{dx}+\left(\frac{dw_2^t}{dx}-\frac{dw_1^t}{dx}\right)H(x-x_0)+\Delta^T\delta(x-x_0) \tag{97}$$

$$\frac{\bar{d}^2w^T}{dx^2}=\frac{d^2w_1^T}{dx^2}+\left(\frac{d^2w_2^t}{dx^2}-\frac{d^2w_1^t}{dx^2}\right)H(x-x_0)+\left[\frac{K_t\Delta^T}{GA'}\left(\frac{1}{\beta}-1\right)-\theta^T\right]+\Delta^T\delta^{(1)}(x-x_0) \tag{98}$$

and

$$\frac{\bar{d}\phi}{dx}=\frac{d\phi_1}{dx}+\left(\frac{d\phi_2}{dx}-\frac{d\phi_1}{dx}\right)H(x-x_0)+\theta^T\delta(x-x_0) \tag{99}$$

$$\frac{\overline{d}^2\phi}{dx^2}=\frac{d^2\phi_1}{dx^2}+\left(\frac{d^2\phi_2}{dx^2}-\frac{d^2\phi_1}{dx^2}\right)H(x-x_0)+\frac{K_r\theta^T}{EI}\left(\frac{1}{\alpha}-1\right)+\theta^T\delta^{(1)}(x-x_0) \tag{100}$$

The deflection and rotation of each beam segment have continuous derivatives and hence they are governed by Equation (84), thus

$$\Omega_1\frac{dw_1^T}{dx}-\frac{d^2\phi_1}{dx^2}+\Omega_1\phi_1=0 \tag{101}$$

$$\frac{d\phi_1}{dx}+\frac{d^2w_1^T}{dx^2}+\frac{q(x)}{GA'}=0 \tag{102}$$

and

$$\Omega_2\frac{dw_2^T}{dx}-\frac{d^2\phi_2}{dx^2}+\Omega_2\phi_2=0 \tag{103}$$

$$\frac{d\phi_2}{dx}+\frac{d^2w_2^T}{dx^2}+\frac{q(x)}{\beta GA'}=0 \tag{104}$$

Now if we let $\Omega_1=\Omega$ and $\Omega_2=\frac{\beta}{\alpha}\Omega$ we can obtain the governing system of equi-

librium Equations for the beam from Equation (97) through Equation (104),

$$\Omega\frac{\overline{d}w^T}{dx}-\frac{\overline{d}^2\phi}{dx^2}+\Omega\phi+\left(\frac{\beta}{\alpha}-1\right)\Omega\frac{dw_2}{dx}H(x-x_0)+\left(\frac{\beta}{\alpha}-1\right)\Omega\phi_2H(x-x_0)$$
$$=\left[\Omega\Delta^T-\frac{K_r\theta^T}{EI}\left(\frac{1}{\alpha}-1\right)\right]\delta(x-x_0)-\theta^T\delta^{(1)}(x-x_0) \tag{105}$$

5.1. Auxiliary Beam Method

This section shows the similarities between the Euler-Bernoulli method and the Timoshenko methods by providing a Timoshenko method example. You can compare it to the Euler-Bernoulli method from earlier. The auxiliary beam is defined for a Timoshenko beam with internal jump discontinuities. The deflection and rotation of the beam are defined below:

$$\bar{w}^T(x) = w^T(x) - \Delta^T H(x-x_0) - \left[\frac{K_t\Delta^T\left(\frac{1}{\beta}-1\right)}{GA'} - \theta^T\right](x-x_0)H(x-x_0)$$

$$\bar{\phi}^T(x) = \phi^T(x) - \theta^T H(x-x_0) - \left[\frac{K_r\theta^T\left(\frac{1}{\alpha}-1\right)}{GA'} - \theta^T\right](x-x_0)H(x-x_0) \tag{106}$$

Substitution Equation (106) into Equation (105) yields,

$$\Omega\frac{\mathrm{d}\bar{w}^T}{\mathrm{d}x} - \frac{\mathrm{d}^2\bar{\phi}}{\mathrm{d}x^2} + \Omega\bar{\phi} = \left[\frac{K_T\Delta^T}{GA'}\left(\frac{1}{\beta}-1\right) - \frac{K_r\Delta^T}{GA'}\left(\frac{1}{\alpha}-1\right)(x-x_0)\right]H(x-x_0)$$

$$\frac{\mathrm{d}^2\bar{w}^T}{\mathrm{d}x^2} + \frac{\mathrm{d}\bar{\phi}}{\mathrm{d}x} + \frac{q(x)}{GA'} + \frac{q(x)}{GA'}\left(\frac{1}{\beta}-1\right)H(x-x_0) = \frac{-K_r\theta^T}{EI}\left(\frac{1}{\alpha}-1\right)H(x-x_0) \tag{107}$$

The boundary conditions for this beam can be found by using the relations shown below:

$$\bar{w}^T(0) = w^T(0), \quad \bar{w}^T(L) = w^T(L) - \Delta^T - \left[\frac{K_t\Delta^T}{GA'}\left(\frac{1}{\beta}-1\right) - \theta^T\right](L-x_0)$$

$$\frac{\mathrm{d}\bar{w}^T(0)}{\mathrm{d}x} = \frac{\mathrm{d}w^T(0)}{\mathrm{d}x}, \quad \frac{\mathrm{d}\bar{w}^T(L)}{\mathrm{d}x} = \frac{\mathrm{d}w^T(L)}{\mathrm{d}x} - \left[\frac{K_t\Delta^T}{GA'}\left(\frac{1}{\beta}-1\right) - \theta^T\right] \tag{108}$$

and

$$\bar{\phi}(0) = \phi(0), \quad \bar{\phi}(L) = \phi(L) - \Theta^T - \left[\frac{K_r\theta^T}{GA'}\left(\frac{1}{\alpha}-1\right) - \theta^T\right](L-x_0)$$

$$\frac{\mathrm{d}\bar{\phi}(0)}{\mathrm{d}x}=\frac{\mathrm{d}\phi(0)}{\mathrm{d}x},\quad \frac{\mathrm{d}\bar{\phi}(L)}{\mathrm{d}x}=\frac{\mathrm{d}\phi^{T}(L)}{\mathrm{d}x}-\left[\frac{K_r\theta^{T}}{GA'}\left(\frac{1}{\alpha}-1\right)-\theta^{T}\right] \tag{109}$$

The continuity condition may be expressed as

$$\frac{\mathrm{d}\bar{\phi}(x_0)}{\mathrm{d}x}=\frac{K_r\theta^{T}}{EI},\quad \frac{\mathrm{d}\bar{w}(x_0)}{\mathrm{d}x}+\bar{\phi}(x_0)=\frac{K_t\Delta^{T}}{GA'} \tag{110}$$

The next section will provide an example to clarify the method shown.

5.2. Timoshenko Beam Example

Suppose the beam used in Example 1 is now a Timoshenko beam, from Equation (107) and Equation (37) we obtain,

$$\Omega\frac{\mathrm{d}\bar{w}}{\mathrm{d}x}-\frac{\mathrm{d}^2\phi}{\mathrm{d}x^2}+\Omega\phi=0 \tag{111}$$

$$\frac{\mathrm{d}^2\bar{w}}{\mathrm{d}x^2}-\frac{\mathrm{d}\phi}{\mathrm{d}x}+\frac{q_0}{GA'}=0 \tag{112}$$

The boundary and continuity conditions for this beam can be written as,

$$\bar{w}^{T}(0)=0,\quad \bar{w}^{T}(L)=\theta^{T}L(1-\lambda),\quad \bar{\phi}(0)=0,\quad \frac{\mathrm{d}\bar{\phi}(L)}{\mathrm{d}x}=0$$

$$\frac{\mathrm{d}\bar{\phi}(\lambda L)}{\mathrm{d}x}=0 \tag{113}$$

After solving this system and applying the boundary and continuity conditions we obtain

$$\bar{\phi}(x)=\frac{q_0}{12EI}\left[2x^3-3(1+\lambda)Lx^2+6\lambda^2x\right] \tag{114}$$

$$\bar{w}^T(x) = \frac{q_0}{24EI}\left[x^4 - 2(1-\lambda)Lx^3 + 6\lambda L^2x^2 - \frac{12}{\Omega}\left[x^2 - (1+\lambda)Lx\right]\right] \tag{115}$$

$$\theta^T = -\frac{L^3 q_0}{24(1-\lambda)EI}\left(-1 + 4\lambda + \frac{12\lambda}{\Omega}\right) \tag{116}$$

From Equation (106) we obtain

$$w^T(x) = \bar{w}^T(x) - \theta^T(x-\lambda L)H(x-\lambda L) \tag{117}$$

$$\phi(x) = \bar{\phi}(x) + \theta^T H(x-\lambda L) \tag{118}$$

Hence,

$$\begin{aligned} w^T(x) = &-\frac{q_0}{24EI}\left(x^4 - 2(1+\lambda)Lx^3 + 6\lambda^2x^2 - \frac{4\lambda-1}{1-\lambda}L^3(x-\lambda L)H(x-\lambda L)\right) \\ &-\frac{12}{\Omega}\left[x^2 - (1+\lambda)Lx - \frac{\lambda L^3}{1-\lambda}(x-\lambda L)H(x-\lambda L)\right] \end{aligned} \tag{119}$$

$$\phi(x) = \frac{q_0}{24EI}\left[4x^3 - 6(1+\lambda)Lx^2 + 12\lambda L^2x - \frac{L^3}{1-\lambda}\left(-1 + 4\lambda + \frac{12\lambda}{\Omega}\right)H(x-\lambda L)\right] \tag{120}$$

As the shear stiffness approaches infinity the effect of shear deformation diminishes until it disappears entirely. From Equation (47) and Equation (116), it is seen that

$$\lim_{\Omega\to\infty}\theta^T = -\theta \tag{121}$$

6. DIRAC-DELTA FUNCTION IN THE STATIC ANALYSIS OF MULTI-CRACKED EULER-BERNOULLI BEAMS

Structural analysis of multi-cracked beams is of greater engineering interest. Research in this area has been mainly concentrated on two classes of problems:

- definition of appropriate linear and non-linear models for representing the effects of cracks under static and dynamical loadings and
- detection of position and severity of the damage by using either static or dynamic tests.

Here we will study an effective and physically based linear modeling of multi-cracked beams subject to the static loading. This result is useful for treating any type of concentrated damage occurring in slender and short beams, e.g. corrosion of steel bars in reinforced concrete members, defects of material and attacks of biotic agents in timber elements etc.

The idea of treating multi-cracked beams with equivalent linear springs at the crack's position is based on the portion of each member into undamaged pieces between two consecutive cracks. For slender Euler-Bernoulli beam, the governing 4th order differential Equation of bending can be written for each subsystem, but it is necessary to impose the pertinent continuity conditions between adjacent subsystems to obtain the static response of the whole beam.

As a result, the computational effort increases with the number of cracks, i.e for n cracks along the beam, $4(n+1)$ algebraic Equations have to be solved to compute $4(n+1)$ integration constants. But this method is inefficient for identifications purposes, when analysis are repeated until position and severity of the damage are found. So one can use finite element method (FEM) in which stiffness matrix and load vector of the non-cracked Euler-Bernoulli beam are modified with some dimensionless coefficients with effects of internal cracks.

Here we will be using generalized functions to handle static and kinematical discontinuities along the beam. Here we require the enforcement of continuity conditions at each jump, and hence additional integration constants are needed. This issue can be tracked in the formulation of "rigidity modeling", which consists of singularities in the flectual stiffness represented by Dirac's delta functions, which in turn are equivalent to internal hinges with rotational linear-elastic springs.

In this work, a non-trivial generalization to multiple discontinuities in the curvature and the slope functions of the integration procedure is presented. The case of Euler-Bernoulli beams under static loads is treated; discontinuities in the curvature and the slope function are modeled as unit step distributions and Dirac's delta, respectively, in the flectural stiffness of the beam. Moreover, the presented procedure is also extended to cases of discontinuities in the axial displacement and in the vertical deflection modeled as Dirac's deltas in the axial stiffness and the shear stiffness, respectively.

6.1. Solution of Euler-Bernoulli Beam with a Flectural Stiffness Model with Multiple Singularities

The well known static governing Equations of Euler-Bernoulli beam with variable Young modulus $E(x)$ and moment of inertia $I(x)$ are given by

$$\frac{\mathrm{d}v}{\mathrm{d}x}(x)+q(x)=0; \quad \frac{\mathrm{d}}{\mathrm{d}x}M(x)-V(x)=0, \tag{122}$$

$$\chi(x)=\frac{M(x)}{E(x)I(x)}; \tag{123}$$

$$\chi(x)=\frac{\mathrm{d}}{\mathrm{d}x}\phi(x); \quad \phi(x)=-\frac{\mathrm{d}}{\mathrm{d}x}u(x); \tag{124}$$

where $q(x)$ is the external load, $V(x)$ and $M(x)$ are the shear force and the bonding moment, respectively, $u(x)$, $\phi(x)$, and $\chi(x)$ are the deflection, slopes and curvature functions, respectively, and prime denotes differentiation where the spatial coordinate x spanning from 0 to the length L of the beam. Since Equation (124),

$$\phi(x)=-\frac{\mathrm{d}^2}{\mathrm{d}x^2}u(x) \tag{125}$$

By Equation (123),

$$\frac{M(x)}{E(x)I(x)} = -\frac{d^2}{dx^2}u(x)$$

$$m(x) = -E(x)I(x)\frac{d^2}{dx^2}u(x) \tag{126}$$

Also, from Equation (123),

$$\frac{d}{dx}M(x) = V(x)$$

$$\frac{d^2}{dx^2}M(x) = \frac{d}{dx}V(x) \tag{127}$$

and by Equation (122) we get,

$$\frac{d}{dx}V(x) = -q(x) \tag{128}$$

By combining Equation (127) and Equation (128), we get

$$\frac{d^2}{dx^2}m(x) = \frac{d^2}{dx^2}\left[-E(x)I(x)\frac{d^2}{dx^2}u(x)\right] = -q(x)$$

$$\Rightarrow -\frac{d^2}{dx^2}\left[E(x)I(x)\frac{d^2}{dx^2}u(x)\right] = -q(x)$$

$$\Rightarrow \frac{d^2}{dx^2}\left[E(x)I(x)\frac{d^2}{dx^2}u(x)\right] = q(x) \tag{129}$$

The flexural stiffness model with a single singularity described by means of a suitable distribution as follows:

$$E(x)I(x)=E_0I_0\left[1-\sum_{i=1}^{n}\alpha_iD(x-x_{0,i})\right] \tag{130}$$

The same model is reconsidered and extended to the case of multiple singularities.

Equation (130) describes a constant flexural stiffness E_0I_0 with n-variations of intensity α_i and abscissas $x_{0,i}$, modeled by means of n distributions here indicates as $D(x-x_{0,i})$. Two types of distribution, such as unit step distribution $U(x-x_{0,i})$ and Dirac's delta $\delta(x-x_{0,i})$ are considered in Equation (130) for the Euler-Bernoulli beam, which leads to two different models as follows

$$E(x)I(x)=E_0I_0\left[1-\sum_{i=1}^{n}r_iU(x-x_{r,i})\right]$$

$$E(x)I(x)=E_0I_0\left[1-\sum_{j=1}^{m}\beta_jU(x-x_{\beta,j})\right] \tag{131}$$

The flexural stiffness provided by Equation (131) describes a beam showing concentrated jumps in the Young modulus $E(x)$ and the inertia moment $I(x)$ of the cross section (Figure 7). In this case, for the flexural stiffness $E(x)I(x)$ to be non-negative, the only constraints to be imposed on jump intensities r_i for the model described by Equation (131) are $\sum_{i=1}^{k}r_i\leq 1; k=1,2,\cdots,n$. On the other hand, if the flexural stiffness is given by Equation (131) for the case of a single Dirac's delta, the slope function will show n concentrated jumps induced by the presence of internal hinges endowed with rotational springs, as depicted in (Figure 8(a)). For this case, constraints on β_j will be discussed later.

Now we will study two different flexural stiffness models given by Equation (131) using the theory of distributions. Further, we present the model with the presence of two different singularities.

6.2. Solution of Euler-Bernoulli Beams in Presence of Multiple Curvature Discontinuities

Here we study the case of multiple jump discontinuities in the flexural stiffness, provided by Equation (131). Such case has been discussed for single (Yavari et al., 2000) [7] and double jumps (Yavari and Sarkani, 2001) [8] . But in those procedures enforcement of continuity conditions (where jumps appear) is required. We provide closed form solution for any number and position of the discontinuities.

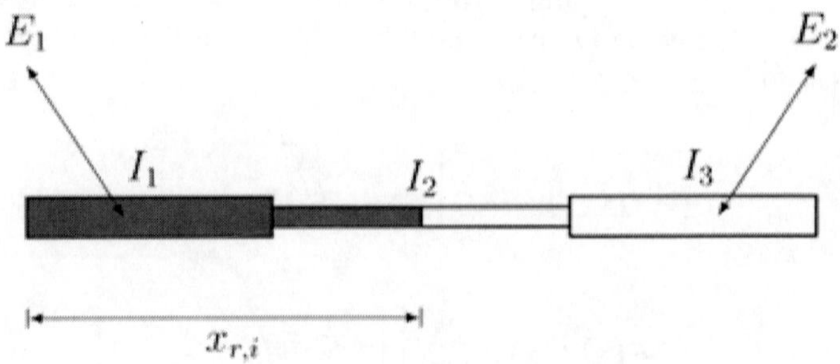

Figure 7. Beam with discontinuities in the Young modulus $E(x)$ and in the moment of inertia $I(x)$.

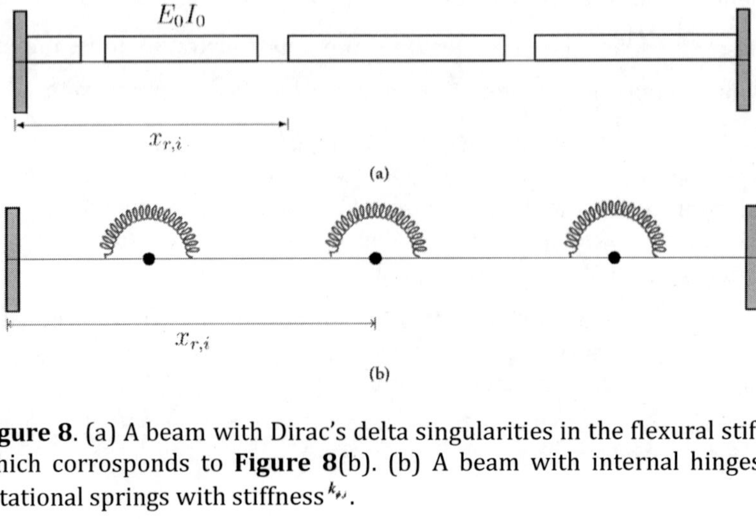

Figure 8. (a) A beam with Dirac's delta singularities in the flexural stiffness which corrosponds to **Figure 8**(b). (b) A beam with internal hinges and rotational springs with stiffness $k_{r,i}$.

In the case of flexural stiffness, Equation (131), the governing Equation (129) takes the following form,

$$\left[E_0 I_0\left(1-\sum_{i=1}^{n} r_i U\left(x-x_{r,i}\right) u''(x)\right)\right]'' = q(x)$$

$$\left[E_0 I_0\left(1-\sum_{i=1}^{n} r_i U\left(x-x_{r,i}\right) u''(x)\right)\right] = q^{(2)}(x) + b_2 x + b_1$$

$$\Rightarrow u''(x) = \frac{b_1 + b_2 x + q^{(2)}(x)}{E_0 I_0\left(1-\sum_{i=1}^{n} r_i U\left(x-x_{r,i}\right)\right)} \tag{132}$$

where b_1 and b_2 are constants of integration and $q^{(k)}(x)$ indicates a primitive of order k of the external load function $q(x)$.

Using the properties of unit step function, Equation (132) can be rewritten as,

$$u''(x) E_0 I_0\left(1-\sum_{i=1}^{n} r_i U\left(x-x_{r,i}\right)\right) = \frac{2b_1}{2E_0 I_0} + \frac{6b_2 x}{6E_0 I_0} + \frac{q^{(2)}(x)}{E_0 I_0}$$
$$= 2c_3 + 6c_4 + \frac{q^{(2)}(x)}{E_0 I_0}$$

$$\Rightarrow u''(x) = \left(2c_3 + 6c_4 + \frac{q^{(2)}(x)}{E_0 I_0}\right)\left[1+\sum_{i=1}^{n} r_i H\left(x-x_{r,i}\right)\right]$$

$$\chi(x) = -u''(x) = -\left(2c_3 + 6c_4 + \frac{q^{(2)}(x)}{E_0 I_0}\right)\left[1+\sum_{i=1}^{n} r_i \mu_i \mu_{i+1} H\left(x-x_{r,i}\right)\right] \tag{133}$$

where

$$c_3 = \frac{b_1}{2E_0 I_0}; \quad c_4 = \frac{b_2}{6E_0 I_0}; \quad \mu_i = \frac{1}{1-\sum_{k=1}^{i-1} r_k} \tag{134}$$

Equation (133) show that the flexural stiffness model given by Equation (131) provides a curvature function with jump discontinuities at $x_{r,i}$; $i=1,2,\cdots,n$ of the curvature function are dependent on the discontinuity intensities at $x_{r,k}$; $k=1,2,\cdots,i-1$.

Integration of Equation (131) provides the slope function as follows:

$$\begin{aligned}\phi(x)=-u'(x)=&-c_2-2c_3\left[x+\sum_{i=1}^{n}r_i\mu_i\mu_{i+1}\left(x-x_{r,i}\right)H\left(x-x_{r,i}\right)\right]\\&-3c_4\left[x^2+\sum_{i=1}^{n}r_i\mu_i\mu_{i+1}\left(x-x_{r,i}\right)H\left(x-x_{r,i}\right)\right]\\&-\frac{q^{(3)}(x)}{E_0I_0}-\sum_{i=1}^{n}r_i\mu_i\mu_{i+1}\left(\frac{q^{(3)}(x)-q^{(3)}\left(x_{r,i}\right)}{E_0I_0}\right)H\left(x-x_{r,i}\right)\end{aligned}\tag{135}$$

Integration of Equation (135) provides the following closed form expression for the deflection function

$$\begin{aligned}u(x)=&c_1+c_2x+c_3\left[x^2+\sum_{i=1}^{n}r_i\mu_i\mu_{i+1}\left(x-x_{r,i}\right)^2H\left(x-x_{r,i}\right)\right]\\&+c_4\left[x^3+\sum_{i=1}^{n}r_i\mu_i\mu_{i+1}\left(x-x_{r,i}\right)^3H\left(x-x_{r,i}\right)\right]\\&+\frac{q^{(4)}(x)}{E_0I_0}+\sum_{i=1}^{n}r_i\mu_i\mu_{i+1}\left(\frac{q^{(4)}(x)-q^{(4)}\left(x_{r,i}\right)-q^{(3)}\left(x_{r,i}\right)\left(x-x_{r,i}\right)}{E_0I_0}\right)\end{aligned}\tag{136}$$

Equation (135) and Equation (136) represents the generalization of multiple flexural stiffness discontinuities, of the type Equation (131), of the closed form expressions for a single discontinuity.

- The bending moment function is obtained by multiplying the curvature function, Equation (133) by Equation (131), as follows:

$$M(x)=E(x)I(x)\chi(x)=-E_0I_0\left(2c_3+6c_4x+\frac{q^{(2)}(x)}{E_0I_0}\right)\tag{137}$$

- The shearing force function is obtained by differentiating Equation (137) as follows:

$$V(x) = M'(x) = -E_0 I_0 \left(6c_4 + \frac{q^{(1)}(x)}{E_0 I_0} \right) \tag{138}$$

Equation (137) and Equation (138) are for a single singularities and show that the flexural stiffness discontinuities do not appear explicitly. In fact, it is expected that bending moment $M(x)$ and shear force $V(x)$ are independent of the flexural stiffness (for statically determinate beams). But, the discontinuity intensities r_i and positions $x_{r,j}$ appear explicitly in the integration constants c_3, c_4 (for statically determinate beams.)

7. SOLUTIONS OF EULER-BERNOULLI BEAMS IN PRESENCE OF MULTIPLE SLOPE DISCONTINUITIES

For the case of multiple slope discontinuities, we adopt a new technique of integration.

In the case of flexural stiffness provided by Equation (131), the governing Equation (129) takes the following form:

$$\left[E_0 I_0 \left(1 - \sum_{j=1}^{m} \beta_j \delta\left(x - x_{\beta,j}\right) \right) u''(x) \right] \tag{139}$$

$$\left(1 - \sum_{j=1}^{m} \beta_j \delta\left(x - x_{\beta,j}\right) \right) u''(x) = \frac{q^{(2)}(x) + b_2 x + b_1}{E_0 I_0}$$

$$\begin{aligned} \Rightarrow u''(x) &= \frac{q^{(2)}(x) + b_2 x + b_1}{E_0 I_0} \frac{1}{1 - \sum_{j=1}^{m} \beta_j \delta\left(x - x_{\beta,j}\right)} \\ &= \frac{q^{(2)}(x) + b_2 x + b_1}{E_0 I_0} \left[1 + \sum_{j=1}^{m} \beta_j \delta\left(x - x_{\beta,j}\right) \right] \\ &= \frac{q^{(2)}(x) + b_2 x + b_1}{E_0 I_0} + \sum_{j=1}^{m} \beta_j \delta\left(x - x_{\beta,j}\right) \end{aligned} \tag{140}$$

To solve the product of two Dirac's delta distributions, we will rely on the technique proposed by Bagarello (1995; 2000) [9] . Bagarello indicates that the product of two Dirac's deltas both centered at x_0 can be reduced to a singe Dirac's delta multiplied by a constant A as follows:

$$\delta\left(x-x_{\beta,j}\right)\delta\left(x-x_{\beta,k}\right)=\begin{cases}A\delta\left(x-x_{\beta,j}\right); & j=k\\ 0; & j\neq k\end{cases} \tag{141}$$

where $A_k=\frac{1}{A}\int_{-1}^{1}\frac{\phi(x)}{xk}\mathrm{d}x$; and $\phi(x)$ is a test function.

Hence,

$$\begin{aligned}u''(x)\delta\left(x-x_{\beta,k}\right)&=\frac{b_1+b_2x+q^{(2)}(x)}{E_0I_0}\delta\left(x-x_{\beta,k}\right)+\sum_{j=1}^{m}\beta_j u''(x)A\delta\left(x-x_{\beta,j}\right)\\ &=\frac{b_1+b_2x+q^{(2)}(x)}{E_0I_0}\delta\left(x-x_{\beta,k}\right)\left[1+\sum_{j=1}^{m}\beta_jA+\sum_{j=1}^{m}\beta_jA\right]\end{aligned}$$

$$u''(x)\delta\left(x-x_{\beta,k}\right)=\frac{1}{1-\beta_jA}\frac{b_1+b_2x+q^{(2)}(x)}{E_0I_0}\delta\left(x-x_{\beta,k}\right) \tag{142}$$

Substituting Equation (142) into Equation (140) given the following explicit expression of the curvature for the considered beam model

$$\begin{aligned}\chi(x)=-u''(x)=&-\left[2c_3+6c_4x+\frac{q^{(2)}(x)}{E_0I_0}\right]\\ &-\sum_{j=1}^{m}\left[2c_3+6c_4x_{\beta,j}+\frac{q^{(2)}\left(x_{\beta,j}\right)}{E_0I_0}\right]\frac{\beta_j}{1-\beta_jA}\delta\left(x-x_{\beta,j}\right)\end{aligned} \tag{143}$$

Integration of Equation (143) provides the following slope function showing discontinuities at the abscissa $x_{\beta,j}$

$$\phi(x) = -u'(x) = -c_2 - 2c_3\left(x + \sum_{j=1}^{m}\frac{\beta_j}{1-\beta_j A}H\left(x - x_{\beta,j}\right)\right)$$
$$-3c_4\left(x^2 + \sum_{j=1}^{m}x_{\beta,j}\frac{\beta_j}{1-\beta_j A}H\left(x - x_{\beta,j}\right)\right)$$
$$-\frac{q^{(3)}(x)}{E_0 I_0} - \sum_{j=1}^{m}\frac{\beta_j}{1-\beta_j A}H\left(x - x_{\beta,j}\right)\frac{q^{(2)}\left(x_{\beta,j}\right)}{E_0 I_0} \tag{144}$$

Further integration of Equation (144) provides the following closed form expression for the deflection function of the beam

$$u(x) = c_1 + c_2 x + 2c_3\left(x^2 + 2\sum_{j=1}^{m}\frac{\beta_j}{1-\beta_j A}H\left(x - x_{\beta,j}\right)\left(x - x_{\beta,j}\right)^{-1}\right)$$
$$+c_4\left(x^3 + 6\sum_{j=1}^{m}x_{\beta,j}\frac{\beta_j}{1-\beta_j A}H\left(x - x_{\beta,j}\right)\left(x - x_{\beta,j}\right)^{-1}\right)$$
$$+\frac{q^{(4)}(x)}{E_0 I_0} + \sum_{j=1}^{m}\frac{\beta_j}{1-\beta_j A}H\left(x - x_{\beta,j}\right)\frac{q^{(2)}\left(x_{\beta,j}\right)}{E_0 I_0}\left(x - x_{\beta,j}\right)^{-1} \tag{145}$$

The bending moment and shear force function formally coincide with Equation (126) and Equation (128), respectively. In fact, for statically determine beams, $m(x)$ and $v(x)$ should not depend on the adopted flexural stiffness. On the contrary, for statically indeterminate beams, the adopted flexural stiffness model will affect the expressions of the constants c_3, c_4.

The slope function defined in Equation (144) presents jump discontinuities $\Delta\phi\left(x_{\beta,j}\right)$ at $x_{\beta,j}$ that are explicitly evaluated as follows:

$$\Delta\phi\left(x_{\beta,j}\right) = -\frac{\beta_j}{1-\beta_j A}\left(2c_3 + 6c_4 x_{\beta,j} + \frac{q^{(2)}\left(x_{\beta,j}\right)}{E_0 I_0}\right); \quad j = 1, 2, \cdots, m \tag{146}$$

and comparison of Equation (146) with the bending moment given by Equation (126) evaluated at $x_{\beta,j}$ leads to

$$\Delta\phi\left(x_{\beta,j}\right)=\frac{\beta_j}{1-\beta_j A}\frac{M\left(x_{\beta,j}\right)}{E_0 I_0};\quad j=1,2,\cdots,m \tag{147}$$

Equation (147) corresponds to the presence of internal hinges at $x_{\beta,j}$; $j=1,2,\cdots,n$, endowed with rotational springs with stiffness $k_{\phi,j}$ as shown in Figure 8(a), given as

$$k_{\phi,j}=\frac{1-\beta_j A}{\beta_j}E_0 I_0;\quad j=1,2,\cdots,m \tag{148}$$

Since rotational spring stiffness can take values from zero (no rotational spring) up to infinity (continuous beam at $x_{\beta,j}$, with no internal hinges), discontinuity values β_j, in view of Equation (148), can take values from 0 up to 1.

However, if rotational spring stiffness $k_{\phi,j}$, are assigned to the related value β_j have to be obtained by Equation (148) for a value of the quantity A among those proposed by Bagereuo (1995) [9] .

REFERENCES

1. Timoshenko, S. and Woinowsky-Kreiger, S. (1959) Theory of Plates and Shells. 2nd Edition, McGraw-Hill, New York.
2. Gorman, D.J. (2008) On the Use of the Dirac Delta Function in the Vibration Analysis of Elastic Structures. International Journal of Solids and Structures, 45, 4605-4614.
3. Timoshenko, S.P. (1976) Strength of Materials: Part I. Krieger, Malabar.
4. Shames, I.H. (1989) Introduction to Solid Mechanics. Prentice-Hall, New York.
5. Kanwal, R.P. (1983) Generalized Functions Theory and Applications. Academic Press, New York.
6. Palmeri, A. and Cicirello, A. (2011) Physically-Based Dirac's Delta Functions in the Static Analysis of Multi-Cracked Euler-Bernoulli and Timoshenko Beams. International Journal of Solids and Structures, 48, 2184-2195.

7. Yavari, A., Sarkani, S. and Moyer, E.T. (2000) On Applications of Generalised Functions to Beam Bending Problems. International Journal of Solids and Structures, 38, 8389-8406.
8. Yavari, A. and Sarkani, S. (2001) On Applications of Generalised Functions to the Analysis of Euler-Bernoulli Beam-Columns with Jump Discontinuities. International Journal of Mechanical Sciences, 43, 1543-1562.
9. Bagarello, F. (1995) Multiplication of Distribution in One Dimension: Possible Approaches and Applications to Delta-Function and Its Derivatives. Journal of Mathematical Analysis and Applications, 196, 885-901.

CHAPTER 12

An Integral Collocation Approach Based on Legendre Polynomials for Solving Riccati, Logistic and Delay Differential Equations

M. M. Khader[1,2], A. M. S. Mahdy[3], M. M. Shehata[3]

[1]*Department of Mathematics and Statistics, College of Science, Al-Imam Mohammad Ibn Saud Islamic University (IMSIU), Riyadh, KSA*

[2]*Department of Mathematics, Faculty of Science, Benha University, Benha, Egypt*

[3]*Department of Mathematics, Faculty of Science, Zagazig University, Zagazig, Egypt*

ABSTRACT

In this paper, we propose and analyze some schemes of the integral collocation formulation based on Legendre polynomials. We implement these formulae to solve numerically Riccati, Logistic and delay differential equations with variable coefficients. The properties of the Legendre polynomials are used to reduce the proposed problems to the solution of non-linear system of algebraic equations using Newton iteration method. We give numerical results to satisfy the accuracy and the applicability of the proposed schemes.

Keywords: Integral Collocation Formulation, Spectral Method, Riccati, Logistic and Delay Differential Equations

1. INTRODUCTION

It is well known that the ordinary differential equations (ODEs) have been the focus of many studies due to their frequent appearance in various applications, such as in fluid mechanics, viscoelasticity, biology, physics and engineering applications, for more details, for example [1] -[5] . Consequently, considerable attention has been given to the efficient numerical solutions of ODEs of physical interest, because it is difficult to find exact solutions. Different numerical methods have been proposed in the literature for solving ODEs [6] -[13] .

The Riccati differential equation (RDE) is named after the Italian Nobleman Count Jacopo Francesco Riccati (1676-1754). The book of Reid [14] contains the fundamental theories of Riccati equation, with applications to random processes, optimal control, and diffusion problems. Besides important engineering science applications that today are considered classical, such as stochastic realization theory, optimal control, robust stabilization, and network synthesis, the newer applications include such areas as financial mathematics [15] . The solution of this equation can be reached using classical numerical methods such as, the forward Euler method and RungeKutta method. Bahnasawi et al. [16] presented the usage of Adomian decomposition method to solve the nonlinear RDE in an analytic form. Tan and Abbasbandy [17] employed the analytic technique called homotopy analysis method to solve the quadratic RDE.

The Logistic model can be obtained by applying the derivative operator on the Logistic equation. The model is initially published by Pierre Verhulst in 1838 [18] . The continuous Logistic model is described by first order ODE. The discrete Logistic model is simple iterative equation that reveals the chaotic property in certain regions [19] . There are many variations of the population modeling [20] . The Verhulst model is the classic example to illustrate the periodic doubling and chaotic behavior in dynamical system [19] . The model describs the population growth may be limited by certain factors like population density [20] .

2. APPLICATIONS OF LOGISTIC EQUATION

A typical application of the Logistic equation is a common model of population growth. Let $u(t)$ represents the population size and t represents the time where the constant $\rho > 0$ defines the growth rate.

Another application of Logistic curve is in medicine, where the Logistic differential equation is used to model the growth of tumors. This application can be considered an extension of the above mentioned use in the framework of ecology.

The solution of Logistic equation is explained the constant population growth rate which not includes the limitation on food supply or spread of diseases [21] . The solution curve of the model is increase exponentially from the multiplication factor up to saturation limit which is maximum carrying capacity [21] ,

$$\frac{\mathrm{d}N}{\mathrm{d}t} = \rho N\left(1 - \frac{N}{K}\right)$$

where N is the population with respect to time, ρ is the rate of maximum population growth and K is the carrying capacity. The solution of continuous Logistic equation is in the form of constant growth rate as in formula $N(t) = N_0 \mathrm{e}^{\rho t}$ where N_0 is the initial population [22] .

A delay differential equation (DDE) is a differential equation in which the derivative of the function at any time depends on the solution at previous time. Introduction of delay in the model enriches its dynamics and allows a precise description of the real life phenomena. DDEs are proved useful in control systems [23] , lasers, traffic models [24] , metal cutting, epidemiology, neuroscience, population dynamics [25] , chemical kinetics [26] , etc. In DDE, one has to provide history of the system over the delay interval $[-\tau, 0]$ as the initial condition. Due to this reason delay systems are infinite dimensional in nature. Because of in infinite dimensionality the DDEs are difficult to analyze analytically and hence the numerical solutions play an important role.

In [27] , Mai-Duy, et al. derived an integral collocation approach based on Chebyshev polynomials and used it for solving numerically the bi-harmonic equations. In [28] , Bhrawy and Alofi introduced a new shifted Chebyshev operational matrix of fractional integration of arbitrary order and applied together with spectral tau method for solving linear fractional differential equations. Khader et al. [29] introduced a new approximate formula of the fractional derivative using Legendre series expansion and used it to solve numerically the fractional diffusion equation.

In this article, we extend the previous work and derive some schemes of the integral collocation formulation based on Legendre polynomials. We implement these formulae to solve numerically Riccati, Logistic and delay differential equations.

Our paper is organized as follows: In Section 2, we derive some integration collocation formulations using Legendre series expansion. In Section 3, we give the integral collocation approach for solving Riccati, Logistic and delay differential equations. In Section 4, the paper ends with a brief conclusion and some remarks.

3. INTEGRATION COLLOCATION FORMULATIONS

The well known Legendre polynomials are defined on the interval $[-1,1]$ and can be determined with the aid of the following recurrence formula [30]

$$L_{k+1}(z)=\frac{2k+1}{k+1}zL_k(z)-\frac{k}{k+1}L_{k-1}(z), k=1,2,\cdots,$$

where $L_0(z)=1$ and $L_1(z)=z$. In order to use these polynomials on the interval $[0,1]$, we define the so called shifted Legendre polynomials by introducing the change of variable $z=2x-1$. Let the shifted Legendre polynomials $L_k(2x-1)$ be denoted by $L_k^*(x)$. Then $L_k^*(x)$ can be obtained as follows

$$L_{k+1}^*(x)=\frac{(2k+1)(2x-1)}{(k+1)}L_k^*(x)-\frac{k}{k+1}L_{k-1}^*(x),$$

where $L_0^*(x)=1$ and $L_1^*(x)=2x-1$. The analytic form of the shifted Legendre polynomials $L_k^*(x)$ of degree k is given by

$$L_k^*(x)=\sum_{i=0}^{k}(-1)^{k+i}\frac{(k+i)!}{(k-i)!(i!)^2}x^i, k=2,3,\cdots. \tag{1}$$

Note that $L_k^*(0)=(-1)^k$ and $L_k^*(1)=1$. The orthogonality condition is

$$\int_0^1 L_i^*(x)L_j^*(x)\mathrm{d}x=\begin{cases}\frac{1}{2i+1}, & \text{for} \quad i=j,\\ 0, & \text{for} \quad i\neq j.\end{cases}$$

In this work, to consider the differential equation of n-th order, we build the integration collocation method using the truncated Legendre series of degree m to represent the n-th derivative of the unknown function $u(x)$ in the following manner

$$\frac{\mathrm{d}^n u(x)}{\mathrm{d}x^n}\cong\sum_{k=0}^{m}a_k L_k^*(x)=\sum_{k=0}^{m}a_k I_k^{(n)}(x). \tag{2}$$

Using the integration we can obtain the lower-order derivatives and the function itself as follows

$$\frac{\mathrm{d}^{n-1}u(x)}{\mathrm{d}x^{n-1}}\cong\sum_{k=0}^{m}a_k I_k^{(n-1)}(x)+c_1, \tag{3}$$

$$\frac{\mathrm{d}^{n-2}u(x)}{\mathrm{d}x^{n-2}}\cong\sum_{k=0}^{m}a_k I_k^{(n-2)}(x)+c_1x+c_2,\cdots \tag{4}$$

$$\frac{\mathrm{d}u(x)}{\mathrm{d}x}=\sum_{k=0}^{m}a_k I_k^{(1)}(x)+c_1\frac{x^{n-2}}{(n-2)!}+c_2\frac{x^{n-3}}{(n-3)!}+\cdots+c_{n-2}x+c_{n-1}, \tag{5}$$

$$u(x)=\sum_{k=0}^{m}a_k I_k^{(0)}(x)+c_1\frac{x^{n-1}}{(n-1)!}+c_2\frac{x^{n-2}}{(n-2)!}+\cdots+c_{n-1}x+c_n, \tag{6}$$

from (1) and (2) we have

$$I_k^{(n)}(x)=\sum_{i=0}^{k}\frac{(-1)^{k+i}(k+i)!}{(k-i)!(i!)^2}x^i,$$

$$I_k^{(n-1)}(x)=\int I_k^{(n)}(x)\mathrm{d}x=\sum_{i=0}^{k}\frac{(-1)^{k+i}(k+i)!}{(k-i)!(i!)^2(i+1)}x^{i+1},$$

$$I_k^{(n-2)}(x)=\int I_k^{(n-1)}(x)\mathrm{d}x=\sum_{i=0}^{k}\frac{(-1)^{k+i}(k+i)!}{(k-i)!(i!)^2(i+1)(i+2)}x^{i+2},$$

$$\cdots$$

$$I_k^{(0)}(x)=\int I_k^{(1)}(x)\mathrm{d}x=\sum_{i=0}^{k}\frac{(-1)^{k+i}(k+i)!x^{i+n}}{(k-i)!(i!)^2(i+1)\cdots(i+n-1)(i+n)}. \tag{7}$$

We now collocate Equatuions (2)-(6) at $(m+1)$ points x_p, $p=0,1,\cdots,m$ as

$$\frac{\mathrm{d}^n u(x_p)}{\mathrm{d}x^n}=\Omega^{(n)}\hat{S},\quad \frac{\mathrm{d}^{n-1}u(x_p)}{\mathrm{d}x^{n-1}}=\Omega^{(n-1)}\hat{S},\cdots,$$
$$\frac{\mathrm{d}u(x_p)}{\mathrm{d}x}=\Omega^{(1)}\hat{S},\quad u(x_p)=\Omega^{(0)}\hat{S}, \tag{8}$$

where $\hat{S}=[a_0,a_1,\cdots,a_m,c_1,c_2,\cdots,c_n]^{\mathrm{T}}$, and $\Omega^{(n)},\Omega^{(n-1)},\cdots,\Omega^{(0)}$ are integrated matrices.

4. INTEGRAL COLLOCATION APPROACH FOR SOLVING RICCATI, LOGISTIC AND DELAY DIFFERENTIAL EQUATIONS

In this section, we introduce the integral collocation approach using Legendre expansion for solving the Riccati, Logistic and delay differential equations.

4.1. Model 1: Riccati Differential Equation

$$\frac{du(x)}{dx}+u^2(x)-1=0, \qquad x>0, \tag{9}$$

we also assume an initial condition

$$u(0)=u^0. \tag{10}$$

The exact solution to this problem at $u^0=0$ is

$$u(x)=\frac{e^{2x}-1}{e^{2x}+1}.$$

The procedure of the implementation is given by the following steps:

1) Approximate the function $u(x)$ using Formula (6) and its relevant derivatives with $m=5$, as follows

$$\frac{du(x)}{dx}\cong\sum_{k=0}^{5}a_k I_k^{(1)}(x),$$
$$u(x)\cong\sum_{k=0}^{5}a_k I_k^{(0)}(x)+c_1, \tag{11}$$

where $I_k^{(0)}(x)$ is defined in (7) as

$$I_k^{(0)}(x)=\sum_{i=0}^{k}(-1)^{k+i}\frac{(k+i)!}{(k-i)!(i!)^2(i+1)}x^{i+1},$$
$$I_k^{(1)}(x)=\sum_{i=0}^{k}(-1)^{k+i}\frac{(k+i)!}{(k-i)!(i!)^2}x^{i}.$$

Then the Riccati differential Equation (9) is transformed to the following approximated form

$$\sum_{k=0}^{5} a_k L_k^*(x) + \left(\sum_{k=0}^{5} a_k I_k^{(0)}(x) + c_1 \right)^2 = 1. \tag{12}$$

We now collocate Equation (12) at $(m+1=6)$ points x_p, $p=0,1,2,3,4,5$ as

$$\sum_{k=0}^{5} a_k L_k^*(x_p) + \left(\sum_{k=0}^{5} a_k I_k^{(0)}(x_p) + c_1 \right)^2 = 1. \tag{13}$$

For suitable collocation points we use the roots of shifted Legendre polynomial $L_6^*(x)$.

2) Also, by substituting from the initial condition (10) in (11) we can obtain $(n=1)$ an equation which gives the value of the constant c_1 as follows

$$c_1 = u^0 = 0. \tag{14}$$

Equations (13) and (14) represent a system of non-linear algebraic equations which contains seven equations for the unknowns $a_k, k=0,1,2,3,4,5$ and c_1.

3) Solve the resulting system using the Newton iteration method to obtain the unknowns $a_k, k=0,1,2,3,4,5$ as follows

$$a_0 = 0.76160,\ \ a_1 = -0.31790,\ \ a_2 = -0.0506,$$
$$a_3 = 0.02896,\ \ a_4 = -0.00111,\ \ a_5 = -0.00112.$$

Therefore, from Formula (12) we can obtain the approximate solution in the form

$$u(x) \cong \sum_{k=0}^{5} a_k I_k^{(0)}(x) + c_1$$

$$s = 0.990x + 0.002x^2 - 0.346x^3 + 0.027x^4 + 0.125x^5 - 0.047x^6.$$

The numerical results of the proposed problem (9) are given in Figure 1 with different values of $m\,(m = 5, 7)$ in the interval $[0,1]$ at $u^0 = 0$. From this figure, since the obtained numerical solutions are in excellent agreement with the exact solution, so, we can conclude that the proposed technique is well for solving such class of ODEs.

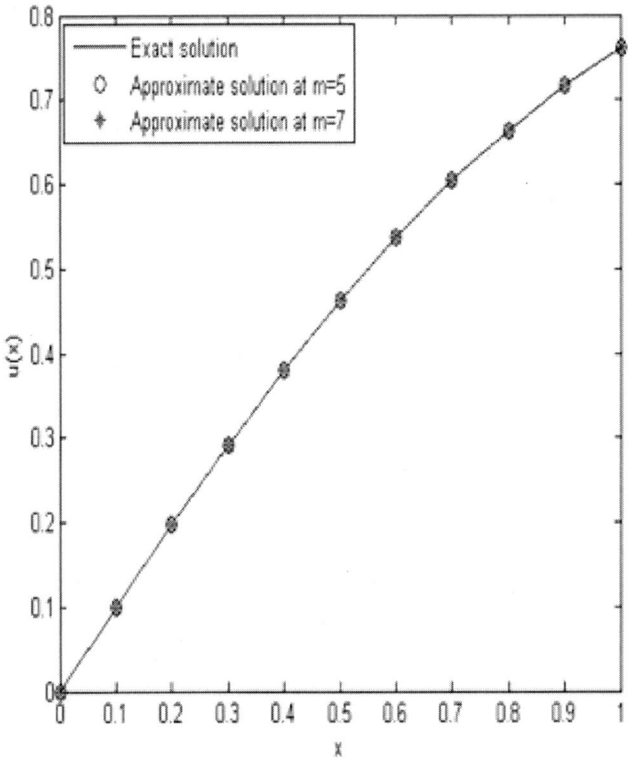

Figure 1. A comparison between the exact solution and the approximate solution at $m = 5$ and $m = 7$.

4.2. Model 2: Logistic Differential Equation

$$\frac{\mathrm{d}u(x)}{\mathrm{d}x} = \rho u(x)(1-u(x)), \quad x>0, \ \rho>0. \tag{15}$$

We also assume an initial condition

$$u(0)=u_0, \qquad u_0>0. \tag{16}$$

The exact solution to this problem is given by

$$u(x)=\frac{u_0}{(1-u_0)\mathrm{e}^{-\rho x}+u_0}.$$

The existence and the uniqueness of the proposed problem (15) are introduced in details in [31] .

The procedure of the implementation is given by the following steps:

1) Approximate the function $u(x)$ and its relevant derivatives with $m=5$, using Formula (11). Then the Logistic differential Equation (15) is transformed to the following approximated form

$$\sum_{k=0}^{5} a_k L_k^*(x) - \rho\left(\sum_{k=0}^{5} a_k I_k^{(0)}(x)+c_1\right)\left(1-\left(\sum_{k=0}^{5} a_k I_k^{(0)}(x)+c_1\right)\right)=0. \tag{17}$$

We now collocate Equation (17) at $(m+1=6)$ points x_p, $p=0,1,2,3,4,5$ as

$$\sum_{k=0}^{5} a_k L_k^*(x_p) - \rho\left(\sum_{k=0}^{5} a_k I_k^{(0)}(x_p)+c_1\right)\left(1-\left(\sum_{k=0}^{5} a_k I_k^{(0)}(x_p)+c_1\right)\right)=0. \tag{18}$$

For suitable collocation points we use roots of shifted Legendre polynomial $L_6^*(x)$ which are

$$x_0 = 0.96623, \quad x_1 = 0.03377,$$
$$x_2 = 0.38069, x_3 = 0.61930,$$
$$x_4 = 0.16931, \quad x_5 = 0.83060. \tag{19}$$

2) Also, by substituting from the initial condition (16) in (11) with $u_0 = 0.85$ we can obtain $(n = 1)$ an equation which gives the value of the constant $c_1 = 0.85$.

Equation (18) represents a system of non-linear algebraic equations which contains six equations for the unknowns $a_k, k = 0,1,2,3,4,5$.

3) Solve the resulting system using the Newton iteration method to obtain the unknowns $a_k, k = 0,1,2,3,4,5$ as follows

$$a_0 = 0.0533, \quad a_1 = -0.0101,$$
$$a_2 = 0.0004, a_3 = 0.00002,$$
$$a_4 = -1.642 \times 10^{-6}, \ a_5 = 5.398 \times 10^{-8}. \tag{20}$$

Therefore, from Formula (11) we can obtain the approximate solution in the form

$$u(x) = \sum_{k=0}^{5} a_k I_k^{(0)}(x) + c_1$$
$$= 1 + x + 0.5x^2 + 0.1667x^3 + 0.0417x^4 + 0.0083x^5 + 0.0014x^6 + 0.0002x^7 + 0.00004x^8.$$

The numerical results of the proposed problem (15) are given in Figure 2 with different values of $m\,(m = 3,5)$ in the interval $[0,1]$. From this figure, since the obtained numerical solutions are in excellent agreement with the exact solution, so, we can conclude that the proposed technique is well for solving such class of ODEs.

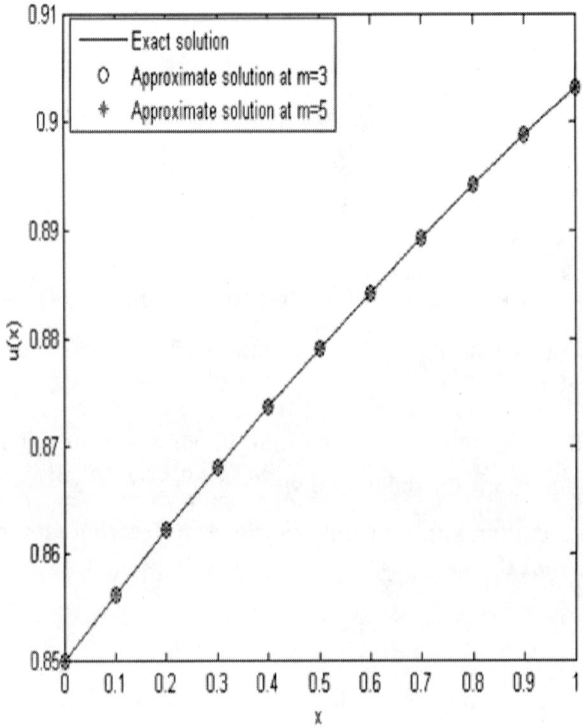

Figure 2. A comparison between the exact solution and the approximate solution at m = 3 and m = 5.

4.3. Model 3: Delay Differential Equation

Consider the linear delay differential equation of third-order

$$\frac{d^3u(x)}{dx^3} = -u(x) - u(x-0.3) + e^{-x+0.3}, 0 \le x \le 1, \quad (21)$$

with the initial conditions

$$u(0) = 1, \frac{du(0)}{dx} = -1, \frac{d^2u(0)}{dx^2} = 1, u(x) = e^{-x}. \quad (22)$$

The exact solution of this model is $u(x)=\mathrm{e}^{-x}$.

The procedure of the implementation is given by the following steps:

1) Approximate the function $u(x)$ using Formula (6) and its relevant derivatives with $m=5$, as follows

$$\frac{\mathrm{d}^3u(x)}{\mathrm{d}x^3}\cong\sum_{k=0}^{5}a_kL_k^*(x)=\sum_{k=0}^{3}a_kI_k^{(3)}(x),$$

$$\frac{\mathrm{d}^2u(x)}{\mathrm{d}x^2}\cong\sum_{k=0}^{5}a_kI_k^{(2)}(x)+c_1,$$

$$\frac{\mathrm{d}u(x)}{\mathrm{d}x}\cong\sum_{k=0}^{5}a_kI_k^{(1)}(x)+c_1x+c_2,$$

$$u(x)\cong\sum_{k=0}^{5}a_kI_k^{(0)}(x)+c_1\frac{x^2}{2!}+c_2\frac{x}{1!}+c_3, \tag{23}$$

where $I_k^{(0)}(x)$, $I_k^{(1)}(x)$ and $I_k^{(s)}(x)$ are defined in (7) as follows

$$I_k^{(0)}(x)=\sum_{i=0}^{k}\frac{(-1)^{k+i}(k+i)!}{(k-i)!(i!)^2(i+1)(i+2)(i+3)}x^{i+3},$$

$$I_k^{(1)}(x)=\sum_{i=0}^{k}\frac{(-1)^{k+i}(k+i)!}{(k-i)!(i!)^2(i+1)(i+2)}x^{i+2},$$

$$I_k^{(2)}(x)=\sum_{i=0}^{k}\frac{(-1)^{k+i}(k+i)!}{(k-i)!(i!)^2(i+1)}x^{i+1}.$$

Then the delay differential Equation (21) is transformed to the following approximated form

$$\sum_{k=0}^{5}a_kL_k^*(x)+\left(\sum_{k=0}^{5}a_kI_k^{(0)}(x)+c_1\frac{x^2}{2!}+c_2\frac{x}{1!}+c_3\right)$$
$$+\left(\sum_{k=0}^{5}a_kI_k^{(0)}(x-0.3)+c_1\frac{(x-0.3)^2}{2!}+c_2\frac{(x-0.3)}{1!}+c_3\right)=\mathrm{e}^{-x+0.3}. \tag{24}$$

We now collocate Equation (24) at $(m+1=6)$ points $x_p,\ p=0,1,2,3,4,5$ as

$$\sum_{k=0}^{5} a_k L_k^*(x_p) + \left(\sum_{k=0}^{5} a_k I_k^{(0)}(x_p) + c_1\frac{x_p^2}{2!} + c_2\frac{x_p}{1!} + c_3\right) + \left(\sum_{k=0}^{5} a_k I_k^{(0)}(x_p)(x_p-0.3) + c_1\frac{(x_p-0.3)^2}{2!} + c_2\frac{(x_p-0.3)}{1!} + c_3\right) = e^{-x_p+0.3}. \quad (25)$$

For suitable collocation points we use the roots of shifted Legendre polynomial $L_6^*(x)$.

2) Also, by substituting from the initial conditions (22) in (23) we can obtain $(n=3)$ equations which give the values of the constants

$$c_1 = 1, \quad c_2 = -1, \quad c_3 = 1. \quad (26)$$

Equations (25) and (26) represent a system of linear algebraic equations which contains nine equations for the unknowns $a_k, k=0,1,2,3,4,5$ and c_1, c_2, c_3.

3) Solve the resulting system using the conjugate gradient method to obtain the unknowns $a_k, k=0,1,2,3,4,5$ as follows

$$\begin{aligned} a_0 &= -0.6321, \quad a_1 = 0.3109, \quad a_2 = -0.0515, \\ a_3 &= 0.0051, \quad a_4 = 0.0004, \quad a_5 = 0.00002. \end{aligned} \quad (27)$$

Therefore, from Formula (23) we can obtain the approximate solution in the form

$$\begin{aligned} u(x) &\cong \sum_{k=0}^{5} a_k I_k^{(0)}(x) + c_1\frac{x^2}{2!} + c_2\frac{x}{1!} + c_3 \\ &= 1 - x + 0.5x^2 - 0.1667x^3 + 0.0417x^4 - 0.0084x^5 + 0.0014x^6 - 0.0002x^7 + 0.00002x^8. \end{aligned}$$

The numerical results of the proposed problem (21) are given in Figure 3 with different values of $m\ (m=5,7)$ in the interval $[0,1]$. From this figure, since the obtained numerical solutions are in excellent agreement with the exact solution, so, we can conclude that the proposed technique is well for solving such class of ODEs.

5. CONCLUSION AND REMARKS

In this article, an integral collocation approach based on Legendre polynomials is introduced for solving M. M. Khader, A. M. S. Mahdy, M. M. Shehata numerically the Riccati, Logistic and delay differential equations. The properties of the Legendre polynomials are used to reduce the proposed problems to system of algebraic equations which are solved by a suitable numerical method. From the obtained numerical results, we can conclude that this method gives results with an excellent agreement with the exact solution. All numerical results are obtained using Matlab program 8.

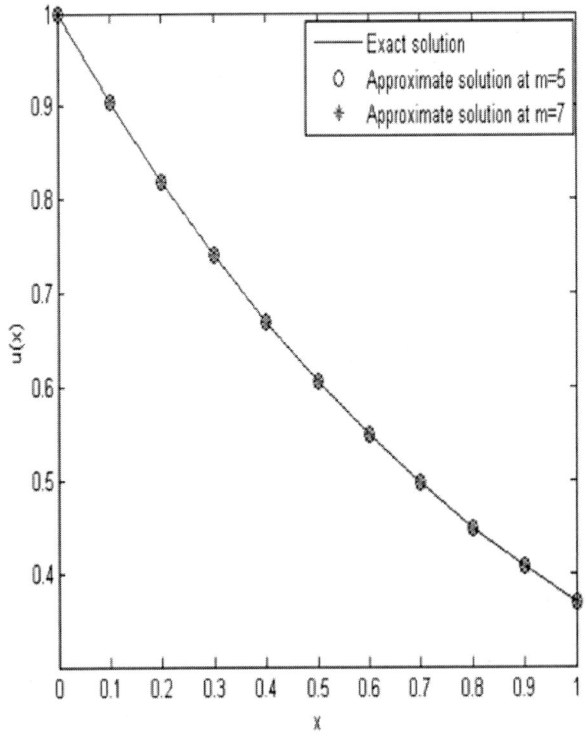

Figure 3. A comparison between the exact solution and the approximate solution at m = 5 and m = 7.

REFERENCES

1. Khader, M.M. (2011) On the Numerical Solutions for the Fractional Diffusion Equation. Communications in Nonlinear Science and Numerical Simulation, 16, 2535-2542.
2. Khader, M.M. (2013) Numerical Treatment for Solving the Perturbed Fractional PDEs Using Hybrid Techniques. Journal of Computational Physics, 250, 565-573.
3. Sweilam, N.H., Khader, M.M. and Mahdy, A.M.S. (2013) Numerical Study for the Fractional Differential Equations Generated by Optimization Problem Using Chebyshev Collocation Method and FDM. Applied Mathematics and Information Science, 7, 2013-2020.
4. Khader, M.M. and Hendy, A.S. (2013) A Numerical Technique for Solving Fractional Variational Problems. Mathematical Methods in Applied Sciences, 36, 1281-1289.
5. Sweilam, N.H., Khader, M.M. and Nagy, A.M. (2011) Numerical Solution of Two-Sided Space-Fractional Wave Equation Using Finite Difference Method. Journal of Computional and Applied Mathematics, 235, 2832-2841.
6. Sweilam, N.H., Khader, M.M. and Adel, M. (2012) On the Stability Analysis of Weighted Average Finite Difference Methods for Fractional Wave Equation. Fractional Differential Calculus, 2, 17-75.
7. Sweilam, N.H. and Khader, M.M. (2010) A Chebyshev Pseudo-Spectral Method for Solving Fractional Order Integro-Differential Equations. ANZIAM, 51, 464-475.
8. Sweilam, N.H., Khader, M.M. and Al-Bar, R.F. (2008) Homotopy Perturbation Method for Linear and Nonlinear System of Fractional Integro-Differential Equations. International Journal of Computational Mathematics and Numerical Simulation, 1, 73-87.
9. Sweilam, N.H., Khader, M.M. and Mahdy, A.M.S. (2012) On the Numerical Solution for the Linear Fractional Klein-Gordon Equation Using Legendre Pseudospectral Method. International Journal of Mathematics and Computer Applications Research, 2, 1-10.
10. Khader, M.M., EL-Danaf, T.S. and Hendy, A.S. (2013) A Computational Matrix Method for Solving Systems of High Order Fractional Differential Equations. Applied Mathematical Modelling, 37, 4035-4050.
11. Khader, M.M., Mahdy, A.M.S. and Shehata, M.M. (2014) Approximate Analytical Solution to the Time-Fractional Biological Population Model Equation. Jokull, 64, 378-394.
12. Sweilam, N.H., Khader, M.M. and Mahdy, A.M.S. (2012) Numerical Studies for Solving Fractional-Order Logistic Equation. International Journal of Pure and Applied Mathematics, 78, 1199-1210.
13. Sweilam, N.H., Khader, M.M. and Mahdy, A.M.S. (2012) Numerical Studies for Fractional-Order Logistic Differential Equation with Two Different Delays. Journal of Applied Mathematics, 2012, Article ID: 764894.
14. Reid, W.T. (1972) Riccati Differential Equations Mathematics in Science and Engineering. Academic Press, New York.
15. Lasiecka, I. and Triggiani, R. (1991) Differential and Algebraic Riccati Equations with Application to Boundary/Point Control Problems: Continuous Theory and Approximation Theory. Lecture Notes in Control and Information Sciences, Springer, Berlin.
16. Bahnasawi, A.A., El-Tawil, M.A. and Abdel-Naby, A. (2004) Solving Riccati Differential Equation Using ADM. Applied Mathematics and Computation, 157, 503-514.
17. Tan, Y. and Abbasbandy, S. (2008) Homotopy Analysis Method for Quadratic Riccati Differential Equation. Communications in Nonlinear Science and Numerical Simulation, 13, 539-546.
18. Cushing, J.M. (1998) An Introduction to Structured Population Dynamics, Society for Industrial and Applied Mathematics.
19. Alligood, K.T., Sauer, T.D. and Yorke, J.A. (1996) An Introduction to Dynamical Systems. Springer, Berlin.
20. Ausloos, M. The Logistic Map and the Route to Chaos: From the Beginnings to Modern Applications XVI. 411, 2006.
21. Pastijn, H. (2006) Chaotic Growth with the Logistic Model of P.-F. Verhulst, Understanding Complex Systems. The Logistic Map and the Route to Chaos. 3-11.

22. Suansook, Y. and Paithoonwattanakij, K. (2009) Dynamic of Logistic Model at Fractional Order. IEEE International Symposium on Industrial Electronics.
23. Fridman, E., Fridman, L. and Shustin, E. (2000) Steady Modes in Relay Control Systems with Time Delay and Periodic Disturbances. Journal of Dynamic Systems, Measurement, and Control, 122, 732-737.
24. Davis, C.L. (2002) Modification of the Optimal Velocity Traffic Model to Include Delay due to Driver Reaction Time. Physica A, 319, 557-567.
25. Kuang, Y. (1993) Delay Differential Equations with Applications in Population Biology. Academic Press, Boston, San Diego, New York.
26. Epstein, I. and Luo, Y. (1991) Differential Delay Equations in Chemical Kinetics: Nonlinear Models: The Cross-Shaped Phase Diagram and the Originator. Journal of Chemical Physics, 95, 244-254.
27. Mai-Duy, N., See, H. and Tran-Cong, T. (2009) A Spectral Collocation Technique Based on Integrated Chebyshev Polynomials for Biharmonic Problems in Irregular Domains. Applied Mathematical Modelling, 33, 284-299.
28. Bhrawy, A.H. and Alofi, A.S. (2013) The Operational Matrix of Fractional Integration for Shifted Chebyshev Polynomials. Applied Mathematics Letters, 26, 26-31.
29. Khader, M.M., Sweilam, N.H. and Mahdy, A.M.S. (2011) An Efficient Numerical Method for Solving the Fractional Diffusion Equation. Journal of Applied Mathematics and Bioinformatics, 1, 1-12.
30. Bell, W.W. (1968) Special Functions for Scientists and Engineers. Great Britain, Butler and Tanner Ltd, Frome and London.
31. El-Sayed, A.M.A., El-Mesiry, A.E.M. and El-Saka, H.A.A. (2007) On the Fractional-Order Logistic Equation. Applied Mathematics Letters, 20, 817-823.

CHAPTER 13

Parabolic Partial Differential Equations as Inverse Moments Problem

Maria B. Pintarelli[1,2]

[1]*Grupo de Aplicaciones Matematicas y Estadisticas de la Facultad de Ingenieria (GAMEFI), Universidad Nacional de La Plata (UNLP), La Plata, Argentina*

[2]*Departamento de Matematica, Facultad de Ciencias Exactas, Universidad Nacional de La Plata (UNLP), La Plata, Argentina*

ABSTRACT

We considerer parabolic partial differential equations $w_t - (w_x)_x = r(x,t)$ under the conditions $w_x(a_1,t) = k_1(t)$ $w_x(b_1,t) = k_2(t)$ $w(x,a_2) = h_1(t)$ $w(x,b_2) = h_2(t)$ on a region $E = (a_1,b_1) \times (a_2,b_2)$. We will see that we can write the equation in partial derivatives as an Fredholm integral equation of first kind and will solve this latter with the techniques of inverse moments problem. We will find an approximated solution and bounds for the error of the estimated solution using the techniques on moments problem. Also we consider the one- dimensional one-phase inverse Stefan problem.

Keywords: Parabolic PDEs, Freholm Integral Equations, Generalized Moment Problem

1. INTRODUCTION

We considerer parabolic partial differential equation of the form:

$$w_t - (w_x)_x = r(x,t) \tag{1}$$

where the unknown function $w(x,t)$ is defined in $E = (a_1, b_1) \times (a_2, b_2)$. $r(x,t)$ is known function. We consider conditions

$$w_x(a_1,t) = k_1(t) \qquad w_x(b_1,t) = k_2(t)$$

$$w(x,a_2) = h_1(t) \qquad w(x,b_2) = h_2(t).$$

This problem was studied under conditions of Cauchy in [1] .

Parabolic differential equations are commonly used in the fields of engineering and science for simulating physical processes. These equations describe various processes in viscous fluid flow, filtration of liquids, gas dynamics, heat conduction, elasticity, biological species, chemical reactions, environmental pollution, etc. [2] [3] .

In a variety of cases, approximations are used to convert parabolic PDEs to ordinary differential equations or even to algebraic equations. The existence and uniqueness properties of this problem are presented in literature. Several numerical methods have been proposed for the solution of this problem [4] -[6] .

Next section is devoted to showing how the differential equation (1) is transformed into integral equation of first kind that can be seen as generalized moments problem. In Section 3 there we present a theorem that guarantees under certain conditions the stability and convergence of the finite generalized moment problem. In Section 4 we exemplify the general method by applying it to some parabolic PDEs of the form (1). Finally in Section 5 the method is applied to solve the one-dimensional one-phase inverse Stefan problem.

The Stefan problem consists of finding w y s such that

$$\frac{\partial w}{\partial t} = \frac{\partial^2 w}{\partial x^2} \quad \text{in} \left\{(x,t); 0 < x < s(t); t > 0\right\}$$

$$-\frac{\partial w}{\partial x}(0,t) = f(t) \qquad t > 0$$

$$w(s(t),t) = 0 \qquad t > 0$$

$$\frac{ds}{dt} = -\frac{\partial w}{\partial x}(s(t),t) \qquad t > 0$$

$$w(x,0) = w_0(x) \qquad x \geq 0$$

$$s(0) = a.$$

The classical Stefan problem is a nonlinear initial value problem with a moving boundary whose position is unknown a priori and it must be determined as part of the solution. The differential equations of parabolic type governing heat diffusion with phase change are an important class of Stefan problems.

The direct Stefan problem requires determining both the temperature and the moving boundary interface when the initial and boundary conditions, and the thermal properties of the heat conducting body are known. Conversely, inverse Stefan problems require determining the initial and/or boundary conditions, and/or thermal properties from additional information which may involve the partial knowledge or measurement of the moving boundary interface position, its velocity in a normal direction, or the temperature at selected interior thermo- couples of the domain.

In this paper we solve the inverse Stefan problem: find $f(t)$ with $s(t)$ known such that the above con- ditions are met.

The d-dimensional generalized moment problem [7] -[10] can be posed as follows: find a function u on a domain $\Omega \subset \mathbf{R}^d$ satisfying the sequence of equations

$$\int_\Omega u(x) g_n(x) \mathrm{d}x = \mu_n \qquad n \in \mathbf{N} \tag{2}$$

where (g_n) is a given sequence of functions lying in $\mathbf{L}^2(\Omega)$ linearly independent.

1. Introduction

Many inverse problems can be formulated as an integral equation of the first kind, namely,

$$\int_a^b K(x,y)u(y)\mathrm{d}y = f(x) \quad x \in (a,b).$$

$K(x,y)$ and $f(x)$ are given functions and $u(y)$ is a solution to be determined; $f(x)$ is a result of experimental measurements and hence is given only at finite set of points. It follows that the above integral equation is equivalent to the following moment problem

$$\int_a^b K(x_n,y)u(y)\mathrm{d}y = f(x_n) \qquad n = 1,2,\cdots.$$

Also we considerer the multidimensional moment problems

$$\int_\Omega K(x_n,y)u(y)\mathrm{d}y = f(x_n) \qquad n = 1,2,\cdots,\Omega \subset \mathbf{R}^d.$$

Moment problems are usually ill-posed. There are various methods of constructing regularized solutions, that is, a approximate solution stable with respect to the given data. One of them is the method of truncated expan- sion.

The method of truncated expansion consists in approximating (2) by finite moment problems

$$\int_\Omega u(x)g_i(x)\mathrm{d}x = \mu_i \qquad i = 1,2,\cdots,n. \tag{3}$$

Solved in the subspace $\langle g_1,g_2,\cdots,g_n\rangle$ generated by $g_1,g_2,\cdots,g_n$ (3) is stable.

Considering the case where the data $\mu = (\mu_1,\mu_2,\cdots,\mu_n)$ are inexact, we apply some convergence theorems and error estimates for the regularized solutions [9] [11] .

2. PARABOLIC PARTIAL DIFFERENTIAL EQUATIONS AS INTEGRAL EQUATIONS OF FIRST KIND

Let $F(w(x,t)) = r(x,t)$ be a partial differential equations such as (1). The solution $w(x,t)$ is defined on the region $E = (a_1,b_1)\times(a_2,b_2)$ and verifies conditiones on the boundary $C = \partial E$:

$$w_x(a_1,t) = k_1(t) \qquad w_x(b_1,t) = k_2(t)$$

$$w(x,a_2) = h_1(t) \qquad w(x,b_2) = h_2(t).$$

We apply the technique used in [1] . Let $F^* = (F_1(w), F_2(w))$ be a vectorial field such that w verifies $div(F^*) = h^*(w)$ with h^* a known function and, reciprocally, if w verifies $div(F^*) = h^*(w)$ then

$$F(w(x,t)) = r(x,t).$$

Specifically in this case $F(w(x,t)) = w_t - (w_x)_x$ and we take

$$F^* = (F_1(w), F_2(w)) = (w_x, -w) \tag{4}$$

$$h^*(w) = -r(x,t). \tag{5}$$

Let $u(x,t,\tau,\xi)$ be the auxiliary function

$$u(x,t,\tau,\xi) = \mathrm{e}^{-t\xi}\cos\left(\frac{x\pi}{b_1}\tau\right). \tag{6}$$

Since

$$udiv\left(F^{*}\right)=uh^{*}\left(w\right)$$

we have

$$\iint_{E}udiv\left(F^{*}\right)\mathrm{d}A=\iint_{E}uh^{*}\left(w\right)\mathrm{d}A.$$

Moreover, as

$$udiv\left(F^{*}\right)=div\left(uF^{*}\right)-F^{*}\cdot\nabla u$$

and

$$\iint_{E}udiv\left(F^{*}\right)\mathrm{d}A=\underbrace{\iint_{E}div\left(uF^{*}\right)\mathrm{d}A}_{=\int_{C}\left(uF^{*}\right)\cdot n\mathrm{d}s}-\iint_{E}F^{*}\cdot\nabla u\mathrm{d}A$$

we obtain

$$\iint_{E}uh^{*}\left(w\right)\mathrm{d}A=\int_{C}\left(uF^{*}\right)\cdot n\mathrm{d}s-\iint_{E}F^{*}\cdot\nabla u\mathrm{d}A \tag{7}$$

where $\nabla u=\left(u_{\tau},u_{\xi}\right)$.

We consider the integral

$$\iint_{E}F^{*}\cdot\nabla u\mathrm{d}A=\iint_{E}F_{1}u_{\tau}+F_{2}u_{\xi}\mathrm{d}A. \tag{8}$$

Integrating by parts:

$$\begin{aligned}\iint_{E}F_{1}u_{\tau}\mathrm{d}A=\int_{a_2}^{b_2}\int_{a_1}^{b_1}F_{1}u_{\tau}\mathrm{d}\tau\mathrm{d}\xi&=\int_{a_2}^{b_2}\left(w\left(b_1,\xi\right)u_{\tau}\left(x,t,b_1,\xi\right)-w\left(a_1,\xi\right)u_{\tau}\left(x,t,a_1,\xi\right)\right)\mathrm{d}\xi\\&-\int_{a_2}^{b_2}\int_{a_1}^{b_1}w\left(\tau,\xi\right)u_{\tau\tau}\mathrm{d}\tau\mathrm{d}\xi.\end{aligned} \tag{9}$$

Note that in (9) if x is a natural number then

$$u_\tau\left(x,t,b_1,\xi\right)=-\mathrm{e}^{-t\xi}\frac{x\pi}{b_1}\sin\left(\frac{x\pi}{b_1}b_1\right)=0 \tag{10}$$

and if $a_1=0$ then

$$u_\tau\left(x,t,a_1,\xi\right)=-\mathrm{e}^{-t\xi}\frac{x\pi}{b_1}\sin\left(\frac{x\pi}{b_1}a_1\right)=0. \tag{11}$$

Thus if $x\in N$ and $a_1=0$:

$$\iint_E F^*\cdot\nabla u dA=-\iint_E\left(wu_{\tau\tau}-wu_\xi\right)\mathrm{d}A=-\int_{a_2}^{b_2}\int_{a_1}^{b_1}wu\left(\left(\frac{x\pi}{b_1}\right)^2+t\right)\mathrm{d}\tau\mathrm{d}\xi. \tag{12}$$

Also if we write $C=\partial E=C_1\cup C_2\cup C_3\cup C_4$ (see Figure 1) then

$$\int_C\left(uF^*\right)\cdot n\mathrm{d}s=\int_{C_1}\left(uF^*\right)\cdot n\mathrm{d}s+\int_{C_2}\left(uF^*\right)\cdot n\mathrm{d}s+\int_{C_3}\left(uF^*\right)\cdot n\mathrm{d}s+\int_{C_4}\left(uF^*\right)\cdot n\mathrm{d}s$$

$$\int_{C_1}\left(uF^*\right)\cdot n\mathrm{d}s=\int_{a_1}^{b_1}u\left(x,t,\tau,a_2\right)\underbrace{w\left(\tau,a_2\right)}_{h_1(\tau)}\mathrm{d}\tau$$

$$\int_{C_3}\left(uF^*\right)\cdot n\mathrm{d}s=-\int_{a_1}^{b_1}u\left(x,t,\tau,b_2\right)\underbrace{w\left(\tau,b_2\right)}_{h_2(\tau)}\mathrm{d}\tau$$

$$\int_{C_2}\left(uF^*\right)\cdot n\mathrm{d}s=\int_{a_2}^{b_2}u\left(x,t,b_1,\xi\right)\underbrace{w_\tau\left(b_1,\xi\right)}_{k_2(\xi)}\mathrm{d}\xi$$

$$\int_{C_4}\left(uF^*\right)\cdot n\mathrm{d}s=-\int_{a_2}^{b_2}u\left(x,t,a_1,\xi\right)\underbrace{w_\tau\left(a_1,\xi\right)}_{k_1(\xi)}\mathrm{d}\xi.$$

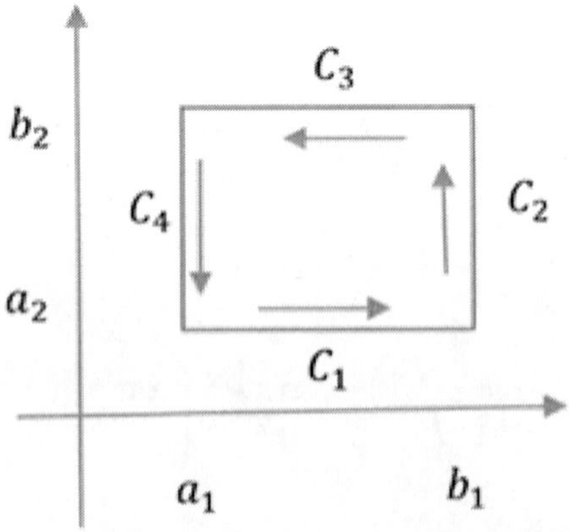

Figure 1. Domain E and contour $\partial E = C$.

We write

$$G(x,t) = \int_{C_1} (uF^*) \cdot n ds + \int_{C_2} (uF^*) \cdot n ds + \int_{C_3} (uF^*) \cdot n ds + \int_{C_4} (uF^*) \cdot n ds$$

$$R(x,t) = \iint_E r(\tau,\xi) u(x,t,\tau,\xi) \, dA$$

finally, if $x \in N$ y $a_1 = 0$ we get:

$$\int_{a_2}^{b_2} \int_{a_1}^{b_1} w(\tau,\xi) u(x,t,\tau,\xi) \, d\tau d\xi = \frac{G(x,t) - R(x,t)}{\left(\left(\frac{x\pi}{b_1}\right)^2 + t\right)} = \varphi(x,t). \tag{13}$$

If $E = \{(\tau,\xi) / \tau \geq 0; a_2 \leq \xi \leq b_2\}$ then you can take

$$u(x,t,\tau,\xi) = e^{-t\xi} \cos(x\tau) \tag{14}$$

and we must have $w(\tau,\xi) \to 0$ when $\tau \to \infty$.

3. SOLUTION OF GENERALIZED MOMENT PROBLEMS

Equation (13) is of the form:

$$\int_{a_1}^{b_1}\int_{a_2}^{b_2} w(\tau,\xi)K(x,t,\tau,\xi)\,\mathrm{d}\tau\,\mathrm{d}\xi = \varphi(x,t).$$

We assign natural values to x and t: $x = i, i = 0,1,\cdots,m$ and $t = j, j = 0,1,\cdots,n$ and we consider the corre- sponding generalized finite moment problem bi-dimensional [12] [13]

$$\int_{a_1}^{b_1}\int_{a_2}^{b_2} w(\tau,\xi)K_{ij}(\tau,\xi)\,\mathrm{d}\tau\,\mathrm{d}\xi = \varphi(i,j) = \mu_{ij} \quad i = 0,1,\cdots,m \quad j = 0,1,\cdots,n.$$

To obtain a numerical approximation of the solution $w(\tau,\xi)$ the truncated expansion method is applied [9] [11] .

We considerer the basis $\{\phi_i(\tau,\xi)\}_{i=0}^{\infty}$ obtained from the sequence $\{K_{ij}(\tau,\xi)\}$ with $i = 0,1,\cdots,m;\ j = 0,1,\cdots,n$ by Gram-Schmidt method and addition of the necessary functions in order to have an orthonormal basis.

To facilitate the calculations we write $\{K_i(\tau,\xi)\}_{i=0}^{r}$ and $\{\mu_i\}_{i=0}^{r}$ with $r = m\cdot n.$

We then approximate the solution $w(\tau,\xi)$ with

$$p_r(\tau,\xi) = \sum_{i=0}^{r}\lambda_i\phi_i(\tau,\xi)$$

and

$$\lambda_i = \sum_{j=0}^{i} C_{ij}\mu_j \qquad i = 0,1,\cdots,r$$

where the coefficients C_{ij} verifies

$$C_{ij} = \left(\sum_{k=j}^{i-1} (-1) \frac{\langle K_i(\tau,\xi) | \phi_k(\tau,\xi) \rangle}{\|\phi_k(\tau,\xi)\|^2} C_{kj} \right) \cdot \|\phi_i(\tau,\xi)\|^{-1} \quad 1 < i \le r;\ 1 \le j < i \tag{15}$$

$$C_{ii} = \|\phi_i(\tau,\xi)\|^{-1} \qquad i = 0,1,\cdots,r. \tag{16}$$

The proof of the following theorem is in [14] .

Theorem 1. Let $\{\mu_m\}_{m=0}^{r}$ be a set of real numbers and let ε and E be two positive numbers such that

$$\sum_{m=0}^{r} \left| \int_{a_2}^{b_2} \int_{a_1}^{b_1} K_m(\tau,\xi) w(\tau,\xi) \mathrm{d}\tau \mathrm{d}\xi - \mu_m \right|^2 \le \varepsilon^2 \tag{17}$$

$$\int_{a_2}^{b_2} \int_{a_1}^{b_1} \left[(b_1 - a_1)^2 w_\tau^2 + (b_2 - a_2)^2 w_\xi^2 \right] \mathrm{d}\tau \mathrm{d}\xi \le E^2 \tag{18}$$

then

$$\int_{a_2}^{b_2} \int_{a_1}^{b_1} |w(\tau,\xi)|^2 \mathrm{d}\tau \mathrm{d}\xi \le \min_m \left\{ \|CC^{\mathrm{T}}\| \varepsilon^2 + \frac{E^2}{8(m+1)^2}; m = 0,1,\cdots,r \right\} \tag{19}$$

where C is the triangular matriz with elements c_{ij} $(1 < i \le r; 1 \le j < i)$.

And

$$\int_{a_2}^{b_2} \int_{a_1}^{b_1} |p_r(\tau,\xi) - w(\tau,\xi)|^2 \mathrm{d}\tau \mathrm{d}\xi \le \|CC^{\mathrm{T}}\| \varepsilon^2 + \frac{E^2}{8(r+1)^2}. \tag{20}$$

If $E = (a_1,\infty) \times (a_2,\infty)$, then (18) it is replaced by

$$\int_{a_2}^{\infty} \int_{a_1}^{\infty} \left[\tau w_\tau^2 + \xi w_\xi^2 \right] \mathrm{Exp}[\tau + \xi] \mathrm{d}\tau \mathrm{d}\xi \le E^2$$

and we must have

$$\tau^m w(\tau,\xi) \to 0 \quad \text{if } \tau \to \infty \quad \text{for all } m \in N$$

and

$$\xi^n w(\tau,\xi) \to 0 \quad \text{if } \xi \to \infty \quad \text{for all } n \in N.$$

4. NUMERICAL EXAMPLES

4.1. Example 1

Let considerer the equation

$$w_t - w_{xx} = 0$$

in the domain $E = (0,2)\times(0,2)$

and boundary condition on ∂E

given by

$$w_x(0,t) = e^{-t} \qquad w_x(2,t) = \cos(2)e^{-t}$$

$$w(x,0) = \sin(x) \qquad w(x,2) = \sin(x)e^{-2}.$$

The exact solution is $w(x,t) = \sin(x)e^{-t}$.

In Figure 2 the approximate numerical solution and the exact one are compared.Were taken $r = 9$ moments.

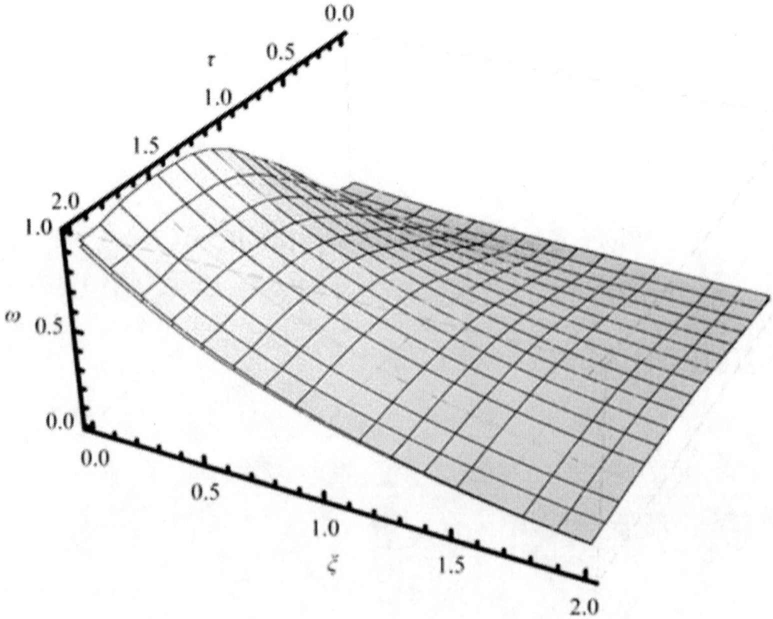

Figure 2. $w(x,t)=\sin(x)e^{-t}$ and $p_9(x,t)$.

The accuracy is, in this case

$$\int_0^2\int_0^2 \left|p_9(x,t)-w(x,t)\right|^2 \mathrm{d}x\mathrm{d}t = 0.0361472.$$

4.2. Example 2

Let considerer the equation

$$w_t - w_{xx} = -4e^{-3t+x}$$

in the domain $E=(0,1)\times(0,\infty)$ and boundary condition on ∂E given by

$$w_x(0,t)=e^{-3t} \qquad w_x(1,t)=e^{1-3t}$$

$$w(x,0) = e^{x}.$$

The exact solution is $w(x,t) = e^{x-3t}$.

In Figure 3 the approximate numerical solution and the exact one are compared.Were taken $r = 9$ moments.

To apply Gram Schmidt to $\{K_{ij}(\tau,\xi)\}$ we consider the inner product

$$\langle f_i(\tau,\xi), f_j(\tau,\xi)\rangle = \int_{a_1}^{b_1}\int_{a_2}^{\infty} f_i(\tau,\xi) f_j(\tau,\xi) e^{-\xi} d\xi d\tau.$$

The accuracy is, with this inner product $\|p_9(x,t) - w(x,t)\|^2 = 0.0250523.$

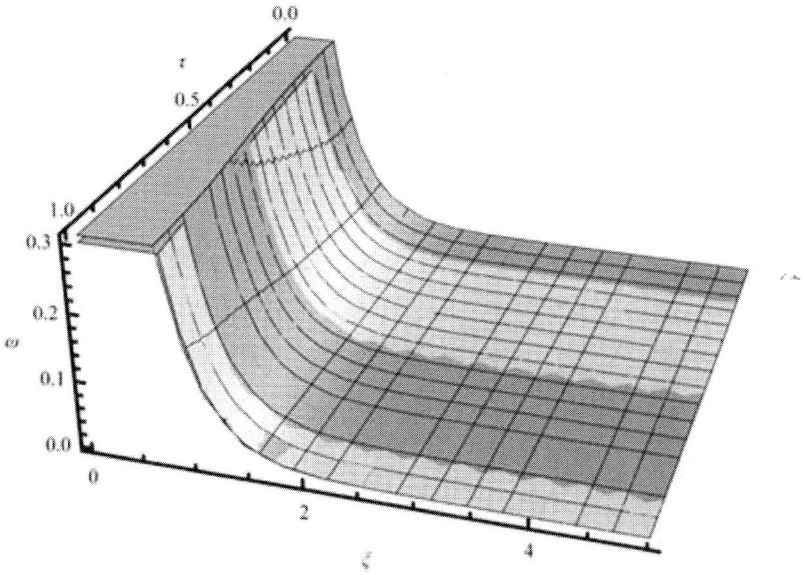

Figure 3. $w(x,t) = e^{x-3t}$ and $p_9(x,t)$.

5. THE ONE-DIMENSIONAL ONE-PHASE INVERSE STEFAN PROBLEM

5.1. The Inverse One-Phase Stefan Problem as Integral Equation

The Stefan problem consists of finding w y s such that

$$\frac{\partial w}{\partial t}=\frac{\partial^2 w}{\partial x^2} \qquad \text{in } \{(x,t);0<x<s(t);t>0\} \tag{21}$$

$$-\frac{\partial w}{\partial x}(0,t)=f(t) \qquad t>0 \tag{22}$$

$$w(s(t),t)=0 \qquad t>0 \tag{23}$$

$$\frac{ds}{dt}=-\frac{\partial w}{\partial x}(s(t),t) \qquad t>0 \tag{24}$$

$$w(x,0)=w_0(x) \qquad x\geq 0 \tag{25}$$

$$s(0)=a. \tag{26}$$

We want to solve the inverse Stefan problem: to find $f(t)$ with $s(t)$ known such that the above con- ditions are met.

We write

$$w_\xi-(w_\tau)_\tau=0. \tag{27}$$

We take the auxiliary function

$$u(x,t,\tau,\xi)=e^{-(t+1)\xi}\cos(x\tau).$$

Therefore

$$\frac{\partial^2 u}{\partial \tau^2} = u_{\tau\tau} = x^2 u \qquad \frac{\partial u}{\partial \xi} = u_\xi = -(t+1)u.$$

We consider the vector field $F^*(w) = (F_1(w), F_2(w))$ with $F_1(w) = w_\tau, F_2(w) = -w$. In this manner $div(F^*(w)) = w_{\tau\tau} - w_\xi$. In consequence if w it is solution of the Equation (27), then $div(F^*(w)) = 0$. Reciprocally, if w satisfies $div(F^*(w)) = 0$, then w it is solution of the Equation (27).

We write $E = \{(\tau,\xi); 0 < \tau < s(\xi); \xi > 0\}$. We use that

$$div(uF^*) = udiv(F^*) - F^* \cdot \nabla u$$

if $\cdot$ it is the scalar product and ∇ is the gradient operator we get

$$\iint_E udiv(F^*)\mathrm{d}A = \iint_E div(uF^*)\mathrm{d}A - \iint_E F^* \cdot \nabla u \mathrm{d}A.$$

By the divergence theorem

$$\iint_E div(uF^*)\mathrm{d}A = \int_C (uF^*) \cdot n\mathrm{d}s$$

with $C = \partial E$, in consequence

$$\iint_E udiv(F^*)\mathrm{d}A = \int_C (uF^*) \cdot n\mathrm{d}s - \iint_E F^* \cdot \nabla u \mathrm{d}A.$$

We calculate $\iint_E F^* \cdot \nabla u \mathrm{d}A$:

First we write

$$F^* \cdot \nabla u = (F_1, F_2) \cdot (u_\tau, u_\xi) = F_1 u_\tau + F_2 u_\xi$$

and $E_T = \{(\tau,\xi); 0 < \tau < s(\xi); 0 < \xi < T\}$, (Figure 4) then we take $T \to \infty$.

If $F_1^p(w) = \int F_1(w)\,\mathrm{d}\tau = w$ then

$$\iint_{E_T} F_1 u_\tau \mathrm{d}A = \int_0^T \int_0^{s(\xi)} F_1 u_\tau \mathrm{d}\tau \mathrm{d}\xi$$
$$= \int_0^T \left[F_1^p\left(w\left(s(\xi),\xi\right)\right) u_\tau\left(x,t,s(\xi),\xi\right) - F_1^p(0,\xi) u_\tau(x,t,0,\xi) \right] \mathrm{d}\xi - \int_0^T \int_0^{s(\xi)} u_{\tau\tau} F_1^p \mathrm{d}\tau \mathrm{d}\xi.$$

As $w\left(s(\xi),\xi\right) = 0$ by (23) and $u_\tau(x,t,0,\xi) = x\sin(x\cdot 0)\mathrm{e}^{-(t+1)\xi} = 0$ then

$$\iint_{E_T} F^* \cdot \nabla u \mathrm{d}A = -\int_0^T \int_0^{s(\xi)} \left(u_{\tau\tau} F_1^p(w) - F_2(w) u_\xi\right) \mathrm{d}\tau \mathrm{d}\xi = \int_0^T \int_0^{s(\xi)} uw\left(x^2 + (t+1)\right) \mathrm{d}\tau \mathrm{d}\xi. \qquad (28)$$

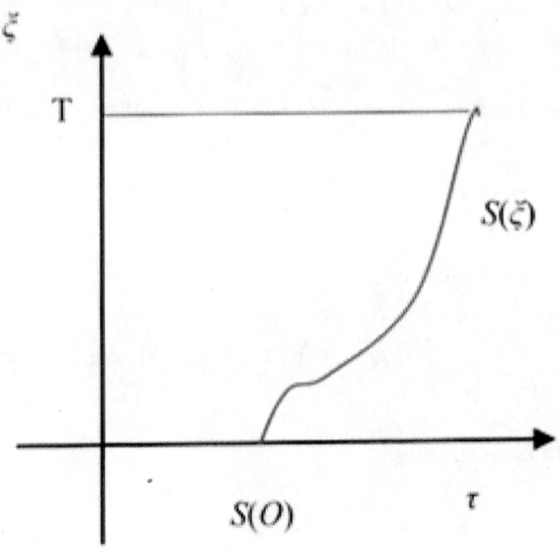

Figure 4. Domain E_T.

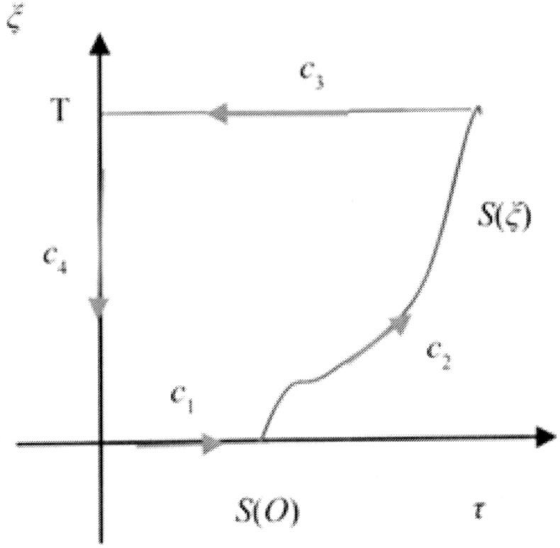

Figure 5. Domain E_T.

Now we developed $\int_C (uF^*)\cdot n ds$. Observe the Figure 5 and:

$$\int_{c1}(uF^*)\cdot n ds=\int_0^{s(0)}u(x,t,\tau,0)\,w(\tau,0)\,d\tau=\int_0^{s(0)}u(x,t,\tau,0)\,w_0(\tau)\,d\tau \quad (29)$$

$$\int_{c3}(uF^*)\cdot n ds=\int_0^{s(T)}u(x,t,\tau,T)\,w(\tau,T)\,dx\xrightarrow[T\to\infty]{}0 \quad (30)$$

$$\int_{c4}(uF^*)\cdot n ds=-\int_0^{T}u(x,t,0,\xi)\,w_\tau(0,\xi)\,d\xi \quad (31)$$

$$\int_{c2}(uF^*)\cdot n ds=-\int_0^{T}u(x,t,s(\xi),\xi)\,s_\xi(\xi)\,d\xi. \quad (32)$$

Then

$$\int_0^\infty\int_0^{s(\xi)}uw\left(x^2+(t+1)\right)d\tau d\xi$$
$$=\int_0^{s(0)}\cos(x\tau)\,w(\tau,0)\,d\tau-\int_0^\infty\cos(xs(\xi))\,e^{-(t+1)\xi}s_\xi(\xi)\,d\xi-\int_0^\infty e^{-(t+1)\xi}w_\tau(0,\xi)\,d\xi. \quad (33)$$

To solve the inverse problem, where $s(\xi)$ is known and $w_\tau(0,\xi)$ is unknown we do $x^2+t+1=0$.

In this manner:

$$\int_0^\infty e^{-(t+1)\xi} w_\tau(0,\xi)\,d\xi$$
$$= \int_0^{s(0)} \cos\left(\sqrt{-t-1}\tau\right) w(\tau,0)\,d\tau - \int_0^\infty \cos\left(\sqrt{-t-1}s(\xi)\right) e^{-(t+1)\xi} s_\xi(\xi)\,d\xi = \varphi(t). \quad (34)$$

We assign values to t: $t=0,1,2,\cdots$

$$\int_0^\infty e^{-(t+1)\xi} w_\tau(0,\xi)\,d\xi = \varphi(t) = \mu_t. \quad (35)$$

We can interpret (35) as a one-dimensional generalized moments problem.

We solve the problem numerically considering the finite generalized moments problem

$$\int_0^\infty e^{-(i+1)\xi} w_\tau(0,\xi)\,d\xi = \varphi(i) = \mu_i \qquad i=0,1,\cdots,n \quad (36)$$

and we apply the truncated expansion method.

5.2. Numerical Approximation to the Solution of the Inverse Stefan Problem

To obtain a numerical approximation of the solution the procedure is analogous to that presented in Section 3. To approximate $w_\tau(0,\xi)$ is taken a base $\phi_i(\xi), i=0,1,2,\cdots,n$ of $L^2(0,\infty)$ obtained from the sequence $g_i(\xi)=e^{-(i+1)\xi}$ by Gram-Schmidt method and necessary functions are added in order to have an orthonormal basis. We then approximate the solution $w_\tau(0,\xi)$ with:

$$p_n(\xi) = \sum_{i=0}^{n} \lambda_i \phi_i(\xi)$$

where

$$\lambda_i = \sum_{j=0}^{i} C_{ij}\mu_j \qquad i = 0,1,\cdots,n$$

and the coefficients C_{ij} verifies

$$C_{ij} = \left(\sum_{r=j}^{i-1} (-1) \frac{\langle g_i(\xi) \mid \phi_r(\xi) \rangle}{\left\| \phi_r(\xi) \right\|^2} C_{rj} \right) \cdot \left\| \phi_i(\xi) \right\|^{-1} \quad 1 < i \le n; 1 \le j < i \tag{37}$$

$$C_{ii} = \left\| \phi_i(\xi) \right\|^{-1} \qquad i = 0,1,\cdots,n. \tag{38}$$

The following theorem is the one-dimensional version of Theorem 1. In [15] is the demonstration when the domain is bounded.

We present here the demonstration when the domain is the interval $(0,\infty)$.

Theorem 2. Let $\{\mu_i\}_{i=0}^{n}$ be a set of real numbers and let ε and M be two positive numbers such that

$$\sum_{i=0}^{n} \left| \int_0^\infty g_i(\xi) x(\xi) \mathrm{d}\xi - \mu_i \right|^2 \le \varepsilon^2 \tag{39}$$

$$\int_0^\infty \left[\xi e^\xi x_\xi^2(\xi) \right] \mathrm{d}\xi \le M^2 \tag{40}$$

then if $\lim_{\xi \to +\infty} \xi^r x(\xi) = 0, \forall r \in N$

$$\int_0^\infty \left| x(\xi) \right|^2 \mathrm{d}\xi \le \min_n \left\{ \left\| C^T C \right\| \varepsilon^2 + \frac{M^2}{(i+1)}; i = 0,1,\cdots,n \right\} \tag{41}$$

and

$$\int_0^\infty |x(\xi)-p_n(\xi)|^2 \, d\xi \le \|C^T C\| \varepsilon^2 + \frac{M^2}{(n+1)} \tag{42}$$

where C is the triangular matriz with elements $C_{ij} \ (1<i\le n; 1\le j<i)$.

Proof. Since the problem is linear we can assume $\mu_i = 0, i = 0,1,\cdots,n$.

We applied Gram-Scmidt method on $\{g_i(\xi)\}_{i=0}^n$ in $L^2(0,\infty)$ and we get $\{\phi_i(\xi)\}_{i=0}^\infty$ then add the resulting set of necessary functions to obtain an orthonormal basis.

We write $x(\xi)$ as

$$x(\xi) = h_n(\xi) + f_n(\xi)$$

where $h_n(\xi)$ it is the orthogonal projection of $x(\xi)$ on the linear space generated by the set $\{g_i(\xi)\}_{i=0}^n$ and $f_n(\xi) = x(\xi) - h_n(\xi)$ it is the orthogonal projection of $x(\xi)$ on the orthogonal complement. Here the underlying structure is the space $L^2(0,\infty)$. We can write

$$h_n(\xi) = \sum_{i=0}^{n} \lambda_i \phi_i(\xi) \qquad f_n(\xi) = \sum_{i=n+1}^{\infty} \lambda_i \phi_i(\xi) \tag{43}$$

where $\lambda_i = \int_0^\infty \phi_i(\xi) x(\xi) \, d\xi, i = 0,1,\cdots$ are the Fourier coefficients in the expansion of $x(\xi)$.

To estimate $h_n(\xi)$ we consider the relationship between the Fourier coefficients λ_i and the moments

$$\mu_i = \int_0^\infty g_i(\xi) x(\xi) \, dt$$

:

$$\lambda_i = \sum_{j=0}^{i} C_{ij}\mu_j \qquad i = 0,1,\cdots,n \tag{44}$$

where C_{ij} they are given in (37) y (38).

In matrix notation

$$\lambda = \begin{pmatrix} \lambda_0 \\ \lambda_1 \\ \vdots \\ \lambda_n \end{pmatrix} \qquad \mu = \begin{pmatrix} \mu_0 \\ \mu_1 \\ \vdots \\ \mu_n \end{pmatrix}. \tag{45}$$

Then

$$\sum_{j=0}^{i} C_{ij} g_j(\xi) = \phi_i(\xi) \qquad i = 0,1,\cdots,n. \tag{46}$$

By (43) until (46) we can write

$$\int_0^\infty |h_n(\xi)|^2 \,\mathrm{d}\xi = \langle \lambda | \lambda \rangle = \langle C^{\mathrm{T}} C\mu | \mu \rangle \le \|C^{\mathrm{T}} C\| \|\mu\|^2 \le \|C^{\mathrm{T}} C\| \varepsilon^2. \tag{47}$$

To estimate f_n we see that each element of the orthonormal set $\phi_i(t)$ can be expanded in terms of the elements other orthonormal basis, in particular the base $\{l_i(\xi) = L_i(\xi)\mathrm{e}^{-\xi}\}_{i=0}^{\infty}$, where $L_i(\xi)$ it represents the Laguerre polynomial of degree i.

These polynomials satisfy

$$\xi \frac{\mathrm{d}^2}{\mathrm{d}\xi^2}\left[L_i(\xi)\right] + (1-\xi)\frac{\mathrm{d}}{\mathrm{d}\xi}\left[L_i(\xi)\right] + iL_i(\xi) = 0 \tag{48}$$

or also

$$-\frac{\mathrm{d}}{\mathrm{d}\xi}\left[\xi \frac{\mathrm{d}}{\mathrm{d}\xi}\left[L_i(\xi)\right]\right] + \xi \frac{\mathrm{d}}{\mathrm{d}\xi}\left[L_i(\xi)\right] = iL_i(\xi). \tag{49}$$

Then

$$\phi_i(\xi)=\sum_{j=0}^{\infty}\gamma_{ij}l_j(\xi) \qquad i=0,1,2,\cdots,n \tag{50}$$

then using (50)

$$f_n(\xi)=\sum_{i=n+1}^{\infty}\lambda_i\phi_i(\xi)=\sum_{i=n+1}^{\infty}\sum_{j=0}^{\infty}\lambda_i\gamma_{ij}l_j(\xi). \tag{51}$$

After several calculations

$$\int_0^{\infty}\left|f_n(\xi)\right|^2\mathrm{d}\xi=\left\|f_n(\xi)\right\|^2=\sum_{i=n+1}^{\infty}\left|\sum_{j=0}^{\infty}\lambda_j\gamma_{ji}\right|^2 \tag{52}$$

and

$$\int_0^{\infty}\left|f_n(\xi)\right|^2\mathrm{d}\xi\le\frac{1}{n+1}\sum_{i=0}^{\infty}(i+1)\left|\chi(i)\right|^2 \tag{53}$$

where

$$\chi(i)=\sum_{j=0}^{\infty}\lambda_j\gamma_{ji}. \tag{54}$$

Now multiplying by $x(\xi)$ and integrating both sides of the differential Equation (49) and assuming that $\lim_{\xi\to\infty}\xi^r x(\xi)=0, \forall r\in N$ we get:

$$\int_0^{\infty}\xi x_\xi^2(\xi)\mathrm{e}^{\xi}\mathrm{d}\xi=\sum_{i=0}^{\infty}(i+1)\chi^2(i) \tag{55}$$

then by (53) and (55):

$$\int_0^\infty |f_n(\xi)|^2 \,\mathrm{d}\xi \le \frac{1}{n+1}\int_0^\infty \xi x_\xi^2(\xi)\mathrm{e}^\xi \mathrm{d}\xi \le \frac{M^2}{n+1} \tag{56}$$

from (47) and (56):

$$\int_0^\infty |x(\xi)|^2 \,\mathrm{d}\xi \le \left\|C^\mathrm{T} C\right\|\varepsilon^2 + \frac{M^2}{(n+1)}. \tag{57}$$

This inequality remains valid if we replace any integer i between 0 and n for n. Then the result (41) it demon- strated. An analogous demonstration proves inequality (42). □

6. NUMERICAL EXAMPLE

Find $w_\tau(0,\xi)$ such that

$$\frac{\partial w}{\partial \xi} = \frac{\partial^2 w}{\partial \tau^2} \quad \text{in } \{(\tau,\xi); 0<\tau<s(\xi); \xi>0\} \quad \text{with } s(\xi)=\sqrt{1+\xi}$$

$$w(s(\xi),\xi)=0 \qquad \xi>0$$

$$\frac{\mathrm{d}s}{\mathrm{d}\xi} = -\frac{\partial w}{\partial \tau}(s(\xi),\xi) \qquad \xi>0$$

$$w(\tau,0)=w_0(\tau)=0.5\mathrm{e}^{0.25}\sqrt{\pi}\left[erf(0.25)-erf\left(\frac{x}{2}\right)\right] \qquad \tau\ge 0.$$

The exact solution is $w_\tau(0,\xi) = -\frac{\mathrm{e}^{0.25}}{2}\sqrt{\frac{1}{1+\xi}}.$

In Figure 6 the approximate numerical solution and the exact one are compared.Were taken $r=6$ moments.

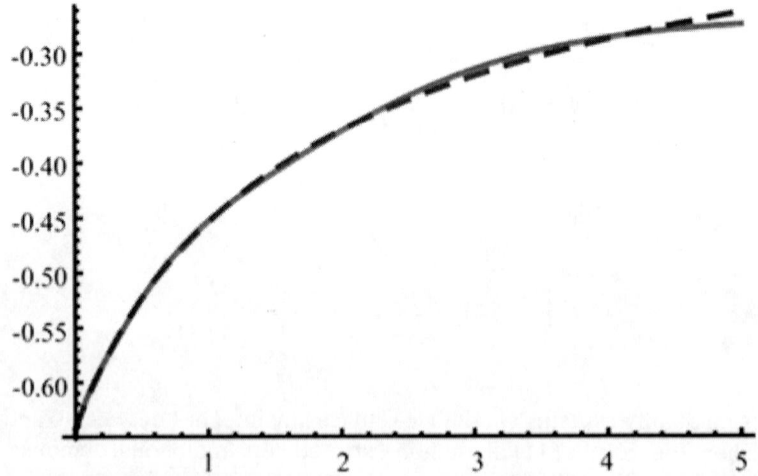

Figure 6. $w_\tau(0,\xi) = -\frac{e^{0.25}}{2}\sqrt{\frac{1}{1+\xi}}$ and $p_6(\tau,\xi)$.

To apply Gram Schmidt to $\{g_i(\xi)\}_{i=0}^{n}$ en $L^2(0,\infty)$ we considerer the inner product

$$\langle f_i(\xi), f_j(\xi)\rangle = \int_0^\infty f_i(\xi) f_j(\xi) e^{-\xi} d\xi.$$

The accuracy is, with this inner product $\|p_6(\xi) - w_\tau(0,\xi)\|^2 = 0.00306424.$

7. CONCLUSIONS

The parabolic partial differential equations

$$w_\xi - (w_\tau)_\tau = r(\tau,\xi)$$

on a region $E = (0,b_1)\times(a_2,b_2)$ can be written as an Fredholm integral equation

$$\int_{a_2}^{b_2}\int_0^{b_1} w(\tau,\xi)u(x,t,\tau,\xi)\mathrm{d}\tau\mathrm{d}\xi = \frac{G(x,t)-R(x,t)}{\left(\left(\frac{x\pi}{b_1}\right)^2+t\right)} = \varphi(x,t) \qquad x\in N.$$

This equation is of the form:

$$\int_0^{b_1}\int_{a_2}^{b_2} w(\tau,\xi)K(x,t,\tau,\xi)\mathrm{d}\tau\mathrm{d}\xi = \varphi(x,t)$$

with

$$K(x,t,\tau,\xi) = \mathrm{e}^{-t\xi}\cos\left(\frac{x\pi}{b_1}\tau\right), x\in N.$$

If $w(\tau,\xi)\in L^2(E)$, then this Fredholm integral equation of first kind can be transformed into a bi-dimen- sional generalized moment problem assigning integer values greater than or equal to zero to variables x and t

$$\int_0^{b_1}\int_{a_2}^{b_2} w(\tau,\xi)K_{ij}(\tau,\xi)\mathrm{d}\tau\mathrm{d}\xi = \mu_{ij} \quad i=0,1,2,\cdots,\ j=0,1,2,\cdots. \tag{58}$$

As the functions $\{K_{ij}(\tau,\xi)\}_{ij}$ are linearly independent then the generalized moment problem defined by (58) can be solved numerically considering the correspondent finite problem.

The inverse Stefan problem which it is to find $w_\tau(0,\xi)$ being $w(\tau,\xi)$ unknown and such that the follow- ing conditions are met

$$\frac{\partial w}{\partial \xi} = \frac{\partial^2 w}{\partial \tau^2} \qquad \text{in } \{(\tau,\xi); 0<\tau<s(\xi); \xi>0\}$$

$$-\frac{\partial w}{\partial \tau}(0,\xi) = f(\xi) \qquad \xi>0$$

$$w(s(\xi),\xi) = 0 \qquad \xi>0$$

$$\frac{\mathrm{d}s}{\mathrm{d}\xi} = -\frac{\partial w}{\partial \tau}(s(\xi),\xi) \qquad \xi>0$$

$$w(\tau,0)=w_0(\tau) \qquad \tau \geq 0$$

$$s(0)=a$$

is equivalent to solve the integral equation

$$\int_0^\infty e^{-(t+1)\xi} w_\tau(0,\xi)\,d\xi$$
$$=\int_0^{s(0)} \cos\left(\sqrt{-t-1}\tau\right) w(\tau,0)\,d\tau - \int_0^\infty \cos\left(\sqrt{-t-1}s(\xi)\right) e^{-(t+1)\xi} s_\xi(\xi)\,d\xi = \varphi(t)$$

which is equivalent to the generalized moments problem

$$\int_0^\infty e^{-(i+1)\xi} w_\tau(0,\xi)\,d\xi = \varphi(i) = \mu_i \qquad i=0,1,2,\cdots$$

and can be solved numerically considering the correspondent finite problem.

REFERENCES

1. Pintarelli, M.B. and Vericat, F. (2014) Partial Differential Equations as Three-Dimensional Inverse Problem of Moments. Journal of Mathematics and System Science, 4, 657-666.
2. Cherniha, R.M. (2011) New Exact Solutions of One Nonlinear Equation in Mathematical Biology and Their Properties. Ukrainian Mathematical Journal, 53, 393-411.
3. Mittal, R.C. and Jiwari, R. (2011) A Higher Order Numerical Scheme for Some Nonlinear Differential Equations: Models in Biology. International Journal for Computational Methods in Engineering Science and Mechanics, 12, 134-140.
4. Forsythe, G.E. and Wasow, W.R. (1960) Finite Difference Methods for Partial Differential Equations. John Wiley and Sons, New York.
5. Zafarullah, A. (1971) Some Stable Implicit Difference Methods for Heat Equation with Derivative Boundary Condition. The Computer Journal, 14, 309-311.
6. Keast, P. and Mitchell, A.R. (1966) On the Instability of the Crank Nicholson Formula under Derivative Boundary Conditions. The Computer Journal, 9, 110-114.
7. Akheizer, N.I. (1965) The Classical Moment Problem. Olivier and Boyd, Edinburgh.
8. Akheizer, N.I. and Krein, M.G. (1962) Some Questions in the Theory of Moment. American Mathematical Society, Providence.
9. Ang, D.D., Gorenflo, R., Le, V.K. and Trong, D.D. (2002) Moment Theory and Some Inverse Problems in Potential Theory and Heat Conduction. Lectures Notes in Mathematics, Springer-Verlag, Berlin.
10. Shohat, J.A. and Tamarkin, J.D. (1943) The Problem of Moments. Mathematics Survey, American Mathematical Society, Providence.
11. Talenti, G. (1987) Recovering a Function from a Finite Number of Moments. Inverse Problems, 3, 501-517.
12. Pintarelli, M.B. and Vericat, F. (2011) Bi-Dimensional Inverse Moment Problems. Far East Journal of Mathematical Sciences, 54, 1-23.

13. Pintarelli, M.B. and Vericat, F. (2012) Klein-Gordon Equation as a Bi-Dimensional Moment Problem. Far East Journal of Mathematical Sciences, 70, 201-225.
14. Pintarelli, M.B. (2015) Linear Partial Differential Equations of First Order as Bi-Dimensional Inverse Moment Problem. Applied Mathematics, 6, 979-989. Pintarelli, M.B. and Vericat, F. (2008) Stability Theorem and Inversion Algorithm for a Generalized Moment Problem. Far East Journal of Mathematical Sciences, 30, 253-274.

CHAPTER 14

On Entropy of Graph Maps That Give Hereditarily Indecomposable Inverse Limits

Jan P. Boro´nski1,2 and Piotr Oprocha2

1. National Supercomputing Centre IT4Innovations, Division of the University of OstravaInstitute for Research and Applications of Fuzzy Modeling Ostrava, Czech Republic

2. Faculty of Applied MathematicsAGH University of Science and Technology, Kraków, Poland

ABSTRACT

We prove that if f:G→G is a map on a topological graph *G* such that the inverse limit lim←−(G,f) is hereditarily indecomposable, and entropy of *f* is positive, then there exists an entropy set with infinite topological entropy. When *G* is the circle and the degree of *f* is positive then the entropy is always infinite and the rotation set of *f* is nondegenerate. This shows that the Anosov-Katok type constructions of the pseudo-circle as a minimal set in volume-preserving smooth dynamical systems, or in complex dynamics, obtained previously by Handel, Herman and Chéritat cannot be modeled on inverse limits. This also extends a previous result of Mouron who proved that if G=[0,1] then h(f)∈{0,∞} and combined with a result of Ito shows that certain dynamical systems on compact finite-dimensional Riemannian manifolds must either have zero entropy on their invariant sets or be non-differentiable.

Keywords: Topological entropy, Attractor, Hereditarily, indecomposable continuum, Pseudo-circle

1 INTRODUCTION

It is well known that hereditary indecomposability can be found in the structure of invariant sets of even very regular dynamical systems. Recall that a nondegenerate compact and connected set, *continuum*, is said to be *indecomposable* if it cannot be decomposed to the union of two proper subcontinua, and it is *hereditarily indecomposable* if every subcontinuum is indecomposable. Handel [26] showed that Bing's pseudo-circle [11], a hereditarily indecomposable cofrontier, can appear as a minimal set for a C∞C∞-smooth volume preserving planar diffeomorphism. A related example was given by Herman [28] in the complex plane. Quite recently Herman's construction has been extended by Chéritat [21] to show that the pseudo-circle can also emerge as the boundary of a Siegel disk. Other prominent examples include those given by Kennedy and Yorke, who exhibited that there exist C∞C∞-smooth dynamical systems in dimensions greater than 2 with uncountably many minimal pseudo-circles, and any small C1C1 perturbation of which manifests the same property (see [33, 34]). Curiously, it is still an open question whether hereditarily indecomposable continua admit dynamical systems with finite, nonzero entropy (see e.g. [41]). Since by a result of Ito ([31], Theorem 1) topological entropy of a diffeomorphism on a compact finite-dimensional Riemannian manifold must be finite, a potential solution could carry important new insights on the interplay between complexity of dynamics and topology of invariant sets. An important facet of this problem appears to be to determine the answer for shift homeomorphisms on inverse limits of graphs. On the one hand side, inverse limits of graphs are often used to construct attractors of dynamical systems on manifolds, where the dynamics on the attractor is conjugate to the shift homeomorphism (see e.g. [4, 6, 7, 19, 38, 43]). On the other hand, famously there are situations where the converse is true: a dynamical system on a manifold can be locally translated to the dynamics of a shift homeomorphism on an inverse system of graphs (see e.g. [9, 14, 48]). Note that, it is well known that many 1-dimensional hereditarily indecomposable continua, such as the pseudo-arc, pseudo-circle or pseudo-solenoids, can be obtained as inverse limits of graphs with a single bonding map (see [32]), and any such bonding map induces a shift homeomorphism on the inverse limit space (see Sect. 3) that, according to [8], extends to a homeomorphism of R3R3. In the case of the pseudo-arc, techniques to construct appropriate bonding maps on the unit interval were first developed by Henderson [27] (a nontransitive map with zero entropy) and Minc and Transue [37] (a transitive map, hence with positive entropy). It had been known for a while that a bonding map that generates the pseudo-arc as the inverse limit must have entropy at least log(2)/2log(2)/2, when positive [16], and recently it has been proved by Mouron [42] that such a positive entropy must be in fact infinite. The main ingredient of Mouron's proof is the notion of stretching and taut cover, especially properties of taut covers (and patterns) of hereditarily indecomposable compacta [44]. In this paper we extend Mouron's result and show that if the entropy is positive then there exists an entropy set with infinite entropy. Our approach is not restricted to interval and works for all inverse limits of topological graphs. Our result is obtained by a careful analysis of the structure of horseshoes in graph maps. The main ingredient of our proof is an old result of Brown [18] which associates a degree of "crookedness" to higher iterates of maps which define hereditarily indecomposable compacta as their inverse limits. Combined with a result of Ito, it also shows that if the attractor constructed this way in a compact finite-dimensional Riemannian manifold is hereditarily indecomposable, then the dynamical system must either have zero entropy, or be non-differentiable. In Sect. 4 we discuss an application of our results to hereditarily indecomposable circle-like continua. We show that when G is the circle and the degree of f is positive then the entropy is always infinite and the rotation set of f is nondegenerate.

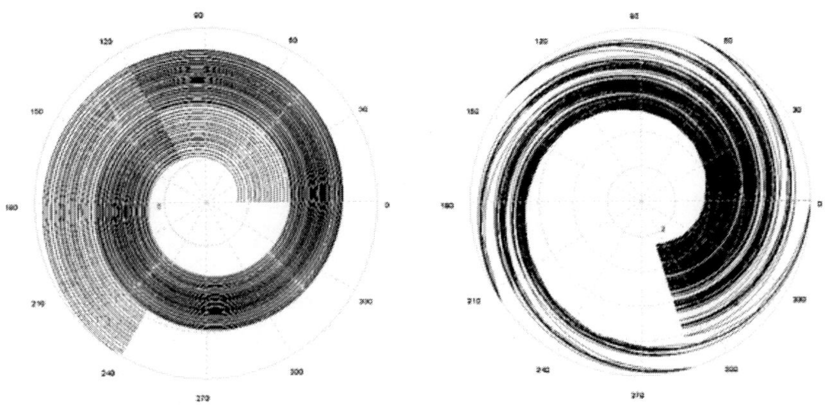

Figure 1. Strange attractors, supporting annulus homeomorphisms with non-unique rotation numbers, obtained by a construction on an inverse limit of circles, as described in [17]

This shows that the Anosov-Katok type constructions of the pseudo-circle as a minimal set in volume-preserving smooth dynamical systems, or in complex dynamics, obtained previously by Handel [26], Herman [28] and Chéritat [21] cannot be modeled on inverse limits. This resembles a known fact for Hénon-type attractors: Williams [48] showed that every hyperbolic, one-dimensional, expanding attractor for a discrete dynamical system is topologically conjugate to the induced map on an inverse limit space based on a branched one-dimensional manifold, but Barge [5] proved that certain dynamical systems with Hénon-type attractors cannot me modeled on inverse limits. In addition, authors' recent example of torus homeomorphism with an attracting pseudo-circle as Birkhoff-type attractor in [17] is non-differentiable and has infinite entropy (see Fig. 1).

2 PRELIMINARIES

2.1 Entropy

Topological entropy is one of the most common measures of complexity of dynamical systems. The reader not familiar with this notion is referred to [1, 47] for more details. Below we recall only the main definitions related to this notion.

Given two finite open covers U,V of X we define the cover U∨V by

$$\mathcal{U} \vee \mathcal{V} = \{U \cap V : U \in \mathcal{U}, V \in \mathcal{V}\}.$$

For each $n \geq 0$ we denote $T^{-n}(\mathcal{U}) = \{T^{-n}(U) : U \in \mathcal{U}\}.$ }. Similarly, for any $0 \leq m \leq n$ define

$$\mathcal{U}_m^n = T^{-m}(\mathcal{U}) \vee T^{-m-1}(\mathcal{U}) \vee \cdots \vee T^{-n}(\mathcal{U}).$$

Denote by $N(\mathcal{U})$ the minimal cardinality among all subcovers $\mathcal{V} \subseteq \mathcal{U}$ and then define

$$h_{\text{top}}(f, \mathcal{U}) = \limsup_{n \to +\infty} \frac{1}{n} \log N(\mathcal{U}_0^{n-1}).$$

Clearly, if VV is a subcover of U $\mathcal{U}$ then $h_{\text{top}}(f, \mathcal{U}) \le h_{\text{top}}(f, \mathcal{V})$. The *topological entropy of (X, f)* is defined by $h_{\text{top}}(f) = \sup_{\mathcal{U}} h_{\text{top}}(f, \mathcal{U})$ where the supremum is taken over all open covers UU of *X*.

For $n \in \mathbb{N}$, the d_n-distance between $x, y \in X$ is defined as

$$d_n(x, y) := \max_{0 \le i \le n-1} \{d(f^i x, f^i y)\}.$$

Fix a set $A \subset X$ and $\varepsilon > 0$. A set $S \subset A$ is (n, ε)-*separated for A* if $d_n(x, y) > \varepsilon$ for any distinct x,y∈Sx,y∈S. Denote $x, y \in S$. Denote $s_n(A, \varepsilon) = \sup\ \{|S| : S \text{ is } (n, \varepsilon) - \text{separated for } A\}$ where |*S*| denotes the cardinality of *S*.

For any compact set A⊂XA⊂X we define its *topological entropy* by

$$h_{\text{top}}(f, A) = \lim_{\varepsilon \to 0} \lim_{n \to \infty} \frac{\log s_n(A, \varepsilon)}{n}.$$

Since *X* is compact, it is well known that $h_{\text{top}}(f) = h_{\text{top}}(f, X)$. If the map *f* is clear from the context we simply write $h_{\text{top}}(f, \mathcal{U}) > 0$.

Entropy pairs were introduced by Blanchard in [12] and later studied by various authors. The main feature related to entropy pairs is that the dynamical system has positive topological entropy if and only if it has an entropy pair. An important generalization of this concept are entropy sets, introduced in [22]. A set K⊂XK⊂X with at least two points is an *entropy set* if for any finite open cover UU of *X* such that *K* is not contained in the closure of any of its elements we have htop(f,U)>0htop(f,U)>0.

It is much easier to detect if a set is an entropy set, if we use the concept IE-tuples (the terminology comes from [35], but the concept appeared earlier in the literature, e.g. [29, Theorem 8.2]). We call a tuple $x = (x_1, \ldots, x_k) \in X^k$ an *IE-tuple* if for every product neighborhood $U_1 \times \ldots \times U_k$ of x, the tuple $(U_1, \ldots, U_k)$ has an independence set T⊂NT⊂N of positive upper density, that is for every finite subset P⊂TP⊂T and function $\{1, 2, \ldots, k\}$ we have $\bigcap_{j \in P} f^{-j}(U_{\zeta(j)}) \neq \emptyset$. Recall that the *upper density* of a set A⊂NA⊂N is the number lim sup $_{n \to \infty} \frac{1}{n} \#(A \cap [0, n))$. It is known (see [13, 29, 35]) that a set *K*, which is not a singleton, is an entropy set iff every finite selection of its points x1,...,xn∈K forms an IE-tuple.

Note that the closure of an entropy set is an entropy set, so in this paper we will consider only closed entropy sets. By the same observation (and Zorn lemma), every entropy set is always contained in a maximal entropy set (in the sense of inclusion).

Positive entropy implies existence of entropy sets. It may happen however that there is no entropy set with entropy close to entropy of the map. The following simple example shows that entropy can be infinite, while every entropy set has finite entropy.

Example 2.1

Call a map g on a nondegenerate interval $[a, b]$ n-fold, n≥2n≥2 if

$$g(a + j(b - a)/n) = a + (b - a)(1 + (-1)^{j+1})/2$$

for j=0,...,nj=0,...,n.

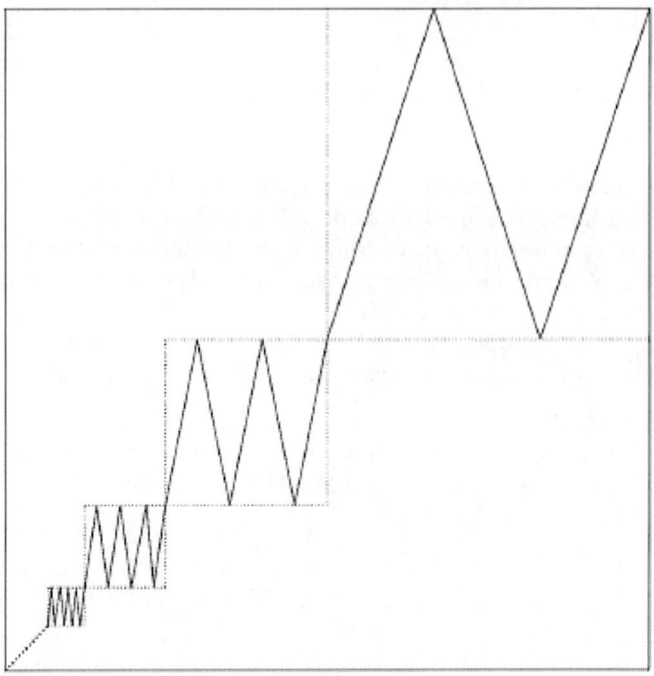

Figure 2. Sketch of the graph of map g in Example 2.1

Put $I_n = [1/2^n, 1/2^{n-1}]$ for $n = 1, 2, \ldots$ and let $f\colon [0,1] \to [0,1]$ be defined $f(0) = 0$ and $f|_{I_n}$ is $(2n+1)$- fold map (see Fig. 2). Observe that *f* is continuous and $h_{\text{top}}(f) = \infty$ since entropy of *n*-fold map is lognlogn.

But every entropy set of *f* must be contained completely in one of the intervals InIn and hence its entropy is finite.

2.2 Graphs and Horseshoes

A *(topological) graph* is a continuum *G* which can be represented as a union of finitely many arcs, any two of them having at most one point in common. Each topological graph is homeomorphic to a simplical complex embedded in the Euclidean space R3R3 (see [40]), hence we may consider each graph *G* endowed with the taxicab metric, that is, the distance between any two points of *G* is equal to the length of the shortest arc in *G* joining these points. A set I⊆G is a *closed interval* if there is a homeomorphism φ:[0,1]→I I such that the set φ((0,1))φis open in *G*. A *path* in a graph *G* is any map ω:[0,1]→G. If ω:[0,1]→G is a homeomorphism on its image and q,r∈[0,1] then we say that ω([q,r]) is an *arc* in *G*and we denote it by ⟨a,b⟩, where a=ω(q)a=ω(q) and b=ω(r)b=ω(r). In such a case, without loss of generality, we shall always assume that ωω is an *isometry* and we shall write t1<t2t1<t2 (resp. t1≤t2t1≤t2) for $t_1 < t_2$ (resp. $t_1 \leq t_2$) for $t_1, t_2 \in \omega([0,1])$ if $\omega^{-1}(t_1) < \omega^{-1}(t_2)$ $\omega^{-1}(t_1) \leq \omega^{-1}(t_2)$).

We say that the set $(a,b) = \langle a,b \rangle \setminus \{a,b\}$ is an *open arc*. Note that sometimes (*a*, *b*) is not open in *G* hence it is not interior of ⟨a,b⟩⟨a,b⟩ in *G* (however it is an open set in ⟨a,b⟩⟨a,b⟩).

Given a map f:G→Gf:G→G and arcs *I*, *J* in *G* we say that *If-coversJ* if there exists an arc K⊆IK⊆Isuch that f(K)=Jf(K)=J. Properties of *f*-covering relation presented below are adapted from [2, p. 590].

Lemma 2.2

Let I,J,K,L⊆GI,J,K,L⊆G be arcs, and let f,g:G→Gf,g:G→G.

1) If *If*-covers *I*, then there exists x∈Ix∈I such that f(x)=xf(x)=x.
2) If I⊆KI⊆K, L⊆JL⊆J and *If*-covers *J*, then *Kf*-covers *L*.
3) If *If*-covers *J* and *Jg*-covers *K*, then *I*(g∘f)(g∘f)-covers *K*.
4) If J⊆f(I)J⊆f(I), and K1,K2⊆JK1,K2⊆J are arcs such that K1∩K2K1∩K2 is at most one point, then *If*-covers K1K1, or *If*-covers K2K2.

Fix an integer s≥2s≥2. An *s-horseshoe* for *f* is an arc I⊆GI⊆G and subarcs J1,...,JsJ1,...,Js of *I* with pairwise disjoint interiors, such that $f(S_j) = I$ for $j = 1, \ldots, s$. An s-horseshoe is *strong* if in addition all the arcs JiJi are contained in the interior of *I* and are pairwise disjoint. It is not hard to show that if *f* has an *s*-horseshoe for s≥4s≥4 then it has an (s−2)(s−2)-strong horseshoe [36].

Definition 2.3

Let f:G→Gf:G→G be a map of a graph *G* with metric *d* and let ε>0ε>0. A path ω:[0,1]→Gω:[0,1]→G is (f,ε)(f,ε)-*crooked,* if there exist *s* and *t* with 0<s≤t<10<s≤t<1 such that $d(f\circ\omega(s), f\circ\omega(1)) \leq \varepsilon$ and $d(f\circ\omega(t), f\circ\omega(0)) \leq \varepsilon$ (see Fig. 3).

Remark 2.4

It is an immediate consequence of the definition, that if for some ε>0and n $n > 0$ the path ω: $[0, 1] \to G$ is (f^n, ε)-*crooked* then it is also (fn+1,ε)(fn+1,ε)-*crooked.*

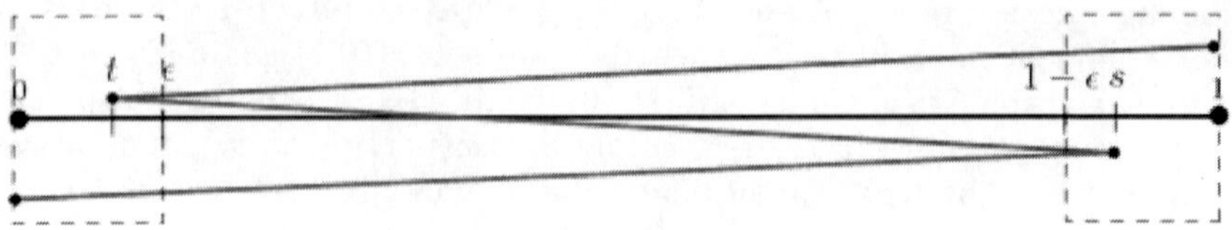

Figure 3. ϵϵ-crooked arc

3 GRAPHS, INVERSE LIMITS, CROOKEDNES AND ENTROPY SETS

Given a map $f : X \to X$ on a metric space *X*, the *inverse limit space* $X_f = \varprojlim(X, f)$ is the space given by

$$X_f = \left\{(x_1, x_2, x_3, \ldots) \in X^{\mathbb{N}} : f(x_{i+1}) = x_i\right\}.$$

The topology of XfXf is induced from the product topology of XNXN, with the basic open sets in XfXfgiven by

$$U_{\leftarrow} = \left(f^{i-1}(U), f^{i-2}(U), \ldots, U, f^{-1}(U), f^{-2}(U), \ldots\right).$$

where U is an open subset of the i-th factor space X. The map f is called a *bonding map* and there is a natural homeomorphism σf: $\sigma_f : X_f \to X_f$, called the *shift homeomorphism*, given by

$$\sigma_f(x_1, x_2, x_3, \ldots) = (f(x_1), f(x_2), f(x_3), \ldots) = (f(x_1), x_1, x_2, \ldots).$$

It is well known that σfσf preserves many dynamical properties of f. In particular, topological entropies of f and σfσf are the same (see [20] for details).

The following result of Brown [18, Lemma 3] will be crucial in the proof of our main results.

Lemma 3.1

Let G be a topological graph. If the inverse limit $\varprojlim(G, f)$ is hereditarily indecomposable then for every δ>0δ>0 there is n>0n>0 such that each path ω $[0, 1] \to G$ *is* (f^n, δ)*-crooked.*

We will also need an important extension of the result of Misiurewicz and Szlenk first proved for interval maps, connecting topological horseshoes with positive topological entropy [36, Theorem B].

Lemma 3.2

If f:G→Gf:G→G has positive topological entropy (i.e. $h_{top}(f) > 0)$ then there exist sequences $\{m_i\}_{i=1}^{\infty}$ *and* $\{k_i\}_{i=1}^{\infty}$ of positive integers such that for each n the map fmnfmn has a knkn-horseshoe and

$$\limsup_{n\to\infty} \frac{1}{m_n} \log(k_n) = h_{top}(f).$$

If $\langle a', b' \rangle$ is an arc, $a' < a < b < b'$ and $\varepsilon > 0$ is such that $d(a, a') \geq \varepsilon$ and $d(b, b') \geq \varepsilon$ then, since distance on G is given by arc length, there are points $a' \leq a_\varepsilon \leq a$ and $b \leq b_\varepsilon \leq b'$, such that $d(a, a_\varepsilon) = \varepsilon$ and $d(b, b_\varepsilon) = \varepsilon$. Therefore we will use the following natural notation $\langle a - \varepsilon, b + \varepsilon \rangle = \langle a_\varepsilon, b_\varepsilon \rangle$.

Lemma 3.3

Let G be a topological graph, $\varepsilon>0\varepsilon>0$ be a number, and let $\langle a-\varepsilon, b+\varepsilon\rangle$ be arcs in G. Fix any even integer $k>0$ *and let* $\delta<\varepsilon/6k$. *If* $g: G \to G$ is a map such that $\langle a-\epsilon, b+\epsilon\rangle \subseteq g(\langle c, d\rangle)$ and each path in G is $(g,\delta)(g,\delta)$-crooked then there are points $p_i, l_i \in \langle c, d\rangle$ such that

$$c \leq l_1 < p_1 < \ldots < l_k < p_k \leq d$$

and such that one of the following conditions holds:

(i) $g(l_i) \in \langle a-\varepsilon, a\rangle,\ g(p_i) \in \langle b, b+\varepsilon\rangle$ *for all* $1 \leq i \leq k$, *or*

(ii) $g(p_i) \in \langle a-\varepsilon, a\rangle,\ g(l_i) \in \langle b, b+\varepsilon\rangle$ *for all* $1 \leq i \leq k$.

Proof

There are $c \leq l_1 < p_k \leq d$ such that $g(l_1) = a-\varepsilon/2,\ g(p_k) = b+\varepsilon/2$ or $g(p_k) = a-\varepsilon/2,\ g(l_1) = b+\varepsilon/2$.

Let us assume that $g(l_1) = a-\varepsilon/2,\ g(p_k) = b+\varepsilon/2$ (for the second case the proof is the same, and corresponds to (ii)). Since any path in G is $(g,\delta)(g,\delta)$-crooked, there are $l1<p1<lk<pkl1<p1<lk<pk$ such that

$$d(g(p_1), b+\varepsilon/2) < \delta \text{ and } d(g(l_k), a-\varepsilon/2) < \delta.$$

Next, let us assume that for some $j\geq 1j\geq 1$, $j<k/2j<k/2$ there are points $c \leq l_1 < p_1 < \ldots < l_j < p_j < l_{k-j+1} < p_{k-j+1} < \ldots < l_k < p_k \leq d$ such that for $i=1,2,\ldots,ji=1,2,\ldots,j$ we have

$$g(p_{k-i}), g(p_i) \in \langle b+\varepsilon/2-2(i-1)\delta, b+\varepsilon/2+2(i-1)\delta\rangle,$$

and

$$g(l_{k-i}), g(l_i) \in \langle a-\varepsilon/2-2(i-1)\delta, a-\varepsilon/2+2(i-1)\delta\rangle.$$

Using again the fact that g is $\delta\delta$-crooked, there are

$$p_i < l_{i+1} < p_{k-i} < l_{k-i+1}$$

such that

$$\begin{aligned} g(l_{i+1}) &\in \langle g(l_i) - \delta, g(l_i) + \delta\rangle \\ &\subseteq \langle a - \varepsilon/2 - (2i-1)\delta, a - \varepsilon/2 + (2i-1)\delta\rangle , \end{aligned}$$

and

$$\begin{aligned} g(p_{k-i}) &\in \langle g(p_{k-i+1}) - \delta, g(l_{k-i+1}) + \delta\rangle \\ &\subseteq \langle b + \varepsilon/2 - (2i-1)\delta, b + \varepsilon/2 + (2i-1)\delta\rangle . \end{aligned}$$

Similarly, there are

$$l_{i+1} < p_{i+1} < l_{k-i} < p_{k-i}$$

such that

$$\begin{aligned} g(p_{i+1}) &\in \langle g(p_{k-i}) - \delta, g(l_{k-i}) + \delta\rangle \\ &\subseteq \langle b + \varepsilon/2 - 2i\delta, b + \varepsilon/2 + 2i\delta\rangle . \\ g(l_{k-i}) &\in \langle g(l_{i+1}) - \delta, g(l_{i+1}) + \delta\rangle \\ &\subseteq \langle a - \varepsilon/2 - 2i\delta, a - \varepsilon/2 + 2i\delta\rangle , \end{aligned}$$

By induction on j=1,2,...,k/2j=1,2,...,k/2, we eventually obtain that there are $c \le l_1 < p_1 < \quad \ldots < l_k < p_k \le d$ such that for i=1,...,ki=1,...,k we have

$$g(l_i) \in \langle a - \varepsilon/2 - k\delta, a - \varepsilon/2 + k\delta\rangle \subseteq \langle a - \varepsilon, a\rangle$$

and

$$g(p_i) \in \langle b + \varepsilon/2 - k\delta, b + \varepsilon/2 + k\delta\rangle \subseteq \langle b, b + \varepsilon\rangle .$$

The proof is complete.ϒ⌋

We have now collected enough tools to prove that the entropy of maps with hereditarily indecomposable inverse limits cannot be bounded, because the size of horseshoes grows very fast when the iteration increases.

Lemma 3.4

Let G be a topological graph, assume that the inverse limit $\varprojlim(G, f)$ is hereditarily indecomposable. Let J, L be closed intervals such that J⊂intLJ⊂intL and there is n>0n>0 such that *J*fnfn-covers *L*. Then for every α>0α>0 there are m,k>0m,k>0 and pairwise disjoint intervals $J_1, \dots, J_k \subset J$ *such that* J_i f^{mn}*-covers L for each* $i = 0, \dots, k$ and

$$\frac{1}{mn} \log(k) > \alpha.$$

Proof Fix any integer $\tilde{k} > 0$ such that $\frac{1}{n} \log(\tilde{k}) > \alpha$. Denote $g = f^n$ and $r = \tilde{k}^2$. Without loss of generality we may assume that $g(J) = L$. Therefore, there are ε>0ε>0 and a<ba<b such that $J \subseteq (a, b) \subset (a - 2\varepsilon, b + 2\varepsilon) \subset L$. Fix any δ<ε/6rδ<ε/6r and let *s* be provided by Lemma 3.1, that is each path in (g^s, δ) crooked. Arc *J* for *g* can be "iterated" for gsgs, that is we may replace *J*by its subinterval J′J′ in such a way that $g^s(J') = L$.

Let us apply Lemma 3.3 to $\langle c', d' \rangle = J' \subset J = \langle c, d \rangle$ obtaining numbers $c' \leq l_1 < p_1 < \dots < l_r < p_r \leq d'$, and for simplicity of notation suppose that $g^s(l_j) \in \langle a - \varepsilon, a \rangle$ $g^s(p_j) \in \langle b, b + \varepsilon \rangle$ (the arguments for the symmetric case $g^s(p_j) \in \langle a - \varepsilon, a \rangle$ and $g^s(l_j) \in \langle b, b + \varepsilon \rangle$ are the same). Then, since we may view gs:J′→Lgs:J′→Las a map between subintervals of the real line, a simple argument shows that each arc $\langle l_j, p_j \rangle$ g^s - covers *J*. Hence, there are arcs $J_1, \dots, J_r$ such that $g^s(J_i) = J$ and since $g(J) = L$ we obtain that each J_i g^{s+1}-covers L. But each path in G is also (g^{s+1}, δ)- crooked, so we can continue this procedure recursively. After *s* steps, we obtain pairwise disjoint arcs $J_{i_1, i_1, i_2, \dots, i_s} \subseteq J$, $i_1, \dots, i_s \in \{1, \dots, r\}$, such that each arc $J_{i_1, i_2, \dots, i_s}$ g^{2s}-cover L. Then there are at least rsrs pairwise disjoint intervals which g^{2s}-cover $\langle a - \varepsilon, b + \varepsilon \rangle$.

But then easy calculation yields that

$$\frac{1}{n}\frac{1}{2s}\log(r^s) \geq \frac{1}{2n}\log(r) \geq \frac{1}{n}\log(\tilde{k}) > \alpha$$

hence it is enough to put k=rsk=rs and m=2sm=2s to end the proof.□

Theorem 3.5

Let G be a topological graph. If the inverse limit $\varprojlim(G, f)$ is hereditarily indecomposable and $h_{top}(f) > 0$ then there exists an entropy set A such that $h_{top}(A) = \infty$.

Proof

Take any $\alpha > 0$ such that $h_{top}(f) > \alpha$. By Lemma 3.2 there are s,k>0s,k>0 such that $\frac{1}{s}\log(k) > \alpha$ and a strong $(k+1)$-horseshoe $J_0, \ldots, J_k \subset int L$ for f^s.

Next we will present a method for enlarging invariant sets induced by horseshoes. For this purpose, suppose that we have constructed pairwise disjoint intervals $J_0, \ldots, J_k \subset int L$ and there is $s > 0$ such that each J_i f^s-covers L. By Lemma 3.4 for every N>0N>0 there are r,m>0r,m>0and pairwise disjoint intervals $I_0, I_1, \ldots, I_r \subset J_0$ such that I_i f^{sm}-covers L for each i=0,...,ri=0,...,r, and

$$\frac{1}{sm}\log(r) > N\alpha.$$

Furthermore, by the definition of horseshoe, for every $w \in \{1, \ldots, k\}^m$ there is an interval $J(w)$ such that $f^{sj}(J(w)) \subset J_{w(j)}$ for all $0 \leq j < m$ and $f^{sm}(J(w)) = L$. Simply we put $J(w) = \bigcap_{j=0}^{m-1} f^{-sj}(J_{w_j})$. In particular intervals $J(w), w \in \{1, \ldots, k\}^m$ form a (strong) horseshoe for fsmfsm in L. Intervals $I_0, I_1, \ldots, I_r \subset J_0$, hence we can add them to the collection of intervals $J(w)$ obtaining a larger horseshoe (with km+r+1km+r+1 elements).

Now we will present a construction from the proof of Proposition II.15 in [15]. While the construction is standard, we will need to make some essential modifications in two places, so we need to recall this construction briefly. For each a∈{1,...,k}a∈{1,...,k} write J(a)=JaJ(a)=Ja. Since J1,...,JkJ1,...,Jk is a strong horseshoe for fsfs, for every a,b∈{1,...,k}a,b∈{1,...,k} we can find an interval $J(ab) \subset J(a)$ such that $f(J(ab)) = J(b)$.

Next, fix n and assume that for any sequence $a_1, \ldots, a_n \in \{1, \ldots, k\}$ we have already constructed intervals

$$J(a_1 a_2 \ldots a_n) \subset J(a_1 a_2 \ldots a_{n-1}) \subset \ldots \subset J(a_1 a_2) \subset J(a_1) \tag{1}$$

and $f^s(J(a_1 a_2 \ldots a_k)) = J(a_2 \ldots a_k)$ for every $2 \leq k \leq n$

By the above construction, for any sequence $\xi \in \{1, \ldots, k\}^{\mathbb{N}}$ the set $A_\xi = \bigcap_{i=1}^{\infty} J(\xi_1 \xi_2 \ldots \xi_i)$ is a nonempty interval, because it is an intersection of a nested sequence of closed intervals. Let ΛΛbe a set that consists of endpoints of all possible sets AξAξ. That way we obtain an invariant set Λ⊂LΛ⊂L for fsfs and a continuous onto map $\pi : \Lambda \to \Sigma_k$ such that ππ is one-to-one for all but countably many points, $\Sigma_k \to \Sigma_k$ is the full shift on k symbols and π∘fs=σ∘ππ∘fs=σ∘π. Namely, ππ sends endpoints of AξAξ to ξξ but it may happen that some (but at most countably many) intervals AξAξ are nondegenerate.

We have to make one important remark at this point. In the case that AξAξ is an interval, it could happen that one of its endpoints is isolated in ΛΛ. To remedy this situation we can take x∈Λx∈Λsuch that π(x)π(x) has dense orbit in ΣkΣk and π−1π(x)π−1π(x) is a singleton. Then x is recurrent, $\hat{\Lambda} = \overline{\{f^{sn}(x) : n \geq 0\}}$ is a closed fsfs-invariant set without isolated points, and π(Λ^)=Σkπ(Λ^)=Σk. Therefore in what follows, we assume that sets ΛΛ constructed by the use of horseshoe are without isolated point.

Furthermore observe that if we take intervals $J_w = J(w),\ w \in \{1, \ldots, k\}^m$m then we obtain a strong horseshoe for fsmfsm. If we fi f^{sm}. If we fix $w_1, w_2 \in \{1, \ldots, k\}^m$ then by the construction $J(w_1 w_2) \subset J_{w_1}$ and $f^{sm}(J(w_1 w_2)) = J_{w_2}$. Hence, in the construction of invariant set for fsmfsmand horseshoe $J_w,\ w \in \{1, \ldots, k\}^m$ we can take $J((w_1)(w_2)) = J(w_1 w_2)$ and extend it inductively on the longer sequences of symbols $J((w_1)(w_2) \ldots (w_n)) = J(w_1 w_2 \ldots w_n)$. Condition (1) ensures that

$$A_\xi = \bigcap_{i=1}^{\infty} J(\xi_1 \xi_2 \ldots \xi_i) = \bigcap_{i=1}^{\infty} J(\xi_1 \xi_2 \ldots \xi_{ni})$$
$$= \bigcap_{i=1}^{\infty} J((\xi_1 \ldots \xi_n)(\xi_{n+1} \ldots \xi_{2n}) \ldots (\xi_{n(i-1)+1} \ldots \xi_{ni}))$$

using horseshoe $J(w),\ w \in \{1,\ldots,k\}^m$ we obtain the same set ΛΛ, but now it is invariant for fsmfsm and connecting factor map is now $\hat{\pi} : \Lambda \to \Sigma_{k^m}$. This immediately implies, that using the above extension of horseshoe, if ΛΛ was fsfs-invariant set produced using horseshoe $J_1,\ldots,J_k$ and Γ is f^{sm}-invariant set constructed using horseshoe consisting of sets $J(w),\ w \in \{1,\ldots,k\}^m$ m together with $I_i,\ 0 \le i \le r$ then $\Lambda \subset \Gamma$ and we can extend π^π^ to a map $\hat{\pi} : \Gamma \to \Sigma_{k^m+r}$. As before, we may assume that ΓΓ does not have isolated points.

By the construction $h_{top}(\Lambda) \ge \frac{1}{s}\log(k)$ and

$$h_{top}(\Gamma) = h_{\text{top}}(f^{sm}|_\Gamma) = \frac{1}{sm}\log(k^m + r) > \frac{1}{sm}\log(r).$$

Recall that if ππ is a factor map, and x1,...,xnx1,...,xn is an IE-tuple, then the fiber $\pi \times \ldots \times \pi(\{(x_1,\ldots,x_n)\})$ contains at least one IE-*n*-tuple (see elementary properties of entropy tuples, e.g. in [29]). But ππ is one-to-one on all but countably many points and sets Λ,ΓΛ,Γdo not have isolated points, hence it is not hard to see that ΛΛ (the same for ΓΓ) is an entropy set, since ΣkΣk is an entropy set. Namely, in arbitrarily neighborhood of any tuple of distinct points x1,...,xnx1,...,xn in ΛΛ we can find a tuple which is IE *n*-tuple, and hence (x1,...,xn)(x1,...,xn) is IE *n*-tuple itself. Then by the above procedure we can construct a sequence of closed entropy sets $\Lambda_1 \subset \Lambda_2 \subset \ldots$ such that $h_{top}(\Lambda_n) \ge n\alpha$. Now, if we denote $D = \overline{\bigcup_j \Lambda_j}$ then clearly *D* is also an entropy set, but now $h_{top}(f) > 0\ then\ h_{top}(f) = \infty.$

As a consequence of Lemma 3.4 we easily obtain the following result. It is a direct consequence of Theorem 3.5, but proving it this way would be overcomplicated.

Corollary 3.6

Let *G* be a topological graph. If the inverse limit lim←−(G,f)lim←(G,f) is hereditarily indecomposable and $h_{top}(f) > 0\ then\ h_{top}(f) = \infty.$

Proof

If $h_{\text{top}}(f) > 0$ then by Lemma 3.2 there is *n* such that fnfn has a 3-horseshoe. In particular, if we denote by *J* the middle interval of three intervals defining 3-horseshoe, then J satisfies assumptions of Lemma 3.4 and hence for every $\alpha > 0$ there are m, k such that

$$h_{\mathrm{top}}(f) = \frac{1}{mn} h_{\mathrm{top}}(f^{mn}) \geq \frac{1}{mn} \log(k) > \alpha.$$

The proof is finished.ϒ⌋

Corollary 3.6 has an important consequence for smooth dynamical systems on compact finite-dimensional Riemannian manifolds, when combined with the following result of Ito.

Theorem 3.7

(Ito [31]) Let (M, g) be a compact n-dimensional Riemannian manifold and $F : M \to M$ *a* C^1*-diffeomorphism*. Then htop(F)htop(F), the topological entropy of *F*, is finite.

Corollary 3.8

Let (M, g) be a compact n-dimensional Riemannian manifold and $\underleftarrow{\lim}(G, f)$ be a homeomorphism with an invariant hereditarily indecomposable continuum $F(X) = X$. *If* $F|X$ is conjugate to a shift homeomorphism on a graph inverse limit $\underleftarrow{\lim}(G, f)$ then either $h_{top}(F) = 0$ *or* F is non-differentiable.

4 CIRCLE-LIKE HEREDITARILY INDECOMPOSABLE CONTINUA AND ENTROPY

There is a well known example of Henderson [27], who proved that shift homeomorphism on the pseudo-arc can have zero topological entropy. For the pseudo-circle, there is also an important example of Handel [26] who constructed a homeomorphism on the pseudo-circle (extendable to the whole plane) with zero topological entropy. Related results in complex dynamics were obtained by Herman [28] and Chéritat [21]. The natural question is whether the example of [27] can be somehow generalized to obtain the pseudo-circle as an inverse limit with one bonding map and with zero entropy. So far such an example was not constructed. We will show, that the reason for lack of such an example is very natural. Simply, such an example does not exist. This situation is in some sense similar to the case of Hénon-type attractors. First, Williams showed that every hyperbolic, one-dimensional, expanding attractor for a discrete dynamical system is topologically conjugate to the induced map on an inverse limit space based on a branched one-manifold [48]. Later Barge showed, however, that certain dynamical systems with Hénon-type attractors cannot me modeled on inverse limits [5].

In this section we are going to show that if $\Lambda_f = \varprojlim(\mathbb{S}^1, f)$ is hereditarily indecomposable, but ΛfΛf is not the pseudo-arc, then $h_{\text{top}}(f) = \infty$. In particular, one cannot construct a finite entropy homeomorphism on the pseudo-circle, or pseudo-solenoids conjugate to the shift homeomorphism on $\varprojlim(\mathbb{S}^1, f)$. Recall that hereditarily indecomposable circle-like continua were classified by Fearnley [25] and Rogers [45]. The classification depends on the degree of the bonding map. According to the classification any continuum in this class is one of the following:

- $(\deg(f) = 0)$ Knaster's *pseudo-arc*: the unique arc-like hereditarily indecomposable continuum.
- $(|\deg(f)| = 1)$ R.H. Bing's *pseudo-circle*: the unique plane separating hereditarily indecomposable circle-like continuum.
- $(|\deg(f)| > 1)$ *Nonplanar pseudosolenoids:* an uncountable family of continua.

Note that every proper subcontinuum of either hereditarily indecomposable circle-like continuum must be arc-like and hereditarily indecomposable so it is homeomorphic to the pseudo-arc.

In what follows we will use some standard fact on rotation numbers and sets for maps on the circle. The reader not familiar with this topic is refereed to Chapter 3 in [1].

We will need the following result of Auslander and Katznelson [3], which extends the classification of periodic-point-free circle homeomorphisms, and implies that every periodic-point-free circle map is semi-conjugate to an irrational rotation.

Theorem 4.1

Suppose $f : \mathbb{S}^1 \to \mathbb{S}^1$ *is such that* $\text{Per}(f) = \emptyset$. *Then*

(1) *for every* $x \in \mathbb{S}^1$ *there is an interval* $J_x = [a_x, b_x]$ *such that* $x \in J_x$,

(2) $f(J_x)$ *is an interval with endpoints* $f(a_x)$ *and* $f(b_x)$,

(3) $f^m(x) \notin f(J_x)$ *for* $m > 1$,

(4) $f^m(J_x) \cap J_{f^m(x)}$ *for* $m = 1, 2, 3, \ldots,$

(5) *intervals* $J_{f^m(x)}$ *for* $m = 0, 1, 2, \ldots$ *are pairwise disjoint,*

(6) *if* $f(x) = f(y)$ *then* $J_x = J_y$,

(7) *the sets* $\mathcal{J} = \{J_x : x \in \mathbb{S}^1\}$ *form a decomposition of* $\mathbb{S}^1$; *i.e.* $J_x = J_y$ *or* $J_x \cap J_y = \emptyset$ *for all* $x, y \in \mathbb{S}^1$,

(8) *for at most countably many x we have* $J_x \neq \{x\}$,

(9) $J_x = J_y$ *if and only if* $J_{f^n(x)} = J_{f^n(y)}$ *for some (every)* $m = 0, 1, 2, \ldots.$

Theorem 4.2

Suppose $f : \mathbb{S}^1 \to \mathbb{S}^1$ *and* $\text{Per}(f) = \emptyset$. *Then* $\Lambda_f = \varprojlim(\mathbb{S}^1, f)$ is decomposable.

Proof

Let *f* be as above and let JJ be the collection of intervals guaranteed by (7) in Theorem 4.1. Without loss of generality we may assume that there is an $x \in \mathbb{S}^1$ such that $J_x \neq \{x\}$, since otherwise by (6) in Theorem 4.1*f* would be a homeomorphism and $\Lambda_f = \mathbb{S}^1$. Additionally observe that $f^{-n}(z) \neq \emptyset$ for every $z \in \mathbb{S}^1$ as otherwise it is not hard to see that *f* has a fixed point.

First we claim that $f^{-n}(J_x)$ is an interval in JJ for every *n*. To see this, fix any n>0n>0 and any $y, z \in f^{-n}(J_x)$. By Theorem 4.1 (4) we have fn $f^n(J_y) \cap J_x \neq \emptyset$ and $f^n(J_z) \cap J_x \neq \emptyset$ which by Theorem 4.1(7) gives $f^{-n}(J_x)$. The proof of the claim is completed by Theorem 4.1 (9).

Put $K_n = f^{-n}(J_x)$. Since K_n is an element of $\mathcal{J}$, without loss of generality we can write $K_n = [a_n, b_n]$ and $f(a_n) = a_{n-1}$ and $f(b_n) = b_{n-1}$. Put $L_n = \{a_n, b_n\} \cup (\mathbb{S}^1 \setminus K_n)$. Then LnLnis an interval,

$$f(L_n) = \{f(a_n), f(b_n)\} \cup (\mathbb{S}^1 \setminus f(K_n)) = \{a_{n-1}, b_{n-1}\} \cup (\mathbb{S}^1 \setminus K_{n-1}) = L_{n-1}$$

and hence the following two maps are onto: $f|_{K_n} : K_n \to K_{n-1}$, $f|_{L_n} : L_n \to L_{n-1}$ for $n = 1, 2, \ldots.$ Therefore the following inverse limits

$$\Gamma = \varprojlim(K_n, f|_{K_n} : n \geq 0),$$
$$\Lambda = \varprojlim(L_n, f|_{L_n} : n \geq 0)$$

are arc-like continua such that $\Lambda_f = \Gamma \cup \Lambda$, in particular ΛfΛf is decomposable.

Although this is not essential for the reasoning in the present paper, note that periodic-point-free circle maps are of degree 1 (see [23]).

Theorem 4.3

If $\Lambda_f = \varprojlim(\mathbb{S}^1, f)$ is hereditarily indecomposable and ΛfΛf is not the pseudo-arc *(i.e.* $\deg(f) \neq 0$*) then* $h_{top}(f) = \infty$.

Proof

Suppose $\Lambda_f = \varprojlim(\mathbb{S}^1, f)$ is hereditarily indecomposable and $||\deg(f)| > 0$. Then ΛfΛf is either the pseudo-circle or it is nonplanar. If ΛfΛf is nonplanar then deg(f)=ndeg(f)=n, for some $|n| > 1$ (see [45, Theorem 9]). Consequently $h_{\text{top}}(f) \geq \log |n|$ by [1] and further, by Theorem 3.6, we obtain that $h_{\text{top}}(f) = \infty$.

Now suppose $\deg(f) = \pm 1$. By Theorem 4.2 we have $\text{Per}(f) \neq \emptyset$. By [30, Theorem 15, p.9] for any integer $m > 0$ we have

$$\varprojlim(\mathbb{S}^1, f) = \varprojlim(\mathbb{S}^1, f^m)$$

and $h_{\text{top}}(f^m) = m \cdot h_{\text{top}}(f)$, hence without loss of generality we may assume that deg $(f) = 1$ and $\text{Fix}(f) \neq \emptyset$. Clearly, for any *m* we have $h_{\text{top}}(f) = \infty$ if and only if $h_{\text{top}}(f^m) = \infty$.

Let $\hat{f} : \mathbb{R} \to \mathbb{R}$ be a lift of *f* to the universal covering (R,τ)(R,τ), where $(\mathbb{R}, \tau)$, where $\tau(t) = (\cos(t), \sin(t))$ and S1S1 is the unit circle in the plane, centered at the origin. It is known that f^f^ can be extended to $\hat{\mathbb{R}} = \mathbb{R} \cup \{+\infty, -\infty\}$, with $\hat{f}(\pm\infty) = \pm\infty$. Because R^R^ is homeomorphic to [0, 1] and $\varprojlim(\mathbb{R}, \hat{f})$ is a universal covering of ΛfΛf with the covering map given by $\hat{\tau}(x_1, x_2, x_3, \ldots) = (\tau(x_1), \tau(x_2), \tau(x_3), \ldots)$

(see [17, Lemma 3.8]), we get that $\varprojlim(\hat{\mathbb{R}}, \hat{f})$ is the pseudo-arc [10]. Since $(f) \neq \emptyset$, $\hat{f}$ can be chosen in such a way so that $\mathrm{Fix}(\hat{f}) \neq \emptyset$. Indeed, any lift of f is uniquely determined by specifying where it sends a single point, and if $c$is a fixed point of f then there is $\hat{c} \in [0, 2\pi]$ such that $\tau(\hat{c}) = c$, hence it is enough to set $\hat{c} = \hat{f}(\hat{c})$.

Set $K_i = \hat{f}^i([0, 2\pi])$ and note that $\hat{c} \in K_i$ for every i. We shall show that the rotation set ρ(f)ρ(f)is nondegenerate. To demonstrate this, we claim that

$$\lim_{i\to\infty} \sup \operatorname{diam}(K_i) = \infty. \tag{2}$$

In order to prove (2), first we claim that

limi→∞supdiam(Ki)≠0.limi→∞supdiam(Ki)≠0.

Recall that f is an onto map, and if B is an interval of length less than 2π2π then τ(B)τ(B) is a proper subset of S1S1. Then it is enough to observe that

$$\tau(K_1) = \tau(\hat{f}([0, 2\pi])) = f(\tau([0, 2\pi])) = f(\mathbb{S}^1) = \mathbb{S}^1,$$

so we must have diam$(K_1) \geq 2\pi$.. Then working by induction, using the relation $K_{i+1} = \hat{f}(K_i)$., we obtain that diam$(K_i) \geq 2\pi$ for every $i > 0$.

Next, let us suppose that $\lim_{i\to\infty} \sup \operatorname{diam}(K_i) < \infty$. Set $K = \bigcup_{i=1}^{\infty} K_i$ and note that K is connected, bounded and $\hat{c} \in K_i$ for every i. Now define sets L_i for $i = 1, 2, \ldots$ as follows. Denote by L1L1 a component of $\hat{f}^{-1}(K)$ such that $\hat{c} \in L_1$, and then for $i > 1$ take as the set LiLi a component of $\hat{f}^{-1}(L_{i-1})$ that contains c^c^. Finally put $L = cl(\bigcup_{i\geq 0} L_i)$. Note that K⊆LK⊆L. It is easy to see that L is connected, as as $\hat{c} \in L_i$ for every i. We claim that L is bounded. Indeed, since $\hat{f}(x + 2k\pi) = \hat{f}(x) + 2k\pi$ we must have $\sup_{x\in[0,2\pi]} \operatorname{diam}(\{\hat{f}^n(x) : n \geq 0\}) = \sup_{x\in\mathbb{R}} \operatorname{diam}(\{\hat{f}^n(x): n \geq 0\})$. Since K is bounded we must have $\sup_{x\in[0,2\pi]} \operatorname{diam}(\{\hat{f}^n(x) : n \geq 0\}) < M$ for some real number M. But now, if p,q∈Lip,q∈Li then

$\mathrm{diam}(K) \geq d(f^i(p), f^i(q)) \geq d(p,q) - 2M$ and hence $(L_i) \leq \mathrm{diam}(K) + 2M$, which gives diam $(L) \leq 2\,\mathrm{diam}(K) + 4M$. By the definition $\hat{f}(L_i) \subset L_{i-1}$ for $i > 1$ and for every n we have

$$\bigcup_{i=n+1}^{\infty} K_i \subset \hat{f}^n(K) \subset \hat{f}^n(L) \subset \hat{f}^{n-1}(L) \subset \ldots \subset L.$$

If we denote $K_i = [p_i, q_i]$ and let $p = \liminf_{i\to\infty} p_i, q = \limsup_{i\to\infty} q_i$. Then

$$\hat{c} \in [p,q] \subset \bigcup_{i=n+1}^{\infty} K_i$$

for every $n > 0$ and $q - p \geq 2\pi$.

Then if we denote $S = \bigcap_{i=0}^{\infty} \hat{f}(L)$ then S is a continuum such that f $\hat{f}(S) = S$ and $[p,q] \subset S$, so S, so S is a closed nondegenerate interval in RR. Now let $\Lambda_S = \varprojlim(S, \hat{f}|_S)$. Since $\Lambda_S \subseteq \varprojlim(\mathbb{R}, \hat{f})$, the continuum ΛSΛS is the psuedo-arc. But

$$\mathbb{S}^1 = \tau([0,2\pi]) \subseteq \tau([p,q]) \subseteq \tau(S)$$

and so $\hat{\tau}(x_1, x_2, x_3, \ldots) = (\tau(x_1), \tau(x_2), \tau(x_3), \ldots)$ defines a continuous onto map from the pseudo-arc ΛSΛS onto the pseudo-circle ΛfΛf. But the pseudo-circle cannot be obtained as the image of the pseudo-arc via a continuous map [46], which is a contradiction. This proves that $\lim_{i\to\infty} \sup \mathrm{diam}(K_i) = \infty$, completing the proof of (2).

Because f^f^ has a fixed point in [0,2π][0,2π] we get by (2) that the rotation set ρ(f^)ρ(f^) is nondegenerate. This implies in turn that $h_{\mathrm{top}}(f) = \infty$ (see [39]) and so, again by Theorem 3.6, we obtain that $h_{\mathrm{top}}(f) = \infty$ which ends the proof.□⌋

The following theorem follows from the above proof.

Theorem 4.4

Suppose f:S1→S1f:S1→S1 is a map with $f:\mathbb{S}^1\to\mathbb{S}^1$ *is a map with* $\deg(f)=1$. *If* $\Lambda_f=\varprojlim(\mathbb{S}^1,f)$ is hereditarily indecomposable then the rotation set ρ(f)ρ(f) is nondegenerate.

In [17] the authors showed that there exists a 2-torus homeomorphism h homotopic to the identity with an attracting R.H. Bing's pseudocircle C such that the rotation set of $h|C$ is not a unique vector. It follows from Theorem 4.3 that the topological entropy of this homeomorphism is infinite.

In [46] Rogers considered a special class of circle-like continua which he called *self-entwined*. Among the many results, he showed that the pseudo-circle is self-entwined, any continuum in this class is indecomposable and no self-entwined continuum is a continuous image of the pseudo-arc. Adapting the proof of Theorem 4.3 and properties of self-entwined continua sketched above, we get the following.

Theorem 4.5

Suppose $\Lambda_f=\varprojlim(\mathbb{S}^1,f)$ *is self-entwined. Then* $h(f)>0$.

Proof

Since ΛfΛf is indecomposable, by Theorem 4.2, f must have a periodic point. Again, without loss of generality assuming that f has a fixed point we can consider f^:R→Rf^:R→R, a lift of f to the universal covering $(\mathbb{R},\tau)$. Then $\hat f$ can be extended to $\hat{\mathbb{R}}=\mathbb{R}\cup\{+\infty,-\infty\}$, $\hat f(\pm\infty)=\pm\infty$.

Because R^R^ is homeomorphic to [0, 1] and $\varprojlim(\mathbb{R},\hat f)$ is a universal covering of ΛfΛf with the covering map given by $\hat\tau(x_1,x_2,x_3,\dots)=(\tau(x_1),\tau(x_2),\tau(x_3),\dots)$ (see [17, Lemma 3.8]), we get that $A=\varprojlim(\hat{\mathbb{R}},\hat f)$ is an arc-like continuum. Now if we assume by contradiction that $h(f)=0$ then f has a one-point rotation set by [39]. Therefore, repeating arguments from the proof of Theorem 4.3, there is a closed f^f^-invariant arc S⊂RS⊂R such that arc-like continuum $\Lambda_S=\varprojlim(S,\hat f|_S)\subset A$ satisfies $\hat\tau(\Lambda_S)=\Lambda_f$. In this case ΛfΛf would be a continuous image of an arc-like continuum. But since every arc-like continuum is a continuous image of the pseudo-arc [24] this leads to a contradiction with Roger's result that no self-entwined continuum is such an image and the proof is complete.□

ACKNOWLEDGMENTS

The authors are grateful to the anonymous referee for his careful reading of this paper and many valuable suggestions. The authors express many thanks to Henk Bruin for fruitful discussions during Workshops on Complexity and Dimension theory of Skew Products Systems held at the Erwin Schrödinger International Institute for Mathematical Physics (ESI) in Vienna, in September of 2013, as well as Henk Bruin's visit at the AGH University of Science and Technology in Kraków, in November of 2013. The authors are grateful for the kind hospitality of the institute during their stay in Vienna. The authors acknowledge financial support of the ESI, the DFG network Grant Oe 538/3-1, OeAD (PL 02/2013) and MNiSW (AT 2/2013-15).

Boroński's work was supported by the European Regional Development Fund in the IT4Innovations Center of Excellence Project (CZ.1.05/1.1.00/02.0070). Boroński also gratefully acknowledges the partial support from the MSK DT1 Support of Science and Research in the Moravian-Silesian Region 2013 (RRC/05/2013). The research of Oprocha was supported by the Polish Ministry of Science and Higher Education from sources for science in the years 2013-2014, Grant no. IP2012 0 04272.

REFERENCES

1. Alsedá, L., Libre, J.L., Misiurewicz, M.: Combinatorial Dynamics and Entropy in Dimension One, Advanced Series in Nonlinear Dynamics 5. World Scientific, Singapore (1993)
2. Alsedà, L., del Río, M.A., Rodríguez, J.A.: Transitivity and dense periodicity for graph maps. J. Differ. Equ. Appl. **9**, 577–598 (2003) MATH
3. Auslander, J., Katznelson, Y.: Continuous maps of the circle without periodic points. Israel J. Math. **32**(4), 375–381 (1979)
4. Barge, M.: Horseshoe maps and inverse limits. Pacific J. Math. **121**(1), 29–39 (1986)
5. Barge, M.: Homoclinic intersections and indecomposability. Proc. Am. Math. Soc. **101**, 541–544 (1987)
6. Barge, M.: A method for constructing attractors. Ergodic Theory Dyn. Syst. **8**(3), 331–349 (1988)
7. Barge, M., Martin, J.: The construction of global attractors. Proc. Am. Math. Soc. **110**(2), 523–525 (1990)
8. Barge, M., Diamond, B.: The dynamics of continuous maps of finite graphs through inverse limits. Trans. Am. Math. Soc. **344**(2), 773–790 (1994)
9. Barge, M., Holte, S.: Nearly one-dimensional Hénon attractors and inverse limits. Nonlinearity **8**(1), 29–42 (1995)
10. Bellamy, D.P., Lewis, W.: An orientation reversing homeomorphism of the plane with invariant pseudo-arc. Proc. Am. Math. Soc. **114**(4), 1145–1149 (1992)
11. Bing, R.H.: Concerning hereditarily indecomposable continua. Pacific J. Math. **1**, 43–51 (1951)

12. Blanchard, F.: A disjointness theorem involving topological entropy. Bull. S.M.F. **121**, 465–478 (1993)
13. Blanchard, F., Huang, W.: Entropy sets, weakly mixing sets and entropy capacity. Discret. Contin. Dyn. Syst. **20**, 275–311 (2008)
14. Block, L.: Diffeomorphisms obtained from endomorphisms. Trans. Am. Math. Soc. **214**, 403–413 (1975)
15. Block, L.S., Coppel, W.A.: Lecture Notes in Mathematics. Dynamics in one dimension, vol. 1513. Springer, Berlin (1992)
16. Block, L., Keesling, J., Uspenskij, V.V.: Inverse limits which are the pseudoarc. Houston J. Math. **26**, 629–638 (2000)
17. Boroński, J.P., Oprocha, P.: Rotational chaos and strange attractors on the 2-torus. Math. Z. **279**, 689–702 (2015)
18. Brown, M.: On the inverse limit of Euclidean NN-Spheres. Trans. Am. Math. Soc. **96**, 129–134 (1960)
19. Bruin, H.: Planar embeddings of inverse limit spaces of unimodal maps. Topology Appl. **96**(3), 191–208 (1999)
20. Chen, L., Li, S.H.: Lecture Notes in Pure and Applied Mathematics. Dynamical connections between a continuous map and its inverse limit space. Continuum theory and dynamical systems, vol. 149, pp. 89–97. Dekker, New York (1993)
21. Chéritat, M.: Relatively compact Siegel disks with non-locally connected boundaries. Math. Ann. **349**(3), 529–542 (2011)
22. Dou, D., Ye, X., Zhang, G.: Entropy sequence and maximal entropy sets. Nonlinearity **19**, 53–74 (2006)
23. Efremova, L.S.: Periodic orbits and the degree of a continuous map of a circle (in Russian). Diff. Integer. Equ. (Gor'kii) **2**, 109–115 (1978)
24. Fearnley, L.: Characterizations of the continuous images of the pseudo-arc. Trans. Am. Math. Soc. **111**, 380–399 (1964)
25. Fearnley, L.: Classification of all hereditarily indecomposable circularly chainable continua. Trans. Am. Math. Soc. **168**, 387–401 (1972)
26. Handel, M.: A pathological area preserving C∞C∞ diffeomorphism of the plane. Proc. Am. Math. Soc. **86**, 163–168 (1982)
27. Henderson, G.W.: The pseudo-arc as an inverse limit with one binding map. Duke Math. J. **31**, 421–425 (1964)
28. Herman, M.: Construction of some curious diffeomorphisms of the Riemann sphere. J. Lond. Math. Soc. **34**(2), 375–384 (1986)MATH
29. Huang, W., Ye, X.: A local variational relation and applications. Israel J. Math. **151**, 237–280 (2006)MATH
30. Ingram, W.T., Mahavier, W.S.: Inverse Limits. From Continua to Chaos. Developments in mathematics, vol. 25. Springer, New York (2012)
31. Ito, S.: An estimate from above for the entropy and the topological entropy of a C1C1-diffeomorphism. Proc. Jpn. Acad. **46**, 226–230 (1970)
32. Kawamura, K., Tuncali, M., Tymchatyn, E.D.: Hereditarily indecomposable inverse limits of graphs. Fund. Math. **185**, 195–210 (2005)
33. Kennedy, J., Yorke, J.A.: Pseudocircles in dynamical systems. Trans. Am. Math. Soc. **343**, 349–366 (1994)
34. Kennedy, J., Yorke, J.A.: Bizarre topology is natural in dynamical systems. Bull. Am. Math. Soc. (N.S.) **32**, 309–316 (1995)
35. Kerr, D., Li, H.: Independence in topological and C∗C∗-dynamics. Math. Ann. **338**, 869–926 (2007)
36. Llibre, J., Misiurewicz, M.: Horseshoes, entropy and periods for graph maps. Topology **32**, 649–664 (1993)
37. Minc, P., Transue, W.R.R.: A transitive map on [0,1][0,1] whose inverse limit is the pseudoarc. Proc. Am. Math. Soc. **111**(4), 1165–1170 (1991)
38. Misiurewicz, M.: Embedding inverse limits of interval maps as attractors. Fund. Math. **125**(1), 23–40 (1985)
39. Miyazawa, M.: Chaos and entropy for circle maps. Tokyo J. Math. **25**(2), 453–458 (2002)

40. Moise, E.E.: Graduate Texts in Mathematics. Geometric topology in dimensions 22 and 33, vol. 47. Springer, New York (1977)
41. Mouron, C.: A chainable continuum that admits a homeomorphism with entropy of arbitrary value. Houston J. Math. **35**(4), 1079–1090 (2009)
42. Mouron, C.: Entropy of shift maps of the pseudo-arc. Topology Appl. **159**, 34–39 (2012)
43. Szczechla, W.: Inverse limits of certain interval mappings as attractors in two dimensions. Fund. Math. **133**(1), 1–23 (1989)
44. Oversteegen, L., Tymchatyn, E.: On hereditarily indecomposable compacta. Geometric and Algebraic Topology, vol. 18, pp. 407–417. Banach Center Publications, Warsaw (1986)
45. Rogers Jr, J.T.: Pseudo-circles and universal circularly chainable continua. Ill. J. Math. **14**, 222–237 (1970)
46. Rogers Jr, J.T.: Mapping the pseudo-arc onto circle-like, self-entwined continua. Mich. Math. J. **17**, 91–96 (1970)
47. Walters, P.: An Introduction to Ergodic Theory. Springer, Berlin (2001)
48. Williams, R.F.: One-dimensional non-wandering sets. Topology **6**, 473–487 (1967)

CHAPTER 15

Stationary Solutions and Connecting Orbits for p-Laplace Equation

Aleksander Cwiszewski1 · Mateusz Maciejewski1

Faculty of Mathematics and Computer Science, Nicolaus Copernicus University, Toruń, Poland

ABSTRACT

We deal with one dimensional p-Laplace equation of the form $u_t = (|u_x|^{p-2}u_x)_x + f(x, u),\ x \in (0, l),\ t > 0,$ under Dirichlet boundary condition, where $p > 2$ and $f\colon [0, l] \times \mathbb{R} \to \mathbb{R}$ is a continuous function with $f(x, 0) = 0.$ We will prove that if there is at least one eigenvalue of the p-Laplace operator between $\lim_{u\to 0} f(x, u)/|u|^{p-2}u$ and $\lim_{|u|\to+\infty} f(x, u)/|u|^{p-2}u,$, then there exists a nontrivial stationary solution. Moreover we show the existence of a connecting orbit between stationary solutions. The results are based on Conley index and detect stationary states even when those based on fixed point theory do not apply. In order to compute the Conley index for nonlinear semiflows deformation along p is used.

1. INTRODUCTION

We shall study a nonlinear p-Laplace equation

$$\begin{cases} u_t(t,x) = \left(|u_x(t,x)|^{p-2} u_x(t,x)\right)_x + f(x, u(t,x)), \ x \in (0,l), \ t \in \mathbb{R}, \\ u(t,0) = u(t,l) = 0, \ t \in \mathbb{R}, \end{cases} \tag{1}$$

with $p > 2, l > 0$ a continuous $f\colon [0,l] \times \mathbb{R} \to \mathbb{R}$, which is locally Lipschitz with respect to the second variable, i.e.

$$\text{for any } R > 0 \text{ there is } L > 0 \text{ such that } |f(x,u) - f(x,v)| \le L|u-v| \text{ for all } x \in [0,l], \ u, v \in [-R, R]. \tag{2}$$

The stationary version of (1), i.e. the elliptic problem

$$\begin{cases} -(|u'(x)|^{p-2}u'(x))' = f(x, u(x)), \ x \in (0,l), \\ u(0) = u(l) = 0, \end{cases} \tag{3}$$

is subject of extensive studies by many authors—see earlier papers [8, 9, 10, 14] as well as more recent examples [4, 5, 12, 15, 17] or [13]. Usually topological degree/index techniques or variational approach are applied. Here we use a dynamical system approach based on the Conley type index from [20] and [18] and Rybakowski's techniques from [19]. In the case of semiflows determined by nonlinear equations with p-Laplace operator, the main difficulty is to compute the Conley index. This task reduces to studying the nonlinear semiflow generated by $u_t = (|u_x|^{p-2}u_x)_x + \lambda|u|^{p-2}u$ with real λ. In order to compute its Conley index, inspired by [8], we use deformation along p. As a result we shall prove the following existence criterion.

Theorem 1.1

Suppose that $f\colon [0,l] \times \mathbb{R} \to \mathbb{R}$ is locally Lipschitz with respect to the second variable, $f(x,0) = 0 \ for \ x \in (0,l)$ and

$$\lim_{u \to 0} \frac{f(x,u)}{|u|^{p-2}u} = f_0'(x) \tag{4}$$

and

$$\lim_{|u|\to+\infty} \frac{f(x,u)}{|u|^{p-2}u} = f'_\infty(x) \tag{5}$$

for some $f'_0, f'_\infty \in C([0,l])$ uniformly with respect to x∈[0,l]x∈[0,l]. Suppose there are $k_0, k_\infty \in$ Nsuch that $\lambda_{k_0}^{(p)} \le f'_0(x) \le \lambda_{k_0+1}^{(p)}$ *and* $\lambda_{k_\infty}^{(p)} \le f'_\infty(x) \le \lambda_{k_\infty+1}^{(p)}$, *for all* $x \in (0,l)$, for all x∈(0,l)x∈(0,l), with the strict inequalities on set of positive measure. *If* $k_0 \ne k_\infty$, then there exists a nontrivial solution u $\bar{u} \in C^1([0,l])$ of(3).

Moreover, there exists a connecting orbit between $u(t_n,\cdot) \to \bar{u}$,, i.e. a full solution *u* of (1) such that either $u(t_n,\cdot) \to \bar{u}$ *for some* $t_n \to +\infty$ *and* $u(t,\cdot) \to 0$ *as* $t \to -\infty$ *or* $u(t_n,\cdot) \to \bar{u}$ *for some* $t_n \to -\infty$ *and* $u(t,\cdot) \to 0$ *as* $t \to +\infty$ (with respect to the max norm of the space $C(0,l)$).

Here recall that λ∈Rλ∈R, for which the problem

$$\begin{cases} -(|u'(x)|^{p-2}u'(x))' = \lambda|u(x)|^{p-2}u(x), & x \in (0,l), \\ u(0) = u(l) = 0, \end{cases}$$

has nonzero solutions, make a sequence of positive numbers $\lambda_n^{(p)}, n \ge 1$, such that $\lambda_n^{(p)} \to +\infty$ and, for any $n \ge 1$, $\lambda_n^{(q)} \to \lambda_n^{(p)}$ whenever $q \to p$ put $\lambda_0^{(p)} := -\infty$.

In this paper we consider a local semiflow (a sort of dynamical system) $\Phi^{(p,f)}$ on the space $X = C_0(0,l) := \{u \in C(0,l) \mid u(0) = u(l) = 0\}$ associated with the equation (1). To find a stationary solution and related connecting trajectory we use the theory of irreducible sets due to Rybakowski [20], where we need to find Conley indices of the zero $K_0 := \{0\}$ and the set K∞K∞ made by all full bounded trajectories of (1). The main difficulty lies in the fact that both the differential operator and continuous term are nonlinear. In order to consider and compute Conley index (due to Rybakowski) for $\Phi^{(p,f)}$ we need to study the existence, compactness and continuity properties of solutions. We shall also exploit the Lyapunov function for the problem and related regularity to find stationary solutions at the ends of full trajectories. What we gain by use of Conley index and what we could not obtain with topological degree techniques is that we show the existence under the condition $k_0 \ne k_\infty$ ∞ while topological degree works in the case where k0k0 and k∞k∞ are of different parities. In addition, we have a full trajectory between two stationary solutions.

The paper is organized as follows. In the rest of the section we give some notation and basic preliminaries on Conley index. Section 2 is devoted mainly to continuity and compactness issues for abstract evolution equations governed by perturbations of *m*-accretive operators and subdifferentials of convex functionals. In Sect. 3 we study the existence and regularity of solutions together with Lyapunov function theory. The continuity and compactness properties with respect to *p* and *f*, which are crucial for computing Conley index via its continuation property, are explored in Sect. 4. Finally, we compute the Conley indices of K0K0 and K∞K∞ and prove Theorem 1.1 in Sect. 5.

PRELIMINARIES

Two pointed topological spaces (X,x0)(X,x0) and (Y,y0)(Y,y0) are said to be *homotopically equivalent* or have *the same homotopy type* if and only if there are maps $f: (X, x_0) \to (Y, y_0)$ and $g: (Y, y_0) \to (X, x_0)$ such that f∘gf∘g is homotopic via a mapping keeping y0y0 fixed to the identity of (Y, y_0) and $g \circ$and g∘fg∘f is homotopic via a mapping keeping x0x0 fixed to the identity of (X, x_0). The homotopy class represented by a space (X_0, x_0) is denoted by $[(X, x_0)]$. is a pair of topological spaces with a nonempty and closed $A \subset X$, then X/A denotes the quotient space, obtained by collapsing the subset *A* to a point [*A*].

Consider $C_0(0, l) := \{u : [0, l] \to \mathbb{R} \mid f$ is continuous, $u(0) = u(l) = 0\}$ $(l > 0)$ with the usual norm $\|u\|_\infty := \max_{[0,l]} |u|$. By $L^p(0, l)$ and $W^{k,p}(0, l)$ and $W_0^{1,p}(0, l)$ we denote the standard Lebesgue and Sobolev spaces on the interval (0, *l*) and we put $H_0^1(0, l) :=$ $W_0^{1,2}(0, l)$. In the same way, by $L^p(0, T; X)$ and $W^{1,p}(0, T; X)$ we denote the spaces with values in a Banach space *X*.

CONLEY INDEX DUE TO RYBAKOWSKI

Here we briefly present homotopy index theory from [20] (see also [18]). Let $\Phi: D \to X$, where *D* is an open subset of $[0, \infty) \times X$, be a local semiflow on a metric space *X*. Let $T_{\bar{u}} = \sup\{t > 0 \mid (t, \bar{u}) \in D\}$. A continuous function u: $J \to X$, where $J \subset \mathbb{R}$ is an interval, is called a *solution of* ΦΦ if and only if $u(t+s) = \Phi_t(u(s))$ for any $t \geq 0$ and $s \in J$ If $t + s \in J$. If $u: [a, +\infty) \to X, a \in \mathbb{R}$, is a solution of ΦΦ, then by the ωω-*limit of u* we mean the set

$$\omega(u) := \left\{\bar{u} \in X \mid \exists (t_n) \text{ in } [a, +\infty) \text{ such that } t_n \to +\infty \text{ and } \bar{u} = \lim_{n\to\infty} u(t_n)\right\}.$$

The $\alpha\alpha$-*limit* of a solution $(-\infty, a] \to X$ of Φ, $a \in \mathbb{R}$, is defined by is defined by

$$\alpha(u) := \left\{\bar{u} \in X \mid \exists (t_n) \text{ in } (-\infty, a] \text{ such that } t_n \to -\infty \text{ and } \bar{u} = \lim_{n\to\infty} u(t_n)\right\}.$$

Let N⊂XN⊂X. By the *invariant part*InvΦ(N)InvΦ(N) of *N* we mean

$$\mathrm{Inv}_\Phi(N) := \{\bar{u} \in N \mid \exists \text{ a solution} u : \mathbb{R} \to N of \Phi \text{ with } u(0) = \bar{u}\}.$$

We shall say that $K \subset X$ is a Φ--*invariant set* or *invariant with respect to* ΦΦ provided $\mathrm{Inv}_\Phi(K) = K$. A Φ-invariant set $K \subset X$ *invariant set* if and only if there exists $N \subset X$ such that $K = \mathrm{Inv}_\Phi(N) \subset$. Such *N* is called an *isolating neighborhood ofK*. The following concept of admissibility is crucial in Rybakowski's version of Conley theory on general metric spaces and enables us to construct a Conley type index without local compactness of *X*.

Definition 1.2

(i) $N \subset X$ is said to be Φ-*admissible* if and only if, for any $t_n \to +\infty$ and (v_n) in X such that $\Phi_{[0,t_n]}(v_n) \subset N$, contains a convergent subsequence.

(ii) *N* is said to be *strongly* Φ-*admissible* if *N* is Φ-admissible and Φ does not *explode*in *N*, i.e. N, i.e. $\Phi_{[0,T_{\bar{u}})}(\bar{u}) \subset N$ implies $T_{\bar{u}} = +\infty$.

We also need the notion of admissibility for the families of local semiflows.

Definition 1.3

$N \subset X$ is said to be $\{\Phi^k\}_{k\in K}$-*admissible* if and only if, for $[0, +\infty)$ with $t_n \to +\infty$, (k_n) in K and (v_n) in X such that $\Phi^{k_n}_{[0,t_n]}(v_n) \subset N$, the sequence $\left(\Phi^{k_n}_{t_n}(v_n)\right)$ contains a convergent subsequence. If additionally Φ^k does not explode in *N* for any $k \in K$, then *N* is *strongly*Φ-*admissible.*

Let $\mathcal{I}(X)$ be the family of pairs (Φ, K) where Φ is a local semiflow on a metric space X and K is an isolated invariant set having a strongly Φ-admissible isolating neighborhood of K. The Conley homotopy index h(Φ,K) of K relative to Φ for $(\Phi, K) \in \mathcal{I}(X)$ is defined by $h(\Phi, K) := [(B/B^-, [B^-])]$ where B is an isolating block of K (relative to Φ) with the exit set $B^- \neq \emptyset$ and if $B^- = \emptyset$ we put $h(\Phi, K) = [(B \cup \{a\}, a)]$ where a is an element that does not belong to B. In particular, $h(\Phi, \emptyset) = \overline{0}$ where $\overline{0} := [(\{a\}, a)]$.

The Conley index has the following properties

(H1) For any $(\Phi, K) \in \mathcal{I}(X)$, if $h(\Phi, K) \neq \overline{0}$, then $K \neq \emptyset$.
(H2) If $(\Phi, K_1), (\Phi, K_2) \in \mathcal{I}(X)$ and $K_1 \cap K_2 = \emptyset$, then $(\Phi, K_1 \cup K_2) \in \mathcal{I}(X)$ and $h(\Phi, K_1 \cup K_2) = h(\Phi, K_1) \vee h(\Phi, K_2)$.
(H3) For any $(\Phi_1, K_1) \in \mathcal{I}(X_1)$ and $(\Phi_2, K_2) \in \mathcal{I}(X_2)$, $(\Phi_1 \times \Phi_2, K_1 \times K_2) \in \mathcal{I}(X_1 \times X_2)$ and $h(\Phi_1 \times \Phi_2, K_1 \times K_2) = h(\Phi_1, K_1) \wedge h(\Phi_2, K_2)$.
(H4) If the family of semiflows $\{\Phi^{(\mu)}\}_{\mu \in [0,1]}$ is continuous and there exists V such that V is strongly $\{\Phi^{(\mu)}\}_{\mu \in [0,1]}$-admissible and $K_\mu = \mathrm{Inv}_{\Phi^{(\mu)}}(V) \subset \mathrm{int}\, V$, $\mu \in [0, 1]$, then

Theorem 1.4

(See [20, Ch. I, Th. 11.1]) Suppose that X is a normed space such that $X_- \oplus X_+$ *with* $k := \dim X_+ < +\infty$-semigroup $\{T(t)\colon X \to X\}_{t \geq 0}$ is such that $T(t)(X_+) \subset X_+$ *and* $T(t)(X_-) \subset X_-$, $\|T(t)x\| \leq Ke^{-\alpha t}\|x\|$ *for* $x \in X_-$, $t \geq 0$ *and some* $\alpha > 0$ *and* $\{T(t)\}_{t \geq 0}$ can be extended to a COCO-group $\{T(t)\}_{t \in \mathbb{R}}$ *on* X_+ such that $\alpha > 0$ *and* $\{T(t)\}_{t \geq 0}$ $\|T(t)x\| \leq Ke^{\beta t}\|x\|$ *for all* $x \in X_+$, $t \leq 0$ and some $\beta > 0$. *Then* $\Phi\colon [0, +\infty) \times X \to X$ given by Φ $\Phi(t, x) := T(t)x$ *is a semiflow on* X, , the set {0}{0} is the set of all bounded full solutions of $\Phi(t, x) := T(t)x$ *is a semiflow on* X, and $h(\Phi, \{0\}) = \Sigma^k$ *where* $\Sigma^k = [(S^k, \overline{s})]$, i.e. it is the homotopy type of a k-dimensional sphere with a point.

We shall use the theory of irreducible sets due to Rybakowski [20]. Recall that an isolated invariant set K (relative to a local semiflow ΦΦ) is called *reducible* if there exist isolated invariant sets K1,K2K1,K2 such that K_1, K_2 such that $K = K_1 \cup K_2$, $K_1 \cap K_2 = \emptyset$, $(\Phi, K_1), (\Phi, K_2) \in \mathcal{I}(X)$ and both $h(\Phi, K_1) \neq \overline{0}$ and $h(\Phi, K_2) \neq \overline{0}$.-. We say that K is *irreducible* if it is not reducible. It is known that the set K is irreducible if one of the following conditions is satisfied: K is connected, $h(\Phi, K) = \overline{0}$ or $h(\Phi, K) = \Sigma^k$ (see [20, Ch. I, Th. 11.6]). The concept of irreducible set turns out to be useful due to the following.

Theorem 1.5

(See [20, Ch. I, Th. 11.5]) *If* $K_0 \subset K \subset X$, $(\Phi, K), (\Phi, K_0) \in \mathcal{I}(X)$, K K is irreducible and $\overline{0} \neq h(\Phi, K_0) \neq h(\Phi, K)$, then there exists a full solution $\sigma(\mathbb{R}) \not\subset K_0$ *and either* $\alpha(\sigma) \subset K_0$ *or* $\omega(\sigma) \subset K_0$. .

2 Properties of Abstract Evolution Equations with *m*-Accretive Operators

Let $D(A) \multimap X$ defined in a Banach space X be an *m*-accretive operator, $f: [0, T] \to X, T > 0$, and $\bar{u} \in \overline{D(A)}$., We shall consider the equation

$$\begin{cases} \dot{u}(t) \in -Au(t) + f(t), & t \in [0, T], \\ u(0) = \bar{u}. \end{cases} \tag{6}$$

Definition 2.1

A continuous function u$[0, T] \to X$ is called an *integral solution* of (6) in X if and only if $u(0) = \bar{u}$ and

$$\|u(t) - v\| \leq \|u(s) - v\| + \int_s^t [u(\tau) - v, f(\tau) - w]_s \, d\tau$$

for all $(v, w) \in \mathrm{Gr}(A)$ and $0 \leq s < t \leq T$, where $[x, y]_s = \inf_{h>0}(\|x + hy\| - \|x\|)/h$.

Remark 2.2

(i)It appears that for any $\bar{u} \in \overline{D(A)}$ and $f \in L^1(0, T; X)$ the problem (6) admits a unique integral solution (see e.g. [1]). It may be also shown that this integral solution is also a mild solution (see e.g. [1]), and as such, is a limit of discrete approximations.

(ii)Now consider Banach spaces X,X~ such that X is continuously embedded into X~ Suppose that a *m*-accretive operator A in X has an extension A~A~ in X~ such that it is *m*-accretive in X~ Then, for any $\tilde{X}$. Then, for any $\bar{u} \in X$, $f \in L^1(0, T; X)$ the integral solution of (6) is also an integral solution of (6) in the space X~ It follows immediately by use of discrete approximations (see (i)).

(iii)Let $\Sigma_A(\bar{u}, f)$ denote the integral solution of (6). Then, for any $\bar{u}_1, \bar{u}_2 \in \overline{D(A)}$ and $f_1, f_2 \in L^1(0, T; X)$ one has (see [1])

$$\|\Sigma_A(\bar{u}_1, f_1)(t) - \Sigma_A(\bar{u}_2, f_2)(t)\| \le \|\bar{u}_1 - \bar{u}_2\| + \int_0^t \|f_1(s) - f_2(s)\| \, ds \text{ for } t \in [0, T].$$

(iv) In particular, if we take $f = 0$, then one may define a family of operators $\{S_A(t): \overline{D(A)} \to \overline{D(A)}\}_{t \ge 0}$ by $S_A(t)\bar{u} := u(t)$ where u is the integral solution of $\dot{u}(t) \in -Au(t), t > 0$, with $u(0) = \bar{u}$ (existing due to Remark 2.2 (i)). It appears that $S_A(0)\bar{u} = \bar{u}$, for all $\bar{u} \in \overline{D(A)}$, and $S_A(t) \circ S_A(s) = S_A(t+s)$, for any $t, s \ge 0$, the mapping $[0, +\infty) \times \overline{D(A)} \ni (t, \bar{u}) \mapsto S_A(t)\bar{u} \in X$ is continuous and

$$\|S_A(t)\bar{u} - S_A(t)\bar{v}\| \le \|\bar{u} - \bar{v}\| \text{ for all } \bar{u}, \bar{v} \in \overline{D(A)}.$$

Let An: $D(A_n) \to X, n \ge 0$, be *m*-accretive operators. We say that AnAn*converges* to *Ain the sense of graphs* (and denote $A_n \overset{\text{Gr}}{\to} A$) if and only if GrA⊂lim infGrAnGrA⊂lim infGrAn. This is equivalent by [2, Proposition 4.4] to the convergence

$$\lim_{n \to \infty} (I + \lambda A_n)^{-1}\bar{u} \to (I + \lambda A_0)^{-1}\bar{u} \text{ for all } \bar{u} \in X, \ \lambda > 0.$$

We shall use the following continuity and compactness result.

Proposition 2.3

Let $A_n : D(A_n) \multimap X, n \ge 0$, be *m*-accretive operators.

(i) (Trotter-Kato theorem) *If* $\bar{u}_n \to \bar{u}_0$ *in* X, $\bar{u}_n \in \overline{D(A_n)}$ *for all* $n \ge 0$, $A_n \overset{\text{Gr}}{\to} A$ *and* $f_n \to f_0$ *in* $L^1(0, T; X)$, *then* $\Sigma_{A_n}(\bar{u}_n, f_n) \to \Sigma_{A_0}(\bar{u}_0, f_0)$ *in* $C([0, T], X)$.

(ii) Assume that $\overline{D(A_n)} = \overline{D(A_1)}$ *for any* $n \ge 1$ and the set $B \subset \overline{D(A_1)}$ *and* $t > 0$. *If* $\{\bar{u}_n\}_{n \ge 1} \subset \overline{D(A_1)}$ and t>0t>0. If {u¯n}n≥1⊂D(A1)¯ is bounded and $\{f_n\} \subset L^1(0, T; X)$) is uniformly integrable, then for any t∈[0,T]t∈[0,T]the set $\{\Sigma_{A_n}(\bar{u}_n, f_n)(t)\}_{n \ge 1}$ is relatively compact.

Proof

(i) can be found in [2, Ch. 4, Th. 4.14] and (ii) is proved in [6, Prop. 2.26]. □?

We shall also consider nonlinear problems of the form

$$\begin{cases} \dot{u}(t) \in -Au(t) + F(u(t)), \ t \in [0, T], \\ u(0) = \bar{u} \end{cases} \tag{7}$$

with a locally $F: X \to X$ and $\bar{u} \in \overline{D(A)}$. We shall say that a continuous u: $[0, T] \to X \ (T > 0)$ is an integral solution of (7) if u is an integral solution of (6) with f:[0,T]→Xf:[0,T]→X given by $f: [0, T] \to X$ given by $f(t) := F(u(t))$ for $t \in [0, T]$. that is, if $u = \Sigma_A(\bar{u}, F \circ u)$. Let us state a general existence and uniqueness theorem.

Proposition 2.4

(i)For any $\bar{u} \in \overline{D(A)}$ there exists a continuous $u: [0, T_{\bar{u}}) \to X$ with $T_{\bar{u}} \in (0, +\infty]$, called a maximal integral solution, such that for each $T \in (0, T_{\bar{u}})$ the function $T_{\bar{u}} \in (0, +\infty]$, is an integral solution of (7). Moreover, either $T_{\bar{u}} = +\infty \ or \ T_{\bar{u}} < +\infty \ and \ \limsup_{t \to T_{\bar{u}}^-} \|u(t)\| = +\infty.$

(ii)Let An: $D(A_n) \multimap X, \ n \geq 0,$ be m-accretive operators with $\overline{D(A_n)} = \overline{D(A_1)}$ for all $n \geq 1 \ and \ A_n \overset{Gr}{\to} A. \ Let \ F_n: X \to X$ X have common Lipschitz constants on bounded subsets of X. Assume that $F_n(\bar{u}) \to F_0(\bar{u}) \ for \ any \ \bar{u} \in X. \ Let \ u_n: [0, T_{\bar{u}_n}) \to X \ for$ be the maximal integral solution of

$$\begin{cases} \dot{u}(t) \in -A_n u(t) + F_n(u(t)), \quad t > 0, \\ u(0) = \bar{u}_n \end{cases} \tag{8}$$

where $\in \overline{D(A_n)} \ for \ all \ n \geq 1 \ and \ \bar{u}_n \to \bar{u}_0 \ as \ n \to +\infty.$. Then $\liminf_{n \to +\infty} T_{\bar{u}_n} \geq$ and $u_n \to u_0 \ in \ C([0, T], X) \ for \ any \ T \in (0, T_{\bar{u}_0})$.

(iii)Assume that m-accretive operators $D(A_n) \multimap X$ with $\overline{D(A_n)} = \overline{D(A_1)}, n \geq 1,$ and locally Lipschitz $F_n\colon X \to X, n \geq 1,$, are such that for any bounded B⊂XB⊂X

$$\bigcup_{n\geq 1} S_{A_n}(t)(B \cap \overline{D(A_n)}) \textit{ is relatively compact and } \bigcup_{n\geq 1} F_n(B) \textit{ is bounded.}$$

If un: $[0, T] \to X, n \geq 1,$, are integral solutions of (8) and there exists R>0R>0 such that $\|u_n(t)\| \leq R \textit{ for all } t \in [0, T] \textit{ and } n \geq 1.$ is relatively compact for any $\{u_n(t)\}_{n\geq 1} \in (0, T].$

Proof

(i)is classic (see [1])

(ii) Fix $T \in (0, T_{\bar{u}_0})$. Let $R := \max_{t\in[0,T]} \|u_0(t)\|$ and let L_R be the common Lipschitz constant for $F_n, n \geq 0$, on the ball $B(0, 3R)$. Define

$$\alpha_n := \sup_{0\leq t\leq T} \|\Sigma_{A_n}(\bar{u}_n, F_n \circ u_0)(t) - \Sigma_{A_0}(\bar{u}_0, F_0 \circ u_0)(t)\|.$$

From Proposition 2.3 (i) it follows that $\alpha_n \to 0^+$ and therefore αne $\alpha_n e^{TL_R} < R \text{ for } n \geq n_0$ for some $\in \mathbb{N}$. Fix $n \geq n_0$.. In order to prove the inequality from the conclusion it suffices to show that $T < T_{\bar{u}_n}$.. Suppose, contrary to our claim, that $T_{\bar{u}_n} \leq T$. Therefore $u_n([0, t]) \subset B(0, 3R)$ and $\|u_n(t)\| > 2R$ for some $t < T$, which follows from (i). Using Remark 2.2 (iii) one has, for $\tau \in [0, t]$,

$$\begin{aligned}\|u_n(\tau) - u_0(\tau)\| &\leq \|\Sigma_{A_n}(\bar{u}_n, F_n \circ u_n)(\tau) - \Sigma_{A_n}(\bar{u}_n, F_n \circ u_0)(\tau)\| \\ &\quad + \|\Sigma_{A_n}(\bar{u}_n, F_n \circ u_0)(\tau) - \Sigma_{A_0}(\bar{u}_0, F_0 \circ u_0)(\tau)\| \\ &\leq \alpha_n + L_R \int_0^\tau \|u_n(s) - u_0(s)\|\, ds.\end{aligned}$$

In consequence, by the Gronwall inequality we obtain the estimate

$$\|u_n(\tau) - u_0(\tau)\| \leq \alpha_n e^{TL_R} < R \text{ for } \tau \in [0, t]. \qquad (9)$$

This means that $\|u_n(t)\| \leq \|u_0(t)\| + R \leq 2R$. This contradicts our assumption. By the estimate $\|u_n(t) - u_0(t)\| \leq \alpha_n e^{TL_R}$ for $t \leq T$, the convergence from the second part of the conclusion holds true.

(iii) is a direct consequence of Proposition 2.3 (ii) with $f_n := F_n \circ u_n$.

Remark 2.5

Let *D* be the set of ($(t, \bar{u}) \in [0, +\infty) \times X$ such that the problem (7) has a solution on [0, *t*] and let $\Phi\colon D \to X$ be defined by $\Phi(t, \bar{u}) = \Phi_t(\bar{u}) := u(t)$, where $[0, t] \to X$ is the integral solution of (7). It clearly follows from Proposition 2.4 that ΦΦ is a local semiflow on *X*.

Now suppose that *H* is a Hilbert space with the scalar product ⟨·,·⟩⟨·,·⟩ and the norm ‖·‖‖·‖ and consider a lower semicontinuous convex functional $\varphi\colon H \to [0, +\infty]$ with $D(\varphi) := \{u \in H \mid \varphi(u) < +\infty\}$ and $\varphi(0) = 0$. . Recall that the *subdifferential∂φ*: $D(\partial\varphi) \to H$ of φφ is defined by $D(\partial\varphi) := \{u \in D(\varphi) \mid$ there exists $\xi \in H$ such that $\langle \xi, v - u \rangle \leq$ for all $v \in H\}$ and $\partial\varphi(u) := \{\xi \in H \mid \langle \xi, v - u \rangle \leq \varphi(v) - \varphi(u)$ for $v \in H\}$ and $\partial\varphi(u) := \{\xi \in H \mid \langle \xi, v - u \rangle \leq \varphi(v) - \varphi(u)$ for all $v \in H\}$.

It is known that ∂φ∂φ is a *m*-accretive operator in *H*. Hence, if $f \in L^2(0, T; H)$ and $\bar{u} \in \overline{D(\partial\phi)}$, , one can consider the following problem

$$\begin{cases} \dot{u}(t) + \partial\varphi u(t) \ni f(t), \; t > 0, \\ u(0) = \bar{u}. \end{cases} \tag{10}$$

It appears that integral solutions in this case are more regular and are strong solutions.

Proposition 2.6

([2, Ch. 4, Th. 4.11 and Lem. 4.4]) If $\in L^2(0, T; H)$, $\bar{u} \in \overline{D(\partial\varphi)}$ and $u \in C([0, T], H)$ is an integral solution of (10), then *u* is a.e. differentiable on (0, *T*) and has the following properties

(i) $u(t) \in D(\partial\varphi)$ *and* $\dot{u}(t) + \partial\varphi(u(t)) \ni f(t)$ *for a.e.* $t \in (0, T)$;

(ii) $\left(t \mapsto t^{1/2}\dot{u}(t)\right) \in L^2(0,T;H)$ *and* $\varphi \circ u \in L^1(0,T)$*;*

(iii) if moreover $\bar{u} \in D(\varphi)$, *then* $u(t) \in D(\varphi)$, *for all* $t \in [0,T]$, for all t∈[0,T]t∈[0,T], the function $\varphi \circ u : [0,T] \to$ R is absolutely continuous,

$$\frac{1}{2}\int_0^t \|\dot{u}(s)\|^2 \,\mathrm{d}s + \varphi(u(t)) \le \frac{1}{2}\int_0^t \|f(s)\|^2 \,\mathrm{d}s + \varphi(\bar{u}), \quad \text{for all } t \in [0,T],$$

and

$$\frac{d}{dt}\varphi(u(t)) = -\|\dot{u}(t)\|^2 + (\dot{u}(t), f(t)) \quad \text{for a.e. } t \in [0,T].$$

In particular, for all $\bar{u} \in \overline{D(\partial\varphi)}$ the function φ∘uφ∘u is continuous on (0, T].

In order to estimate time derivative of solutions we shall need the following result.

Proposition 2.7

(See [2, Th. 4.12]) Assume that $\bar{u} \in \overline{D(\partial\varphi)}$ *and* $f \in W^{1,1}([0,T],H)$. If *u* is the integral solution of (10) with u(0)=u¯u(0)=u¯, then, for all $t \in (0,T]$, $u(t) \in D(\partial\varphi)$,

$$\dot{u}_+(t) + (\partial\varphi)(u(t)) \ni f(t)$$

and

$$t^2\|\dot{u}(t)\|^2 \le \int_0^t s\|f(s)\|^2 \,\mathrm{d}s + 2\left(\|\bar{u}\| + \int_0^t \|f(s)\|\,\mathrm{d}s\right)^2 + \frac{t^2}{2}\left(\int_0^t \|\dot{f}(s)\|\,\mathrm{d}s\right)^2. \quad (11)$$

In particular, if $f \equiv 0$, *then* $t\|\dot{u}(t)\| \le \sqrt{2}\|\bar{u}\|$.

We shall need the following compactness criterion being an extension of [1, Ch. 4, Th. 2.4] to a family of semigroups generated by subdifferentials.

Proposition 2.8

Let $\varphi_n : H \to [0, +\infty], n \geq 1,$, are lower semicontinuous convex functions such that $D(\varphi_n) := \{u \in H \mid \varphi_n(u) < +\infty\}$ is dense in H and $\varphi_n(0) = 0, n \geq 1.$. If for any λ>0λ>0 the set $L_\lambda := \bigcup_{n\geq 1}\{u \in H \mid \varphi_n(u) \leq \lambda\}$ is relatively compact, then the family of semigroups $\{S_{\partial\varphi_n}(t) : H \to H\}_{t\geq 0}$ is compact, i.e. the set $\bigcup_{n\geq 1} S_{\partial\varphi_n}(t)(B)$ is relatively compact for all t>0t>0 and any bounded B⊂HB⊂H.

In order to prove Proposition 2.8 we need the following lemma.

Lemma 2.9

Under the assumptions of Proposition 2.8, for all $n \geq 1, t > 0$ *and* $\bar{u} \in H$ we have

$$\varphi_n(S_{\partial\varphi_n}(t)\bar{u}) \leq \max\left\{1, \frac{M^2}{t}\right\} \quad \textit{with } M := \|\bar{u}\| + \sup_{\bar{v}\in L_0} \|\bar{v}\|.$$

Proof

$$n \geq 1, \bar{u} \in H. \text{ Put } u := S_{\partial\varphi_n}(\cdot)(\bar{u}).$$

By use of Proposition 2.6 (iii) we obtain

$$\frac{\mathrm{d}}{\mathrm{d}t}(\varphi_n(u(t))) = -\|\dot{u}(t)\|^2 \leq 0 \text{ for a.e. } t > 0, \tag{12}$$

which implies in particular that $\varphi_n(u(t)) \leq \varphi_n(\bar{u}) \text{ for all } t \geq 0.$. From the fact that

$$-\dot{u}(t) \in \partial\varphi_n(u(t)) \text{ and } \varphi_n(0) = 0 \text{ it follows that } \varphi_n(u(t)) \leq \|\dot{u}(t)\|\|u(t)\|.$$

Using the contractivity of $\{S_{\partial\varphi_n}(t)\}_{t\geq 0}$ one has, for a.e. $t > 0$,

$$\varphi_n(u(t)) \leq \|\dot{u}(t)\|\|u(t)\| \leq \|\dot{u}(t)\| \left(\|S_{\partial\varphi_n}(t)0\| + \|\bar{u}\|\right) \leq M\|\dot{u}(t)\|. \quad (13)$$

Combining (12) and (13) we obtain

$$\frac{\mathrm{d}}{\mathrm{d}t}(\varphi_n(u(t))) \leq -\frac{1}{M^2}(\varphi_n(u(t)))^2 \text{ for a.e. } t > 0. \quad (14)$$

If $\varphi_n(\bar{u}) \leq 1$, then $\varphi_n(u(t)) \leq \varphi_n(\bar{u}) \leq 1$ for all $t > 0$. If $\varphi_n(\bar{u}) > 1$, then

$$\frac{\mathrm{d}}{\mathrm{d}t}\left(\frac{1}{\varphi_n(u(t))}\right) \geq \frac{1}{M^2}$$

for a.e. $t > 0$ as long as $\varphi(u_n(t)) > 0$. This implies

$$\frac{1}{\varphi_n(u(t))} \geq \frac{1}{\varphi_n(\bar{u})} + \frac{t}{M^2} > \frac{t}{M^2},$$

which ends the proof. □

Proof of Proposition 2.8

Fix $t > 0$ and let $B \subset L^2(0, l)$ be bounded. From Lemma 2.9 we obtain $S_{\partial\varphi_n}(t)(B) \subset L_\lambda$ for some λ independent of *n*. The conclusion follows from the relative compactness of LλLλ. □

3 Existence and Regularity for *p*-Laplace Evolution Equation

First we put the equation (1) in the abstract framework to precise the concept of solution and get information on their regularity. To this end, define

$$D(\mathbf{A}_p) := \{u \in C_0(0, l) \cap C^1(0, l) \mid (|u'|^{p-2}u')' \in C_0(0, l)\},$$
$$\mathbf{A}_p u := -(|u'|^{p-2}u')', \quad u \in D(\mathbf{A}_p).$$

Lemma 3.1

The operator ApAp is *m*-accretive.

Proof

It can be demonstrated along the lines of [5, Lemma 6.1]. □

Definition 3.2

Fix u¯∈C0(0,l)u¯∈C0(0,l) and T>0T>0. By a *solution* of (1) on the interval [0, *T*], with the initial condition u(x,0)=u¯(x)u(x,0)=u¯(x) for x∈[0,l]x∈[0,l], we mean the integral solution of

$$\dot{u}(t) = -\mathbf{A}_p u(t) + \mathbf{F}(u(t)), \ t \in [0, T], \tag{15}$$

with $u(0) = \bar{u}$, where $\mathbf{F}\colon C_0(0,l) \to C_0(0,l)$ is given by $\mathbf{F}(u)(x) := f(x, u(x)), x \in [0,l]$, $u \in C_0(0,l)$.

Theorem 3.3

Assume that a continuous $f\colon [0,l] \times \mathbb{R} \to \mathbb{R}$ satisfies the local Lipschitz condition (2) and that $f(x,0) = 0 \; for \; all \; x \in [0,l]$. Then, for any $x \in [0,l]$, there exists a unique function $u \in C([0, T_{\bar{u}}^{(p,f)}), C_0(0,l)) \; with \; T_{\bar{u}}^{(p,f)} \in (0,+\infty]$ Tu¯(p,f)∈(0,+∞] such that u(0)=u¯u(0)=u¯, the function u|[0,T]u|[0,T] is the solution of (1), for all $T \in (0, T_{\bar{u}}^{(p,f)}]$, *and either* $T_{\bar{u}}^{(p,f)} = +\infty$

$$or \; T_{\bar{u}}^{(p,f)} < +\infty \; and \; \limsup_{t \to T_{\bar{u}}^{(p,f)-}} \|u(t)\|_\infty = +\infty.$$

Proof

The proof follows directly from Proposition 2.4 applied to (15). □

So as to study regularity properties of solutions, it will be convenient to use also an L2-extension of Ap. We shall consider $\bar{\mathbf{A}}_p\colon D(\bar{\mathbf{A}}_p) \to L^2(0,l)$

$$D(\bar{\mathbf{A}}_p) := \{u \in W_0^{1,p}(0,l) \mid (|u'|^{p-2}u')' \in L^2(0,l)\} \text{ and } \bar{\mathbf{A}}_p u := -(|u'|^{p-2}u')'.$$

Clearly, $D(\bar{\mathbf{A}}_p) \subset C^1(0,l)$ and $\mathrm{Gr}(\mathbf{A}_p) \subset \mathrm{Gr}(\bar{\mathbf{A}}_p)$ (by the embedding of $C_0(0,l)$ into $L^2(0,l)$).

Lemma 3.4

(i) The operator $\bar{A}_p$ is m-accretive and there exists $c_p>0$ such that

$$(\bar{\mathbf{A}}_p u - \bar{\mathbf{A}}_p v, u - v)_{L^2} \geq c_p \|u - v\|^p_{W_0^{1,p}} \text{ for all } u, v \in D(\bar{\mathbf{A}}_p). \quad (16)$$

(ii) $\bar{\mathbf{A}}_p = \partial\varphi_p$, where $\partial\varphi_p$ is the subdifferential of the lower semicontinuous convex functional $\varphi_p\colon L^2(0,l) \to \mathbb{R} \cup \{+\infty\}$ given by

$$\varphi_p(u) = \begin{cases} \frac{1}{p}\int_0^l |u'(x)|^p \, dx & \text{if } u \in W_0^{1,p}(0,l) \\ +\infty & \text{if } u \in L^2(0,l) \setminus W_0^{1,p}(0,l). \end{cases}$$

(iii) For any $M, T > 0$ there exists $C = C(M,T) > 0$ (independent of $p \geq 2$) such that $\|S_{\bar{A}_p}(t)\bar{u}_1 - S_{\bar{A}_p}(t)\bar{u}_0\|_{W_0^{1,p}} \leq C(T,M) \cdot \|\bar{u}_1 - \bar{u}_0\|_{L^2}^{1/p}$ whenever $\|\bar{u}_0\|_{L^2}, \|\bar{u}_1\|_{L^2} \leq M$.

Proof

(i)If we change the space $L^p(0,l)$ with $L^2(0,l)$ in the proof of [4, Prop. 4.1], we obtain that the operator $\bar{A}_p$ is maximal monotone, which in Hilbert spaces is equivalent to being m-accretive. In particular, we have the estimate (16).

(ii) Take $(u, v) \in \operatorname{Gr} \bar{\mathbf{A}}_p$, which means that $u \in W_0^{1,p}(0,l)$ and $v = -(j'(u'))'$, where j $j(s) = \frac{1}{p}|s|^p$.

Therefore, for all $w \in W_0^{1,p}(0,l)$ we have

$$\begin{aligned}(v, w - u)_{L^2} &= \int_0^l v(w-u) = \int_0^l j'(u')(w'-u') \leq \int_0^l j(w') - \int_0^l j(u') \\ &= \varphi_p(w) - \varphi_p(u),\end{aligned}$$

which shows that $(u, v) \in \mathrm{Gr}(\partial\varphi_p)$. In other words $\mathrm{Gr}(\bar{\mathbf{A}}_p) \subset \mathrm{Gr}(\partial\varphi_p)$. From the general theory of maximal operators, it follows that, since A¯pA¯p is maximal monotone in a Hilbert space, its graph is maximal among graphs of accretive operators. Therefore we get $\mathrm{Gr}(\partial\varphi_p) = \mathrm{Gr}(\bar{\mathbf{A}}_p)$.

(iii)Let $v_i := S_{\bar{\mathbf{A}}_p}(\cdot)\bar{u}_i$ for $i = 0, 1$ and let $w = v_1 - v_0$. From Proposition 2.7 it can be concluded that $w(s) \in W_0^{1,p}(0, l)$ and that $\|\dot{w}(s)\|_{L^2} \leq \sqrt{2}s^{-1} \cdot (\|\bar{u}_0\|_{L^2} + \|\bar{u}_1\|_{L^2})$ for all s>0s>0. From (i) it follows that

$$c_p\|w(t)\|^p_{W_0^{1,p}} \leq \|\dot{w}(t)\|_{L^2} \cdot \|w(t)\|_{L^2} \leq \frac{2\sqrt{2}M}{t}\|\bar{u}_1 - \bar{u}_0\|_{L^2},$$

which gives the conclusion as $c_p = 2^{2-p} \geq 2^{-p}$.

The following result sheds more light on the regularity of solutions.

Theorem 3.5

If $u \in C([0, T], C_0(0, l))$ is a solution of (1), then

$$u \in C((0, T], W_0^{1,p}(0, l)) \cap W^{1,2}((0, T]; L^2(0, l)),$$
$$u(t) \in D(\bar{\mathbf{A}}_p) \text{ and } \dot{u}(t) = -\bar{\mathbf{A}}_p u(t) + \bar{\mathbf{F}}(u(t)), \text{ for a.e. } t \in [0, T],$$

where $\bar{\mathbf{F}}: C_0(0, l) \to L^2(0, l)$ *is given by* $\bar{\mathbf{F}}(u)(x) := f(x, u(x))$ *for a.e.* $x \in (0, l)$ *and*

$$W^{1,2}((0, T]; L^2(0, l)) = \left\{u \in L^2\left(0, T; L^2(0, l)\right) \mid \dot{u} \in L^2_{loc}\left((0, T]; L^2(0, l)\right)\right\}.$$

Moreover, the functional φp, $\varphi_{p,f}: W_0^{1,p}(0, l) \to \mathbb{R}$ *given by*

$$\varphi_{p,f}(u) := \frac{1}{p}\int_0^l |u'(x)|^p \, \mathrm{d}x - \int_0^l \mathcal{F}(x, u(x)) \, \mathrm{d}x, \quad u \in W_0^{1,p}(0, l),$$

where $\mathcal{F}(x, u) := \int_0^u f(x, \tau) \, \mathrm{d}\tau$, is a Lyapunov function for (1), since for any solution $u \in C([0, T], C_0(0, l))$ *and* $0 < s < t < T$ one has

$$\varphi_{p,f}(u(t)) - \varphi_{p,f}(u(s)) = -\int_s^t \int_0^l |\dot{u}(\tau)|^2 \, dx \, d\tau. \tag{17}$$

Proof

By Remark 2.2, the function u may be viewed as an element of $C([0,T], L^2(0,l))$ and the integral solution (in L2(0,l)L2(0,l)) of

$$\dot{u}(t) = -\bar{\mathbf{A}}_p u(t) + \mathbf{f}(t),\ t \in [0,T],$$

with $\mathbf{f} \in L^2(0,T; L^2(0,l))$ given by $\mathbf{f}(t) := \mathbf{F}(u(t))$, $t \in [0,T]$. This together with Lemma 3.4(ii) and Proposition 2.6, gives that $u(t) \in D(\partial\varphi_p) \subset D(\varphi_p) = W_0^{1,p}(0,l)$ and

$$\dot{u}(t) = -\partial\varphi_p(u(t)) + \mathbf{f}(u(t)),$$

for a.e. t∈[0,T]t∈[0,T], and that φp∘uφp∘u is continuous on (0, *T*]. Therefore, the function $(0,T] \ni t \mapsto \|u(t)\|_{W_0^{1,p}}$,p is continuous. Since $u \in C([0,T], L^2(0,l))$, using the uniform convexity of $W_0^{1,p}(0,l)$ we obtain that $u: (0,T] \to W_0^{1,p}(0,l)$ is continuous.

In order to verify that φp,fφp,f is a Lyapunov function observe that, by Proposition 2.6, one has $u \in W^{1,2}((0,T]; L^2(0,l))$ and

$$\frac{d}{dt}(\varphi_p(u(t))) - (\bar{\mathbf{F}}(u(t)), \dot{u}(t))_{L^2} = -\|\dot{u}(t)\|_{L^2}^2 \text{ for a.e. } t \in [0,T]. \tag{18}$$

Now take any $t \in [0,T]$ such that $\dot{u}(t)$ exists (in $L^2(0,l)$) and any sequence (hn)(hn) in R\{0}R\{0}with hn→0hn→0. Passing to a subsequence, if necessary, we may suppose that $(u(t+h_n)-u(t))/h_n \to \dot{u}(t)$ a.e. on $[0,l]$) such that $g \in L^1(0,l)$ By means of Lebesgue's dominated convergence theorem we have

$(\mathcal{F}(x, u(t+h_n)(x)) - \mathcal{F}(x, u(t)(x)))/h_n \to f(x, u(t)(x))\dot{u}(t)(x)$ a.e. on $[0, l]$.

Furthermore, since FF is Lipschitz with respect to the second variable on bounded sets, for *f* is bounded on bounded sets, we can use dominated convergence theorem to get

$$\frac{\mathrm{d}}{\mathrm{d}t}\left(\int_0^l \mathcal{F}(x, u(t)(x))\,\mathrm{d}x\right) = \int_0^l f(x, u(t)(x))\dot{u}(t)(x)\,\mathrm{d}x = (\bar{\mathbf{F}}(u(t)), \dot{u}(t))_{L^2}.$$

This together with (18) ends the proof. □

We shall also need some compactness and continuity of solutions with respect to the initial data.

Theorem 3.6

Suppose that un: $[0, T] \to C_0(0, l)$ *with* $u_n(0) = \bar{u}_n,\ n \geq 0$, are solutions of (1) and there is R>0R>0 such that $\|u_n(t)\|_\infty \leq R$ *for all* $t \in [0, T]$ *and* $n \geq 0$.

(Continuity) *If* $\bar{u}_n \to \bar{u}_0$ *in* $C_0(0, l)$, *then* $u_n \to u_0$ *in* $C([0, T], C_0(0, l))$. If, additionally, $(\bar{u}_n)$ *is bounded in* $W_0^{1,p}(0, l)$, *then* $u_n(t) \to u_0(t)$ *in* $W_0^{1,p}(0, l)$

(Compactness) The set

$$\{u_n(t)\}_{n\geq 1} \text{ is relatively compact in } C_0(0, l) \text{ for any } t \in (0, T].$$

Before proving the general theorem we shall prove it for the contraction semigroup generated by ApAp.

Lemma 3.7

For any t>0t>0 and bounded $B \subset C_0(0, l)$, *the set* $S_{A_p}(t)(B)$ is a relatively compact subset of $W_0^{1,p}(0, l)$. In particular, the semigroup $\{S_{A_p}(t) \colon C_0(0, l) \to C_0(0, l)\}_{t\geq 0}$

Proof

First observe that for any $\lambda \in \mathbb{R}$, $L_\lambda := \{\bar{u} \in L^2(0, l) \mid \varphi_p(\bar{u}) \leq \lambda\}$ is bounded in W1,p0(0,l)W01,p(0,l), which by the Rellich-Kondrachov embedding theorem means that LλLλ is relatively compact in

L2(0,l)L2(0,l). Therefore, due to Proposition 2.8, the semigroup $\{S_{\bar{\mathbf{A}}_p}(t)\colon L^2(0,l) \to L^2(0,l)\}_{t\geq 0}$ is compact.

Now take a bounded set $B \subset C_0(0,l), t > 0$ and any sequence $(\bar{u}_n)$ in B. Put $\alpha = t/3$. Since the set $\{S_{\mathbf{A}_p}(\alpha)\bar{u}_n\}_{n\geq 1}$ is relatively compact in L2(0,l)L2(0,l), without any loss of generality we may assume that $S_{\mathbf{A}_p}(\alpha)\bar{u}_n \to \tilde{v}_0$ in $L^2(0,l)$. Put $\bar{v}_n := S_{\mathbf{A}_p}(2\alpha)\bar{u}_n$ as well as $\bar{v}_0 := S_{\bar{\mathbf{A}}_p}(\alpha)\tilde{v}_0$. Clearly, $\bar{v}_n \to \bar{v}_0$ in $L^2(0,l)$ and, by Lemma 3.4 (ii) and Proposition 2.7, we see that $\bar{v}_0 \in D(\partial\varphi_p) \subset D(\varphi_p) = W_0^{1,p}(0,l)$, which implies $\bar{v}_0 \in C_0(0,l)$.

In view of Lemma 3.4 (iii), one has

$$\|S_{\mathbf{A}_p}(t)\bar{u}_n - S_{\mathbf{A}_p}(\alpha)\bar{v}_0\|^p_{W_0^{1,p}} = \|S_{\mathbf{A}_p}(\alpha)\bar{v}_n - S_{\mathbf{A}_p}(\alpha)\bar{v}_0\|^p_{W_0^{1,p}} \leq C \cdot \|\bar{v}_n - \bar{v}_0\|_{L^2}$$

for some constant $C > 0$. In consequence, $S_{\mathbf{A}_p}(t)\bar{u}_n \to S_{\mathbf{A}_p}(\alpha)\bar{v}_0$ in $W_0^{1,p}(0,l)$,

Proof of Theorem 3.6

(i) First observe that, due to Proposition 2.4 (ii), $u_n \to u_0$ $C([0,T], C_0(0,l))$. To prove the other part of the assertion assume that (u¯n)(u¯n) is bounded in $W_0^{1,p}(0,l)$. Put $\tilde{f}(x,u) = f(x, r(u))$, where $r\colon \mathbb{R} \to [-R,R]$ is a metric projection. Let $\bar{\mathbf{F}}\colon C_0(0,l) \to L^2(0,l)$ and $\tilde{\mathbf{F}}\colon L^2(0,l) \to L^2(0,l)$ be the Nemytskii operators generated by f (as in Theorem 3.5) and f~f~, respectively. Define a bounded sequence of elements $\mathbf{f}_n \in L^2(0,T; L^2(0,l)), n \geq 1$, by $\mathbf{f}_n := \bar{\mathbf{F}} \circ u_n = \tilde{\mathbf{F}} \circ u_n$. Now observe that in view of Proposition 2.6 (iii) and the boundedness of $(\bar{u}_n)$ in $W_0^{1,p}(0,l)$ and $(\mathbf{f}_n)$ in $L^2(0,T; L^2(0,l))$ we have $u_n \in W^{1,2}(0,T; L^2(0,l))$ and there exists R~>0R~>0 such that ‖u˙n‖L2(0,T;L2(0,l))≤R~‖u˙n‖L2(0,T;L2(0,l))≤R~ for all n≥1n≥1.

Since f~f~ is Lipschitz with respect to the second variable uniformly with respect to x, F~F~ is Lipschitz. Denote its Lipschitz constant by L~L~. In consequence, for all $n \geq 1$ and $t, s \in [0,T]$, we get

$$\|\mathbf{f}_n(t) - \mathbf{f}_n(s)\|_{L^2} \leq \tilde{L}\|u_n(t) - u_n(s)\|_{L^2} \leq \tilde{L}\int_s^t \|\dot{u}_n(\tau)\|_{L^2}\, d\tau. \quad (19)$$

By [3, Thm 1.4.40], this implies that $\|\dot{\mathbf{f}}_n\|_{L^2(0,T;L^2(0,l))} \leq \tilde{L}\|\dot{u}_n\|_{L^2(0,T;L^2(0,l))} \leq \tilde{L}\tilde{R}$. Now fix $t \in (0,T]$. By Proposition 2.7, the sequence $(\dot{u}_n(t))$ is bounded in $L^2(0,l)$. Applying (16) we get, for all n,m≥1n,m≥1,

$$\|u_n(t) - u_m(t)\|^p_{W_0^{1,p}} \leq c_p^{-1}\|\partial\varphi_p(u_n(t)) - \partial\varphi_p(u_m(t))\|_{L^2}\|u_n(t) - u_m(t)\|_{L^2}.$$

Since the values

$$\|\partial\varphi_p(u_n(t)) - \partial\varphi_p(u_m(t))\|_{L^2} \leq \|\mathbf{f}_n(t)\|_{L^2} + \|\mathbf{f}_m(t)\|_{L^2} + \|\dot{u}_n(t)\|_{L^2} + \|\dot{u}_m(t)\|_{L^2}$$

are bounded, we see that (un(t))(un(t)) is a Cauchy sequence in $W_0^{1,p}(0,l)$, which means that $u_n(t) \to u_0(t)$ in $W_0^{1,p}(0,l)$.

(ii) follows immediately from Proposition 2.4 (iii). □

We shall summarize the obtained results in the context of dynamical systems. Define $\mathbf{\Phi}^{(p,f)} : \mathbf{D}^{(p,f)} \to \mathbf{X}$ where $\mathbf{X} := C_0(0,l)$, by

$$\mathbf{D}^{(p,f)} := \left\{(t,\bar{u}) \in [0,+\infty) \times \mathbf{X} \mid t < T_{\bar{u}}^{(p,f)}\right\} \text{ and } \mathbf{\Phi}^{(p,f)}(t,\bar{u}) = \mathbf{\Phi}_t^{(p,f)}(\bar{u}) := u(t),$$

where $[0, T_{\bar{u}}^{(p,f)}) \to \mathbf{X}$ is the maximal integral solution of (1) with u(0)=u¯u(0)=u¯.

By Remark 2.5, ΦΦ(p,f) is a local semiflow on XX.

Theorem 3.8

If $u \in C(\mathbb{R}, \mathbf{X})$ is a bounded solution of (1), then α(u)α(u) and ω(u)ω(u) are nonempty, connected and compact in the space C0(0,l)C0(0,l), $\mathbf{\Phi}_t^{(p,f)}(\alpha(u)) = \alpha(u)$ *and* $\mathbf{\Phi}_t^{(p,f)}(\omega(u)) =$ $\omega(u)$ *for all* $t \geq 0$ *and* $\alpha(u) \cup \omega(u) \subset \mathcal{E}$

where EE is the set of all stationary solutions of (1).

Lemma 3.9

If $u \in C(\mathbb{R}, \mathbf{X})$ is a bounded solution of (1), then

$$\sup_{t\in\mathbb{R}} \|u(t)\|_{W_0^{1,p}} < +\infty.$$

Proof

Suppose the contrary. There exists

(t_n) in $\mathbb{R}$ such that $\|u(t_n)\|_{W_0^{1,p}} \to +\infty$ as $n \to +\infty$.

Clearly, due to Theorem 3.5 and the boundedness of $u(\mathbb{R})$ in $\ddot{C}_0^0(0,l)$, we see that φp,f(u(tn))→+∞φp,f(u(tn))→+∞ and tn→-∞tn→-∞, which simply implies

$$\lim_{t\to-\infty} \varphi_{p,f}(u(t)) = +\infty.$$

On the other hand, for a.e. t∈Rt∈R,

$$\begin{aligned}\frac{\mathrm{d}}{\mathrm{d}t}\left(\frac{1}{2}\|u(t)\|_{L^2}^2\right) &= (\dot{u}(t), u(t))_{L^2}\\ &= -p\,\varphi_{p,f}(u(t)) - p\int_0^l \mathcal{F}(x, u(t)(x))\,\mathrm{d}x + \int_0^l f(x, u(t)(x))u(t)(x)\,\mathrm{d}x\\ &\le -p\,\varphi_{p,f}(u(t)) + C\end{aligned}$$

for some constant C>0C>0. Hence there exists t0∈Rt0∈R such that, for all t≤t0t≤t0,

$$\frac{\mathrm{d}}{\mathrm{d}t}\left(\frac{1}{2}\|u(t)\|_{L^2}^2\right) \le -1,$$

which implies

$$\frac{1}{2}\|u(t)\|_{L^2}^2 \ge t_0 - t + \frac{1}{2}\|u(t_0)\|_{L^2}^2 \to +\infty, \quad \text{as} \to t \to -\infty,$$

a contradiction. □

Proof of Theorem 3.8

First observe that u(R)u(R) is a relatively compact subset of C0(0,l)C0(0,l). Indeed, take any (tn)(tn) in RR. Then $u(t_n) = \Phi_1^{(p,f)}(\bar{u}_n)$ where $\bar{u}_n = u(t_n - 1)$ for $n \geq 1$.. The sequence (u¯n)(u¯n) is bounded, therefore, due to Theorem 3.6, $\left(\Phi_1^{(p,f)}(\bar{u}_n)\right)$ contains a convergent subsequence. Hence α(u)α(u)and ω(u)ω(u) are nonempty, compact and connected. The ΦΦ(p,f)ΦΦ(p,f)-invariance follows in a similar manner by the use of the compactness of Φ(p,f)1Φ1(p,f). Furthermore, it follows from Theorem 3.5 that $u \in C(\mathbb{R}, W_0^{1,p}(0,l))$ and φp,f∘uφp,f∘u is non-increasing, which means that the $\lim_{t\to-\infty} \varphi_{p,f}(u(t))$ and $\lim_{t\to+\infty} \varphi_{p,f}(u(t))$ exist and the latter is finite.

Now take $\bar{u} \in \alpha(u) \cup \omega(u)$ and suppose that either $t_n \to +\infty$ or $t_n \to -\infty$ and u(tn $u(t_n) \to \bar{u}$ as $n \to +\infty$.. By the relative compactness of u(R)u(R), for a fixed τ>0τ>0, passing to a subsequence we may assume that $u(t_n - \tau) \to \bar{u}_0$ in $C_0(0,l)$) for some $\bar{u}_0 \in C_0(0,l)$ such that $\bar{u} = \Phi_\tau^{(p,f)}(\bar{u}_0)$.. Since u(R)u(R) is bounded in C0(0,l)C0(0,l) we see that, by Lemma 3.9, that u(R)u(R) is bounded in W1,p0(0,l)W01,p(0,l). Therefore, by use of Theorem 3.6 (i), we have $u(t_n) = \Phi_\tau^{(p,f)}(u(t_n-\tau)) \to \Phi_\tau^{(p,f)}(\bar{u}_0) = \bar{u}$ in $W_0^{1,p}(0,l)$. For the same reason, we have

$$u(t+t_n) = \Phi_t^{(p,f)}(u(t_n)) \to \Phi_t^{(p,f)}(\bar{u}) \text{ in } W_0^{1,p}(0,l) \text{ for any } t > 0.$$

Hence, for any t≥0t≥0,

$$\varphi_{p,f}(\Phi_t^{(p,f)}(\bar{u})) = \lim_{n\to+\infty} \varphi_{p,f}(u(t+t_n)) = \lim_{n\to+\infty} \varphi_{p,f}(u(t_n)) = \varphi_{p,f}(\bar{u}).$$

This together with (17) means that $\Phi_t^{(p,f)}(\bar{u}) = \bar{u}$ for all $t \geq 0$, i.e. $\bar{u} \in \mathcal{E}$.

Proposition 3.10

Any bounded set N⊂XN⊂X is strongly ΦΦ(p,f)ΦΦ(p,f)-admissible *(*in the sense of Definition 1.2*)*.

Proof

Proof Fix (t_n) in $(0, +\infty)$ and $(\bar{u}_n)$ in $\mathbf{N}$ such that $\mathbf{\Phi}^{(p,f)}([0, t_n] \times \{\bar{u}_n\}) \subset \mathbf{N}$. Then observe that $\mathbf{\Phi}_{t_n}^{(p,f)}(\bar{u}_n) = \mathbf{\Phi}_{\tau}^{(p,f)}(\mathbf{\Phi}_{t_n-\tau}^{(p,f)}(\bar{u}_n)) \subset \mathbf{\Phi}_{\tau}^{(p,f)}(\mathbf{N})$ for a fixed $\tau > 0$ and all sufficiently

4 CONTINUITY AND COMPACTNESS ALONG *P*

We start with a fundamental theorem on continuity and compactness of semigroups with respect to *p*.

Theorem 4.1

If $p_n \geq 2$, $n \geq 1$, *and* $p_n \to p$ *as* $n \to +\infty$, *then* $\mathbf{A}_{p_n} \overset{\mathrm{Gr}}{\to} \mathbf{A}_p$ and therefore

$$S_{\mathbf{A}_{p_n}}(t)\bar{u} \to S_{\mathbf{A}_p}(t)\bar{u} \quad as \quad n \to +\infty \quad for\ any\ \bar{u} \in C_0(0, l)\ and\ t \geq 0.$$

For any bounded $B \subset C_0(0, l)$ *and* $t > 0$, *the set* $\bigcup_{q \in [2, p]} S_{\mathbf{A}_q}(t)(B)$ is relatively compact in C0(0,l)C0(0,l).

In the proof we shall need the following convergence properties.

Lemma 4.2

$$If\ p_n \geq 2,\ n \geq 1,\ and\ p_n \to p\ as\ n \to +\infty,\ then\ \bar{\mathbf{A}}_{p_n} \overset{\mathrm{Gr}}{\to} \bar{\mathbf{A}}_p$$

Proof

Fix $u \in L^2(0, l)$ and $\lambda > 0$. Put

$$q := \sup\{p_n \mid n \geq 1\}, \quad q' := \frac{q}{q-1}, \quad v_n := (I + \lambda \bar{\mathbf{A}}_{p_n})^{-1} u, \quad z_n := |v_n'|^{p_n - 2} v_n'.$$

Therefore, for any n≥1n≥1,

$$\bar{\mathbf{A}}_{p_n} v_n = \frac{1}{\lambda}(u - v_n) \text{ and hence } \int_0^l |v_n'|^{p_n}\,\mathrm{d}x = \frac{1}{\lambda}\int_0^l (u - v_n) v_n\,\mathrm{d}x. \tag{20}$$

From the contractiveness of the resolvents of $\bar{\mathbf{A}}_{p_n}$ it follows that $\|v_n\|_{L^2} \leq \|u\|_{L^2}$. Now, by means of the Hölder inequality, one may show that

- (v_n) is bounded in $H_0^1(0,l)$;
- (z_n) is bounded in $L^{q'}(0,l)$;
- (z'_n) is bounded in $L^2(0,l)$ (note that $z'_n = -\bar{\mathbf{A}}_{p_n} v_n$).

that z′n=−A¯pnvnzn′=−A¯pnvn).

As a consequence, passing to a subsequence if necessary, we can assume that

$$v_n \to v \text{ in } L^2(0,l),\ v'_n \rightharpoonup v' \text{ in } L^2(0,l),\ z_n \to z \text{ in } L^{q'}(0,l),\ z'_n \rightharpoonup z' \text{ in } L^2(0,l)$$

and that $z_n \to z$ almost everywhere. This clearly forces $v'_n \to |z|^{1/(p-1)} \operatorname{sgn} z$ z almost everywhere. Therefore $z = |v'|^{p-2}v',\ z' = -\bar{\mathbf{A}}_p v$ and, clearly,

$$\bar{\mathbf{A}}_{p_n} v_n = \frac{1}{\lambda}(u - v_n) \to \frac{1}{\lambda}(u - v) \text{ in } L^2(0,l).$$

On the other other hand

$$\bar{\mathbf{A}}_{p_n} v_n = -z'_n \rightharpoonup -z' = \bar{\mathbf{A}}_p v$$

which gives $v = \left(I + \lambda \bar{\mathbf{A}}_p\right)^{-1} u$ and ends the proof. □

Proof of Theorem 4.1

(i) Let $\bar{u} \in C_0(0,l)$ and put $\bar{u}_n := \left(I + \lambda \mathbf{A}_{p_n}\right)^{-1} \bar{u},\ \bar{u}_0 :=$ Lemma 4.2 implies that $\bar{u}_n \to \bar{u}_0$ in $L^2(0,l)$. By (16) we get for some constant C>0C>0 that

$$C\|\bar{u}_n - \bar{u}_0\|^{p_n}_{H^1_0} \le \left(\bar{\mathbf{A}}_{p_n}\bar{u}_n - \bar{\mathbf{A}}_p\bar{u}_0, \bar{u}_n - \bar{u}_0\right)_{L^2} - \left(\bar{\mathbf{A}}_{p_n}\bar{u}_0 - \bar{\mathbf{A}}_p\bar{u}_0, \bar{u}_n - \bar{u}_0\right)_{L^2}$$
$$\le \|\lambda^{-1}(\bar{u} - \bar{u}_n) - \bar{\mathbf{A}}_p\bar{u}_0\|_{L^2} \cdot \|\bar{u}_n - \bar{u}_0\|_{L^2} + \tag{21}$$
$$+ \left\| |\bar{u}_0'|^{p_n-2}\bar{u}_0' - |\bar{u}_0'|^{p-2}\bar{u}_0' \right\|_{L^2} \cdot \|\bar{u}_n' - \bar{u}_0'\|_{L^2}.$$

The first term of the right-hand side of (21) converges to zero. Since $\bar{u}_0' \in C(0, l)$, the sequence $|\bar{u}_0'|^{p_n-2}\bar{u}_0' - |\bar{u}_0'|^{p-2}\bar{u}_0'$ ' is bounded in $C(0, l)$ and its pointwise limit equals zero. Therefore $\| |\bar{u}_0'|^{p_n-2}\bar{u}_0' - |\bar{u}_0'|^{p-2}\bar{u}_0' \|_{L^2} \to 0$. Observe, that in order to prove that the second term in (21) tends to zero and therefore that u¯n→u¯0u¯n→u¯0 in H10(0,l)H01(0,l), and consequently in C0(0,l)C0(0,l), it suffices to verify the boundedness of $\bar{u}_n \to \bar{u}_0$ in $H^1_0(0, l)$,

Put $\bar{w}_n = |\bar{u}_n'|^{p_n-2}\bar{u}_n'$ Then the sequence $\bar{w}_n' = -\bar{\mathbf{A}}_{p_n}\bar{u}_n = \lambda^{-1}(\bar{u}_n - \bar{u})$ is bounded in $L^1(0, l)$. As $\bar{u}_n \in C_0(0, l)$., there exist points $x_n \in (0, l)$ such that $\bar{u}_n'(x_n) = 0$ and therefore that $\bar{w}_n(x_n) = 0$. Hence, $|\bar{w}_n(x)| \le M$ for some $M > 0$ and all $x \in [0, T]$. We finally obtain $|\bar{u}_n'(x)| \le M^{1/(p_n-1)}$.

(ii) We shall prove the assertion in two steps.

STEP 1. First we prove that for any bounded $B \subset L^2(0, l)$, $p > 2$ and $t > 0$ the set

$$\bigcup_{q \in [2, p]} S_{\bar{\mathbf{A}}_q}(t)(B)$$

is relatively compact in L2(0,l)L2(0,l). In view of Proposition 2.8, it is sufficient to prove that

$$L_\lambda := \left\{u \in L^2(0, l) \mid u \in D(\varphi_q) \text{ and } \varphi_q(u) \le \lambda \text{ for some } 2 \le q \le p\right\}$$

is relatively compact for any $\lambda \ge 0$. To this end take any $u \in L_\lambda$, i.e. $\varphi_q(u) \le \lambda$ for some $q \in [2, p]$. Then $u \in W^{1,q}_0(0, l)$ and $\int_0^l |u'|^q \le p\lambda$. Therefore u∈W1,20(0,l)u∈W01,2(0,l) and

$$\int_0^l |u'|^2 \le \left(\int_0^l |u'|^q\right)^{\frac{2}{q}} \cdot l^{1-\frac{2}{q}} \le (p\lambda)^{\frac{2}{q}} \cdot l^{1-\frac{2}{q}}.$$

This shows that LλLλ is bounded in H10(0,l)H01(0,l), which due to the Rellich-Kondrachov compact embedding theorem means that LλLλ is relatively compact in L2(0,l)L2(0,l), which ends the proof of Step 1.

STEP 2. Now let us take a bounded B⊂C0(0,l)B⊂C0(0,l) and t>0t>0. Take any (pn)(pn) in [2, *p*] and (u¯n)(u¯n) in *B*. Put α=t/3α=t/3 and define un:=SApn(·)u¯nun:=SApn(·)u¯n. We are going to show that the sequence (un(t))(un(t))has a convergent subsequence in C0(0,l)C0(0,l).

By use of Step 1, without loss of generality we may assume that the sequence (v~n):= $(u_n(\alpha))$ converges in $L^2(0,l)$ to some $\tilde{v}_0 \in L^2(0,l)$ and $p_n \to p_0$. Put

$$\bar{v}_n := u_n(2\alpha) = S_{\mathbf{A}_{p_n}}(\alpha)\tilde{v}_n \text{ and } \bar{v}_0 := S_{\tilde{\mathbf{A}}_{p_0}}(\alpha)\tilde{v}_0.$$

By Lemma 4.2 and Proposition 2.3 (i), $\bar{v}_n \to \bar{v}_0$ in $L^2(0,l)$. Moreover, $\bar{v}_0 \in W_0^{1,p_0}(0,l)$ in view of Proposition 2.7, which implies that $\bar{v}_0 \in C_0(0,l)$.. Therefore we can put

$$v_0 := S_{\mathbf{A}_{p_0}}(\cdot)\bar{v}_0 \colon [0,\alpha] \to C_0(0,l), \quad v_n := S_{\mathbf{A}_{p_n}}(\cdot)\bar{v}_n \colon [0,\alpha] \to C_0(0,l).$$

Then, clearly un(t)=vn(α)un(t)=vn(α) for all n≥1n≥1. Further, we note that

$$\|v_n(\alpha) - v_0(\alpha)\|_\infty \le \|S_{\tilde{\mathbf{A}}_{p_n}}(\alpha)\bar{v}_n - S_{\tilde{\mathbf{A}}_{p_n}}(\alpha)\bar{v}_0\|_\infty + \|S_{\mathbf{A}_{p_n}}(\alpha)\bar{v}_0 - S_{\mathbf{A}_{p_0}}(\alpha)\bar{v}_0\|_\infty. \quad (22)$$

By Lemma 3.4 (iii) and due to the continuity of the embedding of $H_0^1(0,l)$ into $C_0(0,l)$ we get

$$\|S_{\mathbf{A}_{p_n}}(\alpha)\bar{v}_n - S_{\mathbf{A}_{p_n}}(\alpha)\bar{v}_0\|_\infty \le C\|S_{\tilde{\mathbf{A}}_{p_n}}(\alpha)\bar{v}_n - S_{\tilde{\mathbf{A}}_{p_n}}(\alpha)\bar{v}_0\|_{H_0^1} \le C'\|\bar{v}_n - \bar{v}_0\|_{L^2}^{1/p_n} \to 0$$

for some constants C,C′>0C,C′>0. Finally by Proposition 2.3 (i), we get $S_{\mathbf{A}_{p_n}}(\alpha)\bar{v}_0 \to S_{\mathbf{A}_{p_0}}(\alpha)\bar{v}_0$ in $C_0(0,l)$. 0,l). This together with (22) proves that $u_n(t) = v_n(\alpha) \to v_0(\alpha)$ in $C_0(0, l)$, which completes the proof. □

We will summarize continuity and compactness properties for the equation (1) in the following

Theorem 4.3

Theorem 4.3 *(i)* (Continuity) *Let $p_n \to p_0$ in $[2, +\infty)$ and $\bar{u}_n \to \bar{u}_0$ in $C_0(0, l)$. Assume that the functions $f_n\colon [0, l] \times \mathbb{R} \to \mathbb{R}$, $n \geq 0$ satisfy (2) with common Lipschitz constants and that $f_n(x, u) \to f_0(x, u)$ uniformly for (x, u) from bounded subsets of $[0, l] \times \mathbb{R}$. Then $\liminf_{n\to+\infty} T_{\bar{u}_n}^{(p_n, f_n)} \geq T_{\bar{u}_0}^{(p_0, f_0)}$. Moreover, if $u_n\colon [0, T] \to C_0(0, l)$, $n \geq 0$, with fixed $T > 0$ are the solutions of (1) with $p = p_n$, $f = f_n$ $u_n(0) = \bar{u}_n$, and there is $R > 0$ such that $\|u_n(t)\| \leq R$ for all $t \in [0, T]$, $n \geq 1$, then $u_n \to u_0$ in $C([0, T], C_0(0, l))$.*

(ii) (Compactness) *Assume that $\{p_n\}_{n\geq 1}$ is bounded in $[2, +\infty)$, $f_n\colon [0, l] \times \mathbb{R} \to \mathbb{R}$ are locally Lipschitz and such that the set $\bigcup_{n\geq 1} f_n([0, l] \times [-r, r])$ is bounded for any $r > 0$. If $u_n \in C([0, T], C_0(0, l))$, $n \geq 1$, are solutions of (1) with $p = p_n$, $f = f_n$ and $\bar{u} = \bar{u}_n$ and there is $R > 0$ such that $\|u_n(t)\|_\infty \leq R$ for all $t \in [0, T]$ and $n \geq 1$, then for any $t \in (0, T]$ the sequence $(u_n(t))$ contains a convergent subsequence in the space $C_0(0, l)$.*

Proof

(i) follows directly from Theorem 4.1 and Proposition 2.4 (ii). The condition (ii) comes from Theorem 4.1 and Proposition 2.4 (iii). □

At the end of this section, we express the obtained results in terms of parameterized semiflows. To this end, let $\mathbf{X} = C_0(0, l)$ and let $f\colon [0, l] \times \mathbb{R} \times [0, 1] \to \mathbb{R}$ be continuous with f $f(x, 0, \mu) = 0$, for all $x \in [0, l]$ and $\mu \in [0, 1]$, and such that for any R>0R>0 there exists $L > 0$ such that $|f(x, u, \mu) - f(x, v, \mu)| \leq L|u - v|$ whenever x $x \in [0, l]$, $|u|, |v| \leq R$ $\mu \in [0, 1]$. Consider

$$\dot{u}(t) = -\mathbf{A}_p u(t) + \mathbf{F}(u(t), \mu), \ t > 0, \tag{23}$$

where $\mathbf{F}\colon \mathbf{X} \times [0, 1] \to \mathbf{X}$ is defined by

$$[\mathbf{F}(u, \mu)](x) = f(x, u(x), \mu), \ x \in [0, l], \ \mu \in [0, 1], \ u \in \mathbf{X}.$$

Define $\mathbf{\Phi}^{(p,\mu)}\colon \mathbf{D}^{(p,\mu)} \to \mathbf{X}$ by

$$\mathbf{D}^{(p,\mu)} := \{(t, \bar{u}) \in [0, +\infty) \times \mathbf{X} \mid t < T_{\bar{u}}^{(p,\mu)}\}$$

And

$$\mathbf{\Phi}^{(p,\mu)}(t, \bar{u}) = \mathbf{\Phi}_t^{(p,\mu)}(\bar{u}) := u(t) \text{ where } u\colon [0, T_{\bar{u}}^{(p,\mu)}) \to \mathbf{X}$$

is the maximal integral solution of (23) with $u(0) = \bar{u}$. The above results imply that this family of semiflows is continuous with respect to *p* and μμ and that bounded sets are strongly admissible.

Proposition 4.4

(i) I $\mathit{If}\ \bar{u}_n \to \bar{u}_0\ \mathit{in}\ C_0(0, l),\ t_n \to t_0\ \mathit{in}\ [0, +\infty),\ p_n \to p_0\ \mathit{in}\ [2, +\infty),\ \mu_n \to \mu_0\ \mathit{in}\ [0, 1]\ \mathit{and}\ (t_0, \bar{u}_0) \in \mathbf{D}^{(p_0,\mu_0)},\ \mathit{then}\ (t_n, \bar{u}_n) \in \mathbf{D}^{(p_n,\mu_n)}$ for large *n* and $\mathbf{\Phi}_{t_n}^{(p_n,\mu_n)}(\bar{u}_n) \to \mathbf{\Phi}_{t_0}^{(p_0,\mu_0)}(\bar{u}_0)\ \mathit{as}\ n \to +\infty.$

(ii) If $\mathbf{N} \subset \mathbf{X}\ \mathit{is\ bounded,\ then,\ for\ any}\ (p_n)\ \mathit{in}\ [2, +\infty)\ \mathit{an}\ (\mu_n)$ NN is strongly $(\mathbf{\Phi}^{(p_n,\mu_n)})$admissible.

Proof

(i) is a straightforward consequence of Theorem 4.3 (i). The proof (ii) goes along the lines of Proposition 3.10 with use of Theorem 4.3 (ii). □

5 Proof of Theorem 1.1

We start with a computation of Conley index at zero and at infinity.

Theorem 5.1

Under the assumptions of Theorem 1.1

(i)K0:={0}K0:={0} is an isolated invariant set with respect to ΦΦ(p,f)ΦΦ(p,f) and

$$h(\mathbf{\Phi}^{(p,f)}, K_0) = \Sigma^{k_0},$$

where k0k0 is such that $\lambda_{k_0} \le f_0'(x) \le \lambda_{k_0+1},\ \mathit{for\ all}\ x \in [0, l],$ for all x∈[0,l]x∈[0,l], and the inequalities are strict on a set of positive measure.

(ii)K∞K∞ consisting of all bounded full orbits for ΦΦ(p,f)ΦΦ(p,f) is a bounded isolated invariant set with a strongly admissible neighborhood and

$$h(\Phi^{(p,f)}, K_\infty) = \Sigma^{k_\infty},$$

where k∞k∞ is such that $\lambda_{k_\infty} \leq f'_\infty(x) \leq \lambda_{k_\infty+1}$, *for all* $x \in [0, l]$, and the inequalities are strict on a set of positive measure.

The computation of the indices of K0K0 and K∞K∞ will be reduced to computing the Conley index of zero in a special case.

Lemma 5.2

Let $g \in C([0, l])$ be a such that for some $k \geq 1$, $\lambda_k^{(p)} \leq g(x) \leq \lambda_{k+1}^{(p)}$ for all x∈[0,l]x∈[0,l] with the strict inequalities on a set of positive measure and let $\{\Phi^{(p,g)}\}_{t\geq 0}$ be the local semiflow generated on $C_0(0, l)$ by the problem

$$\begin{cases} u_t = \left(|u_x|^{p-2} u_x\right)_x + g(x)|u|^{p-2}u, \; x \in (0, l), \; t > 0, \\ u(t, 0) = u(t, l) = 0, \; t > 0. \end{cases} \tag{24}$$

Then u≡0u≡0 is the only full bounded solution of ΦΦ(p,g)ΦΦ(p,g) and, in particular, K:={0}K:={0} is an isolated invariant set relative to $\Phi^{(p,g)}$. *Moreover* $h(\Phi^{(p,g)}, K) = \Sigma^k$.

In the proof we shall use the following lemma.

Lemma 5.3

(See e.g. [7] and [8]) If *g* is as in Lemma 5.2, then the problem

$$\begin{cases} -\left(|u_x|^{p-2} u_x\right)_x = g(x)|u|^{p-2}u, \; x \in (0, l), \\ u(0) = u(l) = 0. \end{cases} \tag{25}$$

has no nontrivial weak solutions.

Proof of Lemma 5.2

Given a full bounded solution $u \in C(\mathbb{R}, C_0(0,l))$ of (24). By Theorem 3.8, α(u)∪ω(u)⊂Eα(u)∪ω(u)⊂E. If *u* was nontrivial, then we would get two different equilibria, since due to Theorem 3.5 the Lyapunov function would change its value along the nontrivial solution. But according to Lemma 5.3, in this case we have E={0}E={0}, which is a contradiction showing that there are no nontrivial bounded solutions of (24), which shows that K={0}K={0} is an isolated invariant set.

Observe that we get

$$h(\Phi^{(p,g)}, K) = h(\Phi^{(p,\lambda)}, K) \quad \text{for any} \quad \lambda \in (\lambda_k^{(p)}, \lambda_{k+1}^{(p)}). \tag{26}$$

Indeed, it is enough to consider a family of semiflows $\Phi^{(p,\tilde{g}(\cdot,\mu))}, \mu \in [0,1]$, with $\tilde{g}: [0,l] \times [0,1] \to [\lambda_k^{(p)}, \lambda_{k+1}^{(p)}]$ given by $\tilde{g}(x,\mu) = \mu g(x) + (1-\mu)\lambda, x \in [0,l], \mu \in [0,1]$, given by μ∈[0,1], and use Proposition 4.4 and the first part of the proof together with the continuation property (*H*4) of Conley index.

Next we consider the family $\{\Phi^{(\tilde{p}(\mu),\tilde{\lambda}(\mu))}\}_{\mu\in[0,1]}$ of local semiflows on C0(0,l)C0(0,l) with continuous

$$\tilde{p}: [0,1] \to [2,p] \text{ and } \tilde{\lambda}: [0,1] \to \mathbb{R} \text{ with } \tilde{p}(0) = p, \tilde{p}(1) = 2$$

and $\lambda_k^{\tilde{p}(\mu)} < \tilde{\lambda}(\mu) < \lambda_{k+1}^{\tilde{p}(\mu)}$ for all $\mu \in [0,1]$. for all μ∈[0,1]μ∈[0,1]. Again, in view of Proposition 4.4 we may apply the continuation property of Conley index and obtain

$$h(\Phi^{(p,\tilde{\lambda}(0))}, K) = h(\Phi^{(2,\tilde{\lambda}(1))}, K). \tag{27}$$

Using the spectral decomposition given by the Laplace operator A2A2 (and A¯2A¯2) together with Theorem 1.4, we get

$$h(\Phi^{(2,\tilde{\lambda}(1))}, K) = \Sigma^k. \tag{28}$$

By combining (26), (27) and (28) we get the assertion. □

Lemma 5.4

Under the assumptions (4) and (5), $\textit{if } \mathbb{R} \ni r_n \to 0^+,\ \mathbb{R} \ni R_n \to \infty$ $\bar{v}_n \to \bar{v}_0 \textit{ in } C_0(0,l), \textit{ then}$ then

$$r_n^{1-p}\mathbf{F}(r_n\bar{v}_n) \to f_0'|\bar{v}_0|^{p-2}\bar{v}_0 \textit{ and } R_n^{1-p}\mathbf{F}(R_n\bar{v}_n) \to f_\infty'|\bar{v}_0|^{p-2}\bar{v}_0 \textit{ in } C_0(0,l).$$

Proof

The first convergence follows directly from (4). We shall prove the second convergence. Fix ε>0ε>0. Put $\text{Fix } \varepsilon > 0. \text{ Put } V := \sup\{\|\bar{v}_n\|^{p-1} \mid n \in \mathbb{N}\}$ 0 be such that

$$\left|\frac{f(x,u)}{|u|^{p-2}u} - f_\infty'(x)\right| < \frac{\varepsilon}{2V} \text{ for all } x \in [0,l] \text{ and } |u| \geq D. \tag{29}$$

For n sufficiently large

$$\left|f_\infty'(x)|\bar{v}_n(x)|^{p-2}\bar{v}_n(x) - f_\infty'(x)|\bar{v}_0(x)|^{p-2}\bar{v}_0(x)\right| < \frac{\varepsilon}{2} \text{ for all } x \in [0,l] \tag{30}$$

and

$$\frac{\sup|f([0,l]\times[-D,D])| + \|f_\infty'\|D^{p-1}}{R_n^{p-1}} < \frac{\varepsilon}{2}. \tag{31}$$

Then

$$\left|\frac{f(x,R_n\bar{v}_n(x))}{R_n^{p-1}} - f_\infty'(x)|\bar{v}_0(x)|^{p-2}\bar{v}_0(x)\right| < \varepsilon \text{ for } x \in [0,l]. \tag{32}$$

Indeed, set $x \in [0, l]$. If $|R_n \bar{v}_n(x)| \geq D$, then (32) follows from (29) and (30). If $|R_n \bar{v}_n(x)| <$ D, then (32) is implied by (30) and (31). □

Proof of Theorem 5.1

(i) Consider $\tilde{f}(x, u, \mu) := \mu f(x, u) + (1 - \mu) f_0'(x)|u|^{p-2}u$, and the family of local semiflows $\{\tilde{\mathbf{\Phi}}^{(\mu)}\}_{\mu \in [0,1]}$ on $\mathbf{X} = C_0(0, l)$ generated by the equations

$$\begin{cases} u_t = \left(|u_x|^{p-2} u_x\right)_x + \tilde{f}(x, u, \mu), \ x \in (0, l), \ t \in \mathbb{R}, \\ u(t, 0) = u(t, l) = 0, \ t \in \mathbb{R}. \end{cases} \tag{33}$$

We claim that there exists $r_0 > 0$ such that for all $r \in (0, r_0]$, $\mathbf{N}_r := B(0, r)$ is an isolating neighborhood of K_0 relative to $\tilde{\mathbf{\Phi}}^{(\mu)}$ for all $\mu \in [0, 1]$. Suppose to the contrary, that there exist (r_n) in $(0, +\infty)$, (μ_n) in $[0, 1]$ with $r_n \to 0^+$ together with full solutions $u_n \in C(\mathbb{R}, \mathbf{X})$ of $\tilde{\mathbf{\Phi}}^{(\mu_n)}$ such that $\|u_n(0)\|_\infty = \sup_{t \in \mathbb{R}} \|u_n(t)\|_\infty = r_n$ for all $n \geq 1$. Define $v_n \in C(\mathbb{R}, \mathbf{X})$ by $v_n(t) := r_n^{-1} u_n(t/r_n^{p-2})$, $t \in \mathbb{R}$, $n \geq 1$. Then $\|v_n(0)\|_\infty = 1$ and v_n is a solution of

$$\begin{cases} u_t = \left(|u_x|^{p-2} u_x\right)_x + f_n(x, u), \ x \in (0, l), \ t \in \mathbb{R}, \\ u(t, 0) = u(t, l) = 0, t \in \mathbb{R}, \end{cases}$$

where fn $f_n \colon [0, l] \times \mathbb{R} \to \mathbb{R}$, $f_n(x, u) := r_n^{-(p-1)} \tilde{f}(x, r_n u, \mu_n)$, :=rn $|f(x, u)| \leq M|u|^{p-1}$ for some $M > 0$. By (4) and (5), $r^{1-p}|f(x, ru)| \leq M|u|^{p-1}$ and consequently

$$\|\mathbf{F}_n(v_n(t))\|_\infty \leq M \text{ for } t \in \mathbb{R}, \tag{34}$$

where $[\mathbf{F}_n(\bar{u})](x) = f_n(x, \bar{u}(x))$ for $\bar{u} \in C_0(0, l)$, $x \in [0, l]$. Thus, we can use Proposition 2.3(ii), which together with Theorem 3.6 and a diagonal argument gives a subsequence (still denoted by (vn)(vn)) converging pointwise to some v0∈C(R,X)v0∈C(R,X) . From Lemma 5.4 and (34) it follows that $\mathbf{F}_n \circ v_n \to f_0'|v_0|^{p-2}v_0$ in $L^1([0, T], C_0(0, l))$. Therefore, Proposition 2.3 (i) gives that v0v0 is a nonzero full bounded integral solution of

$$\begin{cases} v_t = \left(|v_x|^{p-2} v_x\right)_x + f_0'(x)|v|^{p-2}v, \ x \in (0,l), \ t \in \mathbb{R}, \\ v(t,0) = v(t,l) = 0, t \in R. \end{cases}$$

This, in view of Lemma 5.2, gives a contradiction, which proves the existence of a proper isolating neighborhood NrNr of K0K0 relative to the local semiflows $\tilde{\Phi}^{(\mu)}, \mu \in [0,1].$ Now Proposition 4.4, the homotopy invariance of Conley index (H4) and Lemma 5.2 imply the required equality

$$\tilde{f}(x,u,\mu) := \mu f(x,u) + (1-\mu) f_\infty'(x)|u|^{p-2}u, \text{ for } x \in [0,l],$$
$$\{\tilde{\Phi}^{(\mu)}\}_{\mu\in[0,1]} \text{ on } \mathbf{X} = C_0(0,l)$$

(ii) The proof is analogous to that for (i). Here we change properly the definition of $\tilde{f}(x,u,\mu) := \mu f(x,u) + (1-\mu) f_\infty'(x)|u|^{p-2}u, \text{ for } x \in [0,l].$, and consider the family of local semiflows $\{\tilde{\Phi}^{(\mu)}\}_{\mu\in[0,1]} \text{ on } \mathbf{X} = C_0(0,l)$ generated by the equations (33). We can show that there exists $R_0 > 0$ such that for all R>R0R>R0NR:=B(0,R)NR:=B(0,R) is an isolating neighborhood of K∞K∞ relative to ΦΦ~(μ)ΦΦ~(μ) for all μ∈[0,1]μ∈[0,1]. Indeed, on the contrary, suppose that there exist $R_n \to +\infty$ and $\mu_n \in [0,1], n \geq 1$ together with full solutions $u_n \in C(\mathbb{R}, \mathbf{X})$ of $\tilde{\Phi}^{(\mu_n)}$ such that $\|u_n(0)\|_\infty = \sup_{t\in\mathbb{R}} \|u_n(t)\|_\infty = R_n$. Define $v_n \in C(\mathbb{R}, \mathbf{X})$ by $v_n(t) := R_n^{-1} u_n(t/R_n^{p-2}), t \in \mathbb{R}, n \geq 1$. Then $\|v_n(0)\|_\infty = 1$ and v_n and vnvn is an integral solution of

$$\begin{cases} u_t = \left(|u_x|^{p-2} u_x\right)_x + f_n(x,u), \ x \in (0,l), \ t \in \mathbb{R}, \\ u(t,0) = u(t,l) = 0, t \in \mathbb{R}, \end{cases}$$

where

$$f_n : [0,l] \times \mathbb{R} \to \mathbb{R}, \ f_n(x,u) := R_n^{-(p-1)} \tilde{f}(x, R_n u, \mu_n), x \in [0,l], u \in \mathbb{R}, n \geq 1.$$

In a similar manner as in (i), there exists a subsequence of (vn)(vn) converging uniformly on bounded intervals to some bounded v0∈C(R,X)v0∈C(R,X) with ‖v0(0)‖=1‖v0(0)‖=1 that is an integral solution of

$$\begin{cases} v_t = \left(|v_x|^{p-2} v_x\right)_x + f'_\infty(x)|v|^{p-2}v, \ x \in (0,l), \ t \in \mathbb{R}, \\ v(t,0) = v(t,l) = 0, t \in \mathbb{R}, \end{cases}$$

which is impossible. Finally, using the homotopy invariance of Conley index and Lemma 5.2one has

$$h(\Phi^{(p,f)}, K_\infty) = h(\tilde{\Phi}^{(1)}, K_\infty) = h(\tilde{\Phi}^{(0)}, \{0\}) = \Sigma^{k_\infty},$$

which completes the proof. □?

Proof of Theorem 1.1

By Theorem 5.1,

$$(K_0, \Phi^{(p,f)}), (K_\infty, \Phi^{(p,f)}) \in \mathcal{I}(\mathbf{X})$$

and

$$h(\Phi^{(p,f)}, K_0) = \Sigma^{k_0} \quad \text{and} \quad h(\Phi^{(p,f)}, K_\infty) = \Sigma^{k_\infty}.$$

This means that K0K0 and K∞K∞ are irreducible and that $h(\Phi^{(p,f)}, K_0) \neq h(\Phi^{(p,f)}, K_\infty)$. Hence, in view of Theorem 1.5, we get a full solution $u : \mathbb{R} \to \mathbf{X}$ of $\Phi^{(p,f)}$ with $u(\mathbb{R}) \not\subset K_0$ and such that either $\alpha(u) \subset K_0$ or $\omega(u) \subset K_0$. In view of Theorem 3.8, since *u* is a nontrivial solution of ΦΦ(p,f)ΦΦ(p,f), there is a nontrivial stationary solution u¯u¯ of (1) such that either $\lim_{t\to-\infty} u(t) = 0$ $\bar{u} \in \omega(u)$ or $\bar{u} \in \alpha(u)$ and $\lim_{t\to+\infty} u(t) = 0$. Since each solution of $\Phi^{(p,f)}$) is a solution of (1), the proof is completed. □?

ACKNOWLEDGEMENTS

The study of both the authors was supported by the NCN Grant 2013/09/B/ST1/01963.

REFERENCES

1. Barbu, V.: Analysis and Control of Infinite Dimensional Systems, Mathematics in Science and Engineering, vol. 190. Academic Press, Boston (1993)
2. Barbu, V.: Nonlinear Differential Equations of Monotone Types in Banach Spaces. Springer, New York (2010)
3. Cazenave, Th, Haraux, A.: An introduction to Semilinear Evolution Equations. Clarendon Press, Oxford (1998)
4. Ćwiszewski, A., Maciejewski, M.: Positive stationary solutions for pp- Laplacian problems with nonpositive perturbation. J. Differ. Equ. **254**, 1120–1136 (2013)
5. Ćwiszewski, A., Kryszewski, W.: Constrained topological degree and positive solutions of fully nonlinear boundary value problems. J. Differ. Equ. **247**, 2235–2269 (2009)
6. Ćwiszewski, A.: Degree theory for perturbations of m-accretive operators generating compact semigroups with constraints. J. Evol. Equ. **7**, 1–33 (2007)
7. del Pino, M., Manasevich, R., Elgueta, M.: Sturm's comparison theorem for (|u′|p−1u′)′+c(t)(|u|p−1u)(|u′|p−1u′)′+c(t)(|u|p−1u), p>1p>1. Technical Report , Dept. of Math., F.C.F.M., U. of Chile, (1988)
8. del Pino, M., Elgueta, M., Manásevich, R.: A homotopic deformation along p of a Leray-Schauder degree result and existence for (u′|u′|p−2)′+f(x,u)=0(u′|u′|p−2)′+f(x,u)=0, u(0)=u(T)=0u(0)=u(T)=0, p>1p>1. J. Differ. Equ. **80**, 1–13 (1989)
9. del Pino, M., Manásevich, R.: Multiple solutions for the p-Laplacian under global resonance. Proc. Am. Math. Soc. **112**(1), 131–138 (1991)
10. Drabek, P.: Solvability and bifurcations of nonlinear equations, Pitman Research Notes in Mathematics Series 264, Longman (1992)
11. Dunford, N., Schwartz, J.T.: Linear Operators, Parts I and II. Wiley-Interscience, New York (1966)
12. Lan, K., Yang, X., Yang, G.: Positive solutions of one-dimensional p-Laplacian equations and applications to population models of one species. Topol. Methods Nonlinear Anal. **46**(1), 431–445 (2015)
13. Maciejewski, M.: Positive solutions to p-Laplace reaction-diffusion systems with nonpositive right-hand side. Topol. Methods Nonlinear Anal. **46**(2), 731–754 (2015)
14. Mansevich, R., Njoku, F.I., Zanolin, F.: Positive solutions for the one-dimensional p-Laplacian. Differ. Integral Equ. **8**(1), 213–222 (1995)
15. Montreanu, D., Montreanu, V.V., Papageorgiou, N.S.: A multiplicity theorem for problems with the pp-Laplacian. Nonlinear Anal. **68**, 1016–1027 (2008)
16. Ôtani, M.: A remark on certain nonlinear elliptic equations. Proc. Fac. Sci. Tokai Univ. **19**, 23–28 (1984)
17. Precup, R.: Abstract weak Harnack inequality, multiple fixed points and p-Laplace equations. J. Fixed Point Theory Appl. **12**, 193–206 (2012)
18. Rybakowski, K.P.: On the homotopy index for infinite-dimensional semiflows. Trans. Am. Math. Soc. **269**, 351–382 (1982)

19. Rybakowski, K.P.: Trajectories joining critical points of nonlinear parabolic and hyperbolic partial differential equations. J. Differ. Equ. **51**, 182–212 (1984)
20. Rybakowski, K.P.: The Homotopy Index and Partial Differential Equations, Universitext. Springer, Berlin (1987)

INDEX

T

U